RaumFragen: Stadt – Region – Landschaft

Series Editors

Olaf Kühne, Forschungsbereich Geographie, Eberhard Karls Universität Tübingen, Tübingen, Germany

Sebastian Kinder, Forschungsbereich Geographie, Eberhard Karls Universität Tübingen, Tübingen, Germany

Olaf Schnur, Stadt- und Quartiersforschung, Berlin, Germany

RaumFragen: Stadt – Region – Landschaft | SpaceAffairs: City – Region – Landscape

Im Zuge des „spatial turns" der Sozial- und Geisteswissenschaften hat sich die Zahl der wissenschaftlichen Forschungen in diesem Bereich deutlich erhöht. Mit der Reihe „RaumFragen: Stadt – Region – Landschaft" wird Wissenschaftlerinnen und Wissenschaftlern ein Forum angeboten, innovative Ansätze der Anthropogeographie und sozialwissenschaftlichen Raumforschung zu präsentieren. Die Reihe orientiert sich an grundsätzlichen Fragen des gesellschaftlichen Raumverständnisses. Dabei ist es das Ziel, unterschiedliche Theorieansätze der anthropogeographischen und sozialwissenschaftlichen Stadt- und Regionalforschung zu integrieren. Räumliche Bezüge sollen dabei insbesondere auf mikro- und mesoskaliger Ebene liegen. Die Reihe umfasst theoretische sowie theoriegeleitete empirische Arbeiten. Dazu gehören Monographien und Sammelbände, aber auch Einführungen in Teilaspekte der stadt- und regionalbezogenen geographischen und sozialwissenschaftlichen Forschung. Ergänzend werden auch Tagungsbände und Qualifikationsarbeiten (Dissertationen, Habilitationsschriften) publiziert.

In the course of the "spatial Turn" of the social sciences and humanities, the number of scientific researches in this field has increased significantly. With the series "RaumFragen: Stadt – Region – Landschaft" scientists are offered a forum to present innovative approaches in anthropogeography and social space research. The series focuses on fundamental questions of the social understanding of space. The aim is to integrate different theoretical approaches of anthropogeographical and social-scientific urban and regional research. Spatial references should be on a micro- and mesoscale level in particular. The series comprises theoretical and theory-based empirical work. These include monographs and anthologies, but also introductions to some aspects of urban and regional geographical and social science research. In addition, conference proceedings and qualification papers (dissertations, postdoctoral theses) are also published.

More information about this series at http://www.springer.com/series/10584

Olaf Kühne · Corinna Jenal

The Multivillage-Metropolis Baton Rouge

A Neopragmatic Landscape Approach

Springer VS

Olaf Kühne
Geographisches Institut
Universität Tübingen
Tübingen, Germany

Corinna Jenal
Stadt- und Regionalentwicklung
Universität Tübingen
Tübingen, Germany

ISSN 2625-6991 ISSN 2625-7009 (electronic)
RaumFragen: Stadt – Region – Landschaft
ISBN 978-3-658-30715-8 ISBN 978-3-658-30716-5 (eBook)
https://doi.org/10.1007/978-3-658-30716-5

This work contains media enhancements, which are displayed with a "play" icon. Material in the print book can be viewed on a mobile device by downloading the Springer Nature "More Media" app available in the major app stores. The media enhancements in the online version of the work can be accessed directly by authorized users.

Responsible Editor: Cori Antonia Mackrodt
This Springer VS imprint is published by the registered company Springer Fachmedien Wiesbaden GmbH part of Springer Nature.
The registered company address is: Abraham-Lincoln-Str. 46, 65189 Wiesbaden, Germany

Acknowledgements

We would like to take this opportunity to thank all those involved who made this book possible and who not only supported us during our research stay with all their expertise and inspired the project with their suggestions in joint discussions, but who also welcomed us with great hospitality.

We would like to thank all participants in the qualitative interviews for taking the time to contribute to this research project and to share with us insights, assessments, and interpretations of 'their city', thus providing us with a deeper insight into the living environments of Baton Rouge. We would also like to express our sincere thanks to John McEwen for collecting and providing the spectacular aerial photographs of the downtown and Mississippi River areas.

Our sincere thanks go to our colleagues at Louisiana State University, Prof. Kent Mathewson and Prof. Craig Colten, Department of Geography and Anthropology, who have always given us their full support as researchers and towards our research project. They have given us important insights and suggestions about Baton Rouge/Louisiana throughout our research path, and who have reflected critically and constructively with us on the results obtained.

We would also like to thank the student assistants Linda Baum, Anne-Cathrin Kayser, and Lara Koegst, who have always provided reliable support for this project and beyond in our research and in the procurement and administration of literature. We would like to thank Ted Cahill for transcribing the interviews, assisting with the evaluation of the songs on Baton Rouge, and revising the English language manuscript. We would also like to thank our chair assistant Diana Strauß, who coordinates our travel planning and research stays with the equal care and conscientiousness and who always supports us with her immunity against paperwork intolerance during the indignities of the administrative apparatus.

We would like to thank our colleagues Mohammed Al-khanbashi, Karsten Berr, and Timo Sedelmeier for their constructive and inspiring cooperation beyond the scope of this paper, as well as the entire chair team.

Special thanks go to our partners in life who not only tolerate but support our work and enthusiasm for the topic of settlement development, here in the case of Baton Rouge, and patiently endure research-related absences.

May, 2020 The Authors

Contents

List of Figures

List of Tables

List of Textboxes

1 Introduction: Of Biographical Approaches, Neopragmatisms, Urban–Rural Hybrids and… 'Why the Hell Baton Rouge?'

Baton Rouge, Louisiana, is not a focus of world publicity, not even of national publicity. Baton Rouge has not been the focus of much (spatial) scientific interest to date. Nevertheless, Baton Rouge deserves a special (scientific) interest. Baton Rouge shows a development, which on the one hand can be interpreted from regional and national aspects, increasingly also with global influences, yet on the other hand the development of Baton Rouge is characterized by individual events, which in feedback with the supra-local developments have formed a specific biography. More than almost any other city in the United States, it is shaped by political change: Baton Rouge "having flown under ten flags" (Armstrong 2010, p. 9), "the flags of France, England, Spain, the independent republic of West Florida, the United States and the Confederate States of America were flown over it, some of them more than once" (Gleason and Brockway 1991, vii). During this long period of time, Baton Rouge has undergone the development from a small to a medium-sized Southern city, then becoming the capital of Louisiana, home of Louisiana State University and a center of the petrochemical industry – all of the while in constant competition with and mostly in the shadow of Louisiana's global metropolis, New Orleans. In addition to these substantive reasons, there are also formal reasons for investigating Baton Rouge: Baton Rouge, with just under 230,000 inhabitants (metropolitan area just under 600,000 inhabitants), is so highly differentiated socially, culturally, economically, and also in terms of urban development, that different patterns of interpretation and evaluation of the urban biography are to be expected. On the other hand, the complexity of the settlement structure, the social, economic, and cultural interdependencies and demarcations are not so high that only partial aspects of the urban biography can be dealt with (such as in Los Angeles or San Diego; see Kühne 2012a, 2016a; Kühne and Schönwald 2015a, 2015b). Baton Rouge also has a remarkable presence in mass media as well as social media. But while cities with a supraregional presence also experience an intensive scientific investigation (outstanding in the

O. Kühne and C. Jenal, *The Multivillage-Metropolis Baton Rouge*, RaumFragen: Stadt – Region – Landschaft, https://doi.org/10.1007/978-3-658-30716-5_1

United States due to the 'schools' established there, certainly Chicago and Los Angeles; see for instance Bulmer 1984; Dear 2004; Dear 2005; Hennig 2012), the engagement with Baton Rouge, as mentioned above, remained rather restrained (some exceptions: Carleton 1981; Gleason and Brockway 1991; Hendry and Edwards 2009; Rodrigue and Phillips 2011), a special focus was usually relegated to a presentation of the city's history, combined with the presentation of numerous historical photographs. The only existing detailed geographical survey of Baton Rouge dates from 1963 and is, remarkably, written in German (Brill 1963). Due to the author's marked inclination to cartographic surveys, this work provides an essential basis for dealing with the development of Baton Rouge - especially in the second half of the twentieth century and the first two (not quite completed) decades of the twentieth century. Compared to Baton Rouge, New Orleans is once again one of the many (due to its high symbolic charge, especially in terms of culture, but also its ecological vulnerability: Campanella 2002, 2008; Colten 2006a) and the scientific attention paid to Louisiana as a whole can also be characterized as extensive (see e.g. Kniffen and Hilliard 1988; Wilds et al. 1996; Yodis et al. 2016). In this respect, Baton Rouge is also suitable for testing an approach to regional studies that not only synthesizes existing published knowledge, but also empirically generates new knowledge.

Synthetic studies of spatial aspects have largely fallen out of fashion in human geography, as sectoral views of specific spaces dominate. The category of 'space' is no longer unquestionably given; at the end of the twentieth century, constructivist thinking was also received in 'regional geography'. The constructed nature of spaces (as in Hall 1995; Marston 2000; Paasi 1991, 2010; Schultz 1997; van Houtum and Lagendijk 2001) was reflected upon, as well as the interweaving of regional knowledge with non-scientific patterns of interpretation and evaluation, and also the contingency of indicators and scientific perspectives. The 'new regional geography' was born when Gilbert (1988; in addition, especially Thrift 1991) applied various perspectives, Marxist and humanist approaches, and theories of practice, to the concept of 'region' (Entrikin 1996; cf. Gebhardt et al. 2004; Paasi 2002). With the 'new regional geography', the scaled framing also changed; the focus of the considerations shifted to the interdependencies between the different levels of scale (Paasi 2009). With the Constructivist turn, questions such as how regions are discursively generated, why people believe in the existence of regions, what identification function they have, how their boundaries are constructed and, last but not least, how their institutionalizations and meanings change (Paasi 1991; summarizing: Kühne 2019d), came into focus. This shift in focus was also associated with a change in the level of investigation: materialities lost their importance, the observation of the social construction of 'region' gained in importance. The present work - although based on the social constructivist tradition - would like to direct an increased focus on material elements, namely in their recursive feedbacks with social and individual constructions. This also means the integration of research results from a classical positivist perspective, an approach that we theoretically frame as '*neopragmatic*', which will be discussed in more detail in Chapter 2.

We have chosen the term '*biography*' as a metaphor to describe our work on Baton Rouge, as this term links well-documented events of individual (here urban) history with supra-individual events, such as, for example, the individual synchronization of one's own history and its attribution of meaning to supra-individual expectations - including the resulting self-dramatization in feedback with the image conveyed to the individual. Furthermore, the metaphor facilitates the reference of personal biographies to events, interpretations, and evaluations of urban biography (as already shown in Kühne and Schönwald 2015a, 2015b). The fact that regions, urban areas, and landscapes cannot be understood as ambiguously given objects (material or social) demonstrates numerous approaches from regional geography as well as geographical and sociological landscape research (among many: Blotevogel 1996; Chilla et al. 2012; Chilla et al. 2016; Gailing and Leibenath 2015; Jones and Paasi 2015; Kühne 2018d, 2019c; Paasi 1986, 2005, 2008; Weber 2019). In this respect, a theoretically reflected 'urban landscape biography' must always take into account spatial contingencies of varying intensity and bandwidth and not be tempted to subject them to a logic of reinterpretation and binary coding (more on this: Bauman 1990, 1995; Hard 2003; Kühne 2012b; Lyotard 1979). In this respect, we are aware that we are writing just one of many possible 'urban landscape biographies' of Baton Rouge, which is also marked by a specific 'déformation professionelle' (geography, political science, sociology; for more details: Berr et al. 2019; Berr and Kühne 2019; Kühne 2018b).

The metaphor of the biography of a city has also been used in various other places (Caughey and Caughey 1977; Hibbert 1969, 2001; Jones 2006; Large 2002; Larsen et al. 2007; Loew 2011; Logan 2000), but without this metaphor having received any further conceptual support. The approach we have chosen follows the approach of 'landscape biographies', which has been discussed in landscape research for about a decade and which discusses different political, planning, and specifically local influences on the development of a space understood as a 'landscape' (Roymans 1995; Roymans et al. 2009; Vervloet et al. 2010). Our approach to 'urban landscape biography', on the other hand, is expanded in three dimensions: first, it refers specifically to different spatial levels of influence on the urban landscape; second, it refers specifically to social science theory formation in order to frame the connection between material and social developments; third, it expands the existing approach to include the individual experience of urban developments. Since it is difficult to deal with such a wide range of topics with an isolated theoretical approach (such as a positivist or radical constructivist approach), a neopragmatic approach is chosen in this study (Chilla et al. 2015; Eckardt 2014; Kühne 2019c, e). This is based on the assumption that different theoretical focuses (with appropriate justification) can be chosen for complex issues (such as dealing with an urban landscape biography).

The term '*urban–rural hybrid*', as it appears in the title of the book, is intended to describe current spatial developments beyond dichotomous or trichotomous spatial characterizations, as they are found in the conceptual separation of city and country or that of city, suburbia, and country. The term refers multi-dimensionally to different mixing ratios of polarities: social (society-community), natural and cultural, architectural, and also economic (among many: Jenal and Kühne 2018; Kühne 2012a, 2016c; Weber 2018a). Since these compartments do not arrange themselves continuously, 'spatial pastiches' are created, which are characterized by discontinuities, breaks, and different spatial valences (Aitken and Zonn 1994; Kühne 2012a, 2015, 2017). This study is part of a large number of studies that deal with the relationship between supra-local rationales and local specifics – being more economic oriented in the United States than in Western Europe – and their (urban) landscape results (among other): Berking and Löw 2008; Beuka 2004; Clarke 1993; Conzen 2010; Duncan and Duncan 2004; Holzner 1996; Schein 1997; Wilson and Groth 2003; Zapatka 1995).

The objectives of this thesis are divided into both theoretical/methodological and object-related:

1. the approach of 'neopragmatic landscape research' should be operationalized and examined for its sustainability
2. to further develop the approach of 'urban landscape biographies'.
3. to depict and interpret the main lines of development, events, and supra-local influences on the biography of Baton Rouge
4. to emphasize the media presentation of Baton Rouge on the Internet - as an evolving compartment of a landscape biography - in order to capture elements of its image and self-description.
5. to examine the everyday life in Baton Rouge regarding its attributions of meaning and significance.

The downtown area of Baton Rouge borders the eastern bank of the Mississippi River (Fig. 1.1), with the rest of the city extending north, south, and east. On the opposite bank of the Mississippi River is the independent West Baton Rouge Parish town of Port Allen (Baton Rouge itself is part of the West Baton Rouge Parish and its capital). Baton Rouge is surrounded by growing suburban communities, especially to the south and east. The population density generally decreases outwards, although in a south-easterly direction somewhat less rapidly than in the other directions (Antipova 2010). Baton Rouge can be reached from the west, east, and southeast via Interstate Highways. The municipal airport in the north of the city connects Baton Rouge to the hubs of Atlanta, Dallas/Fort Worth, Houston, and Charlotte. Rail transport is limited to freight transport.

In addition to this introduction and the conclusion (Chapter 9), the present study is divided into seven chapters and a Post Scriptum (Chapter 10), which deals ironically with specifics of Baton Rouge. First, however, Chapter 2 sets out the theoretical framework of the work, focusing largely on the neopragmatic approach to regional geography, by explaining how we combine theoretical approaches and related methods, as well. Essential aspects of settlement development that are neither specific to Louisiana nor to Baton Rouge but do follow global or, at least, national logics will be presented in Chapter 3. In Chapter 4, we outline the integration of Baton Rouge into

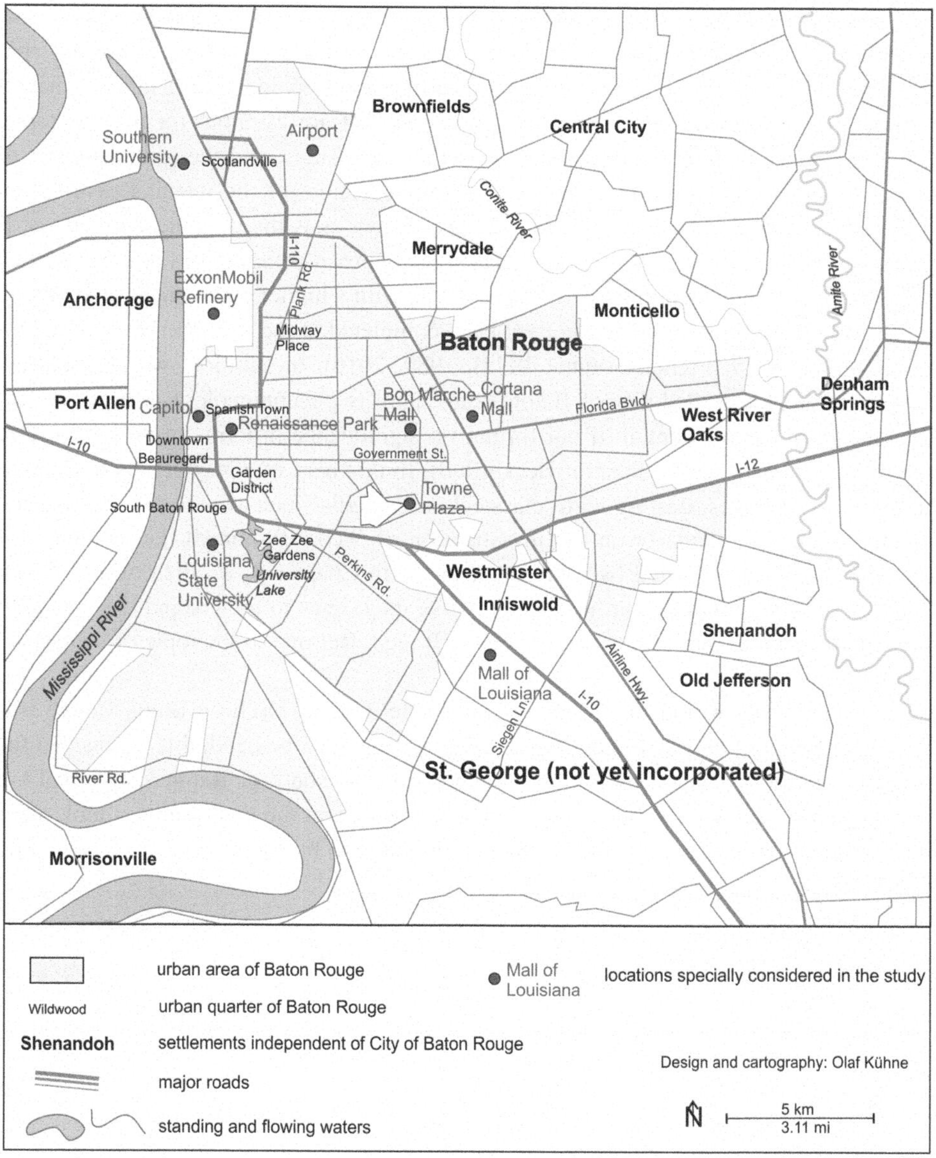

Fig. 1.1 General map of Baton Rouge. In particular, those spatial elements are presented which are given greater importance in the present book

the spatial-historical development of Louisiana as well as natural regional conditions that had and still have an influence on the development of Baton Rouge communities (from the regional deposit of oil to the risk of flooding). Chapter 5 presents essential elements of the biography of Baton Rouge: a topographical description at the beginning, the restrained development as a country town and the elevation to capital of Louisiana, the influence of the petrochemical industry as well as the universities, to current spatial differentiations. We will deal with these differentiations in more detail in Chapter 6, by taking a closer look at certain compartments that are essential for the biography of Baton Rouge, such as the downtown area, the universities, and the spatial impact of the petrochemical industry. In addition, material indicators of the unequal distribution of life chances (Dahrendorf 1979b) are also examined, as is the ongoing effort towards incorporating St. George, which is a collection of suburban communities in the southeast of Baton Rouge, inhabited mostly by white people with a higher endowment of symbolic capital - the incorporation having not yet been completed at the time the book was written (Bourdieu 1979 [French original 1972], 1989, 2016). In addition, we also address the question of the walkability of Baton Rouge and its phenomenological accessibility. Mass media have an essential, if not formative, significance in terms of how we create the world (Luhmann 1996; Schmitt-Beck 1990). In this respect, Chapter 7 is devoted to this complex of topics, dealing with 'classical mass media' such as national newspapers, but our focus is particularly on the investigation of Web 2.0 content, i.e. content that is produced and re-produced by users themselves. In Chapter 8, we then turn to Baton Rouge as a living environment, using qualitative interviews to understand the constructions, interpretations, and evaluations of Baton Rouge. But first to the topic of theoretical framing.

At this point we would like to give a short note: Our research on Baton Rouge was carried out between mid-2018 and November 2019. On the one hand, this means that the processes around the incorporation of St. George in the south of Baton Rouge had not yet been completed. On the other hand, this also means that the incorporation took place before the corona pandemic with its effects impossible to be estimated today (beginning of May 2020).

Theoretical Framework and Methodological Approach

2

In order to structure the understanding of space and landscape, the Three World Theory of Karl Popper (Popper 1973; Popper 2019 [1987]) will be referred to (see e.g. Kühne 2008b, d). Next to the world of material objects (World 1) there is World 2 (of individual consciousness) and the world of social contents (World 3). This also includes objects that are assigned to several worlds, such as those objects that are part of the material world but nevertheless receive World 3 contents because they have been shaped, placed, or somewhat symbolically discharged (Popper 2019 [1987]). With this approach Popper pursued the idea of overcoming the dualistic world view. As a result of the social constructivist framework, we see a constitutive meaning in World 3. In this respect, we see the constitutive dimension of social constructivist landscape research not in the material, but in the social dimension (see Fig. 2.1). In the following, different theoretical approaches to landscape (and space) will be discussed (and assigned to these three worlds).

An 'urban landscape biography', which is ultimately in the tradition of 'regional geography' as well as 'landscape geography', should provide itself with a theoretical framework, on the one hand because of the complexity of the object of research, and on the other hand because of the historical confusion and conflicts (which occurred more eruptively in German geography than in the Anglo-Saxon world). This theoretical framework, a neopragmatic one, will be presented in more detail after we have looked at some aspects of the history of landscape research (here with a strong focus on (West) Germany, we cannot, nor desire to, deny our regional disciplinary socialization at this point). In addition, the concept of urban landscape biography pursued here will be presented. In addition, some current urban landscape developments will be presented, which accompany the content-related discussion of the topic 'Landscape Biography Baton Rouge'. From the basic position of neopragmatism we derive a methodical procedure at the end of the chapter.

O. Kühne and C. Jenal, *The Multivillage-Metropolis Baton Rouge*, RaumFragen: Stadt – Region – Landschaft, https://doi.org/10.1007/978-3-658-30716-5_2

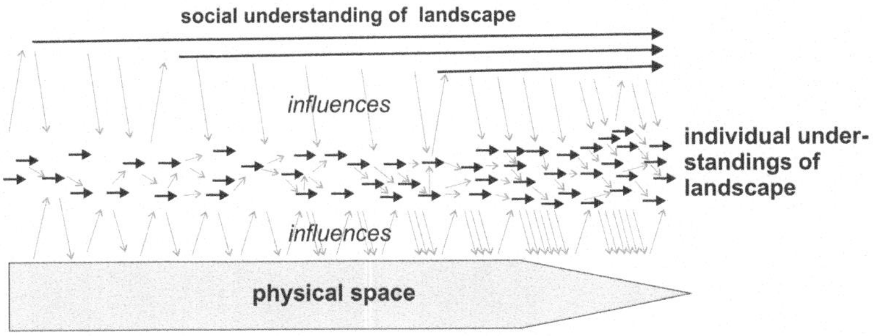

Fig. 2.1 The 'triple landscape change' (Kühne 2019c) from a social constructivist perspective. Constitutive are the societal understandings of landscape, which have an effect on individual landscape understandings, which in turn in special cases also have an effect on societal understandings (especially via the 'expert special knowledge'; see Textbox 1), but also influence each other directly. However, the individual level of landscape is also influenced by physical space, if this provides the basic substrate for symbolization, aesthetic experience, and emotional affection. But it is also transformed by individuals (based on social conventions and in interactions between individuals). (Own picture)

2.1 Developments in Landscape Research - A Brief Outline of Where We Stand Today

Basically, theoretical access to landscape refers to different weightings, understandings, and mutual references of the social, the individual, and materialities. Basic theoretical positions which have had or still have a greater significance in landscape research can be described as essentialist, positivist, constructivist, and over-representational. In the following we will briefly characterize these positions in order to classify them in relation to our work regarding the 'urban landscape biography' of Baton Rouge. The individual approaches have already been dealt with in more detail elsewhere, so we will only present their basic features here (see detailed Kühne 2018d, c; Winchester et al. 2003; Wylie 2007).

In contrast to a geo-deterministic position, an essentialist basic position of 'landscape' developed in the second half of the nineteenth century - according to some predecessors, such as Alexander von Humboldt. The latter assumed that 'landscape' had an independent nature, formed by a centuries-long mutual shaping of 'nature' and 'culture'. Correspondingly, it has 'essential' qualities, i.e. qualities that indicate its 'essence' and 'accidental' elements, i.e. those that are to be found there by chance. In particular, therefore, elements that arise from a global rationality. With regard to landscape, for example, traditional house and field forms or a local dialect are understood as essential, but not prefabricated houses, large-scale agricultural projects, large-scale infrastructure, etc., which is considered accidental (for more details: Albert 2005; Chilla et al. 2015, 2016; Glasze and Mattissek 2009; Trepl 2012). Accordingly, a kind of historically formed 'state of equilibrium' is assumed, which is influenced by accidental elements: "If the change takes place quickly, it initially appears disharmonious, since the equilibrium is disturbed for a longer period of time" (Lautensach 1973, p. 26 f.). From the higher valuation of the 'essential', the norm of its preservation is now derived (e.g. in Quasten 1997; Wöbse 1999, 2002). It is not possible to clearly assign this approach to one or more of the three worlds of Karl Popper, since essentialist theory attempts to fathom the essence 'behind things', a request that eludes empirical operationalization and is also not used in this work due to its speculative implications.

The position of positivist landscape research, on the other hand, is rooted in a tradition of scientific thought. Here, a 'landscape' is understood as an object that can be described empirically by means of measurement, weighing and counting, and inductively deduced by the analytical mind (Eisel 2009; Körner 2005; Tilley 1997, 2012). This research deals notably with World 1, but also with those hybrid areas that can be assigned to both World 1 and World 3, namely material objects that have been shaped, transformed, or placed by humans. The positivist paradigm is also partly found in quantitative social research (as access to World 3), when questionnaires, opinions, and attitudes of 'the population' are to be recorded 'objectively'. With the help of an ever-increasing availability of computer services, the empirically obtained data are modelled and used to make forecasts about landscape conditions. According to the postulate of the freedom of value of modern (positivist) science, however, norms of action are difficult to derive. It was precisely 'traditional landscape geography', popularized by Carl O. Sauer (1889–1975) also in the Anglo-Saxon world, that increasingly combined the aim of understanding the 'essence of landscape' with empirical methods of quantitatively recording the 'essential' features of a landscape, also in order to be able to name (and often declare undesirable) 'accidental' features (Duncan and Duncan 2009; Hard 1970, 1977; Kühne 2019c). This essentialist, positivist understanding of landscape came under criticism not only because of the conservative basic attitude (although it is still represented, albeit implicitly, in landscape planning and regional development; see in this context e.g. Chilla et al. 2016; Gailing 2014) and its focus on material objects and the holistic attitude, but also because of its focus on the rural and the associated disregard for the urban. As well, the growing intertwining of the formerly rural and the urban, such as the focus on the 'exotic', which is often assumed to be particularly close to the 'essence' of a space interpreted as a landscape (e.g. Blotevogel 1996; Duncan 1980, 1990; Mathewson 2009).

In Germany, this criticism culminated in the temporary rejection of the landscape paradigm (in the wake of the Kiel Geographers' Day in 1969; see in detail Berr and Kühne 2020; Berr and Schenk 2019; Schenk 2013), in the Anglo-Saxon world to a turn towards a social-constructivist understanding of landscape, while at the same time using the newly formulated understanding to criticize (often on the basis of neo-Marxist approaches) the spatially shaping power relations in society (among many others: Cosgrove 1984; Duncan and Duncan 2004; Mitchell 2002; Zukin 1993). In German geography, the positivist paradigm was separated from the normative content of essentialism and referred to seemingly more innocuous categories of 'space' or 'region' (Chilla et al. 2016), while 'critical geography' led a marginal existence (Gebhardt 2016).

2.2 Basic Theoretical Positions of the Work - The Neopragmatic Framework

The primary epistemological interest of social constructivist landscape theory lies in how socially shared conventions of landscape come about, how this becomes largely binding in the course of the socialization process and is thus socially 'normalized' (among many others: Greider and Garkovich 1994; Kühne 2019b, f; Stemmer 2016; Weber et al. 2017; on the theoretical foundations: Berger and Luckmann 1966; Schütz 1971 [1962]; Schütz and Luckmann 2003 [1975]; see on the topic of socialization of landscape: Textbox 1). Other constructivist approaches place the discourse in the focus of their own interest or the development and shaping of specific sub-societal system logics in relation to landscape. Discourse-theoretical approaches pose the question of how certain discursive links between statements can become hegemonic, i.e. which interpretations and evaluations of landscape are within and which are outside a discourse about landscape, such as the interpretation of whether or not a certain spatial element, such as a wind turbine, belongs to a landscape (e.g. Otto 2019; Otto and Leibenath 2013; Weber 2018b; on the theoretical foundations: Laclau and Mouffe 1985; Mouffe 2000). The autopoietic systems theory approach to landscape research is based on the sociology of Niklas Luhmann (Luhmann 1984, 1986, 2017), which in turn draws on the approaches of Talcott Parsons' structural-functionalist systems theory (1991 [1951]) and radical constructivism (Maturana and Varela 1987; Varela et al. 1974). Luhmann assumes that modern society organizes itself autopoietically in subsystems, i.e. self-referential, self-organizing, and self-preserving units, which exclusively perform specific tasks for the whole of society (for example, the economy for its supply of goods and services), and which observe their environment according to a specific logic, such as making money in the economy, generating power in politics, or gaining knowledge in science. In terms of landscape this means: Landscape is only communicated within and between systems if they are brought into resonance, for example, if money can be earned with landscape (as in tourism), power can be generated (positioning for or against planned landscape changes), (publishable) knowledge can be generated (science), current information can be disseminated and morally evaluated (mass media; for more details see Kneer and Nassehi 1997; Kühne 2014, 2018a, 2019a; van Assche 2010; van Assche and Verschraegen 2008). While the discourse-theoretical approach and the autopoietic systems theory primarily focus on the structures and processes of World 3, the social constructivist perspective is not limited to these alone, but also deals with the question of how the understanding, interpretation, and valuation of landscape, as laid out in World 3, also influences World 2 and how, in turn, this World 2 is able to influence World 3.

Textbox 1: The Socialization of Landscape - A Differentiation

The individual appropriation of what is commonly understood by landscape, how it is to be interpreted and evaluated, is rather discontinuous and can be divided into phases. In the first phase, the 'normal native landscape' is formed. This can be understood as the result of the appropriation of the living environment during childhood. Parents, if available, (especially older) siblings, but also grandparents, later on the peer group etc. act as mediators also. This 'normal native landscape' is "filled with first memories of regional language, sounds, smells, colors, gestures, moods, and conversational things, all deeply anchored in the memory" (Hüppauf 2007, p. 112). It has a strong positive emotional connotation and the norm of permanence is attached to the material objects it occupies (Kühne 2011; Kühne et al. 2013; Stotten 2019; Weber et al. 2019). The 'stereotypical landscape', on the other hand, is not - like the 'normal landscape of the home country' - based on unquestioned familiarity, but rather on socially shared, especially aesthetic, patterns of interpretation and evaluation (Kühne 2008a, b, c). It is formed expressly on the basis of secondary information, such as through school lessons (textbooks in particular), but also films, books, often illustrated books, calendars, and increasingly through the Internet. In this process, ideas about the individual usefulness of landscape are also formed (for example, as a backdrop for sports or aesthetic considerations). The 'stereotypical landscape' forms the common sense understanding of landscape in a particular social association (e.g. a language community), and is thus marked by social influence to a much greater extent than the 'normal landscape at home', which is shaped by individual experience. In an optional third phase, 'expert special knowledge' (Kühne 2019c; see also Kühne 2008a; Stotten 2013) will be acquired in vocational training, especially in a relevant scientific landscape-related study program. They are generally strongly cognitive and subject to a subject-specific deficit orientation (especially when essentialist patterns of interpretation are conveyed). The 'expert special knowledge stocks' are still changeable in principle, since they are subject to the competing paradigmatic basic orientations described in this chapter (cf. Hokema 2015; Kühne 2008b; Leibenath 2014; Wojtkiewicz 2015).

In recent years, interest (especially in Anglo-Saxon landscape research) has increasingly focused on intermediary approaches. These are characterized by the fact that, on the one hand, they reject the strict subject-object separation, which is represented in particular by a positivist understanding of the world (but also by classical discourse research and autopoietic systems theory), but, on the other hand, they also attach greater importance to non-human living and non-living material objects. These are the actor network theory (ANT), the assemblage theory, and the rediscovery of phenomenology. While the ANT seeks to break down the subject dichotomy with the help of the network concept and also to assign non-human objects the status of 'actants' (Kneer 2009; Latour 1998, 2002 [1999]; Schulz-Schaeffer 2000), assemblage theory - starting from a basic constructivist position - focuses on how material things are discursively negotiated and how they affect society (Färber 2014; Landa 2006; Mattissek and Wiertz 2014). Phenomenological landscape research, conversely, makes the experiencing individual (in Popper's sense, World 2) the starting point for its considerations. Here the question is raised of how material arrangements are experienced, how atmospheres are created between these and individuals (Kazig 2013; McEwen 2014; Wylie 2005). Thus, principally, the reference from World 2 to World 1 is addressed. Although phenomenology and social constructivism go back in part to the same roots, namely the phenomenological sociology of Alfred Schütz, the foci of the approaches are quite different: if social constructivism focuses on the social generation and transmission of knowledge, in which the individual is particularly important as a carrier of social knowledge, phenomenology focuses on the experiencing individual - in this case, landscapes as objects (World 2). The difference between the approaches becomes particularly clear in material objects. Are they of significance for social constructivist landscape research when they are symbolically occupied or, when viewed aesthetically, are assembled to form the social or individual construct 'landscape'? For phenomenological landscape research, they are the constitutive counterpart for the experiencing individual, especially when dealing with atmospheres (Düttmann 2000; for a comparison between social constructivist and phenomenological landscape theory see in more detail: Kühne 2019c). Kazig (2013, p. 221) clarifies the significance of atmospheres for (phenomenological) landscape research: "The concept of atmosphere, which focuses on the dynamic, sensorily mediated relationship between environmental qualities, subjective states, and activities, helps to systematically capture the perceptual dimension of landscaping".

Neopragmatism offers less a 'new' theoretical basis for landscape research, but rather leads to a more object-oriented approach to theoretical foundations (e.g. Chilla et al. 2015, 2016; Eckardt 2014). Philosophical pragmatism, i.e. the truth being seen as constituted by usefulness and usability, has significantly influenced studies by the Chicago School (Joas 1988; Schubert et al. 2010) and has led to a selection of methods based on the object of research (usually on a medium scale, e.g. in the form of an urban district). Neopragmatic landscape research is characterized by a meta-perspective as opposed to the object perspective of pragmatic research. Since complex research questions can hardly be framed with the help of a theoretical perspective, different theories and, also, methods are used in neopragmatic research (Eckardt 2014; Fine 2000; Kühne 2019e). By means of 'theoretical' as well as empirical 'triangulation', a differentiated picture of complex research objects is to be generated, whereby (partial) contradictions between theoretical approaches or the relationship between the various justifiably chosen theories and the equally chosen empirical methods are also accepted. To this extent, a basic social constructivist perspective can be combined with methods of quantitative social research, although these are more likely to be attributed to a positivist world view (Kühne 2018c).

A further difference to 'classical' theoretical approaches that strive for 'purity' is the potential for inter- and, ultimately, transdisciplinarity: if 'world explanation' (particularly as in the context of positivism or essentialism) is regarded as the essential goal of classical theories, especially in the version of one discipline, then neopragmatic approaches cannot only generate cross-disciplinary perspectives, they can also be used to generate instructions for political or administrative practice (Berr et al. 2019; Chilla et al. 2015, 2016; Weber et al. 2016).

The following paragraph explains how this neopragmatic approach can be profitably applied to urban landscape geography research.

2.3 Neopragmatic Urban Landscape Biography Research

The concept of landscape biography was specifically formulated by archaeologists in the Netherlands in the late 1990s (e.g. Kolen 1995; Roymans 1995; Roymans et al. 2009) and became popular outside this discipline in the 2000s (Kolen et al. 2015; Kühne and Schönwald 2015b; Meijles and van Hoven 2010; Roymans et al. 2009; Vervloet et al. 2010). Today, the concept is interdisciplinary and transdisciplinary: archaeology, anthropology, historical geography, geology, and geomorphology all draw on the concept, as does practical spatial planning (Roymans et al. 2009). Accordingly, landscape biography becomes an open concept for many interpretations (Vervloet et al. 2010). The concept of landscape biography is related to the anthropological work of Appadurai (1988, 2015 [1986]) and Kopytoff (2015 [1986]) as well as the geographer Samuels (1979). Samuels distinguishes himself from approaches that viewed landscape as a passive material by-product of anonymous economic and social developments (Roymans et al. 2009). Samuels (1979) placed individual and collective action related to landscapes at the center of his considerations. In addition to 'landscapes of impression', which he understood as inscriptions of ideologies and cultural representations (but also planning concepts), he focused on 'landscapes of expression'. These are the result of the individual and collective actions of people. According to Roymans et al. (2009), the concept of landscape biography, as developed after 2000, goes beyond its historical predecessors in three aspects:

1. If it is geared to the consideration of the longue durée, processes from prehistoric times to the present are taken into account. This is linked to the goal of studying landscape at any point in its history as the result of a long and complex interaction process between mentalities and values, institutional and governmental changes, as well as social and economic developments and ecological dynamics (Roymans et al. 2009).
2. Special focus is placed on the different stages of development of the landscape: in this way, each new 'landscape layer' builds on past ones and is interwoven with the biography of people, creating links between people and generations (Roymans et al. 2009).

3. Does not the concept of landscape biography assume a sharp break between past and present? Thus, current social developments are based on the past, the same applies to different discourses related to landscape.

The concept of landscape biography takes into account both the social dimension of landscape (in terms of values and norms; World 3), individual influences (especially as inscriptions; World 2), and the material level, in the form of a landscape understood as material (World 1; Roymans et al. 2009; van der Valk and van Dijk 2009; Vervloet et al. 2010). The confrontation between local peculiarities with supra-local (today extending to global) influences gives rise to both specific conventions in dealing with the (object) landscape and their inscription in this landscape. The concept of landscape biography shows a fundamentally positivist understanding of landscape, landscape is understood as an object in which social developments are inscribed through individual action (here we find a proximity to the metaphor of the 'palimpsest', which was very popular in the late phase of 'classical landscape geography'; see Duncan and Duncan 2009). On the whole, the scientific-theoretical framework of the concept remains little reflected; a kind of juxtaposition of scientific results takes place without disciplinary traditions and logics being questioned or related to each other. This is also where our further development of the concept begins, but first we will explain the general structure of our approach.

2.4 The Conception of the Present Work as a 'Neopragmatic Landscape Biography'

Our approach of a 'neopragmatic landscape biography', an 'urban' one in the case of this study, is structurally divided into the levels of the 'natural initial substrate' – these being composed of tectonic/geological/geomorphological structures, flora and fauna, atmosphere and hydrosphere, all at different spatial scales (Fig. 2.2). For example, the local geomorphological and geological structure is essentially determined by the integration of tectonic processes, continental uplift and subsidence processes or regional erosion and sedimentation processes. The climate is also involved in processes at different scales (including the global circulation of the atmosphere and global climate change). Flora and fauna, for example, are also not determined by local factors alone, but are involved in the global exchange of species; the same applies to the hydrosphere (World 1; according to Popper).

The 'natural initial substrate' (which is of course also an abstraction, as will be explained in more detail later) is contrasted with a society divided into subsystems, which is becoming increasingly globalized (in the course of its modernization). For example, local policy is influenced by regional (federal) and national legislation, which in turn is incorporated into a framework of international treaties and agreements. The local economy is also involved in supplier and sales relationships, in ideas about the organization of economic activities that are of allochthonous origin, and the same

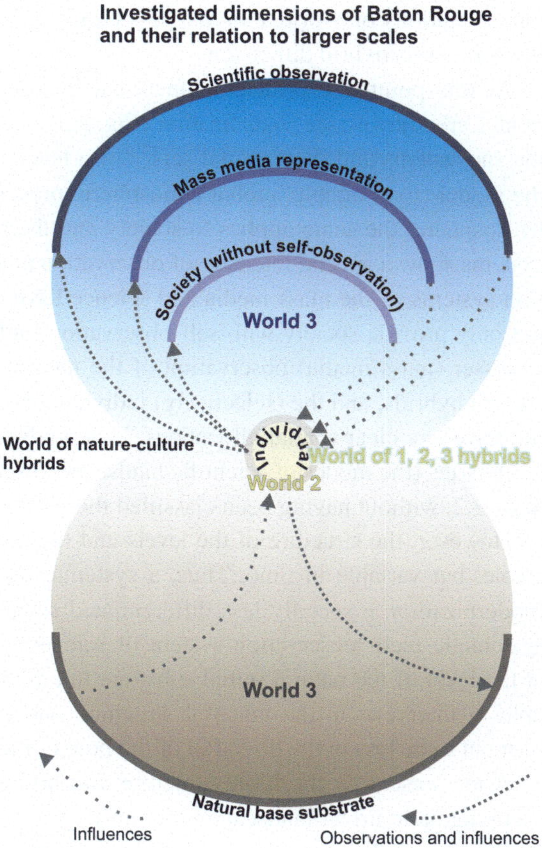

Fig. 2.2 Essential dimensions of the presented understanding of a 'neopragmatic urban landscape biography'. (own presentation)

applies to social roles, norms, and values as well as cultural influences. Although to be understood as parts of society, but highlighted here for analytical reasons, are the social observation systems (especially self-observation systems) of the mass media and science (World 3).

Between society and the natural foundations stands the individual (World 2), who - on the basis of social patterns of interpretation, evaluation and action - intervenes in the natural foundations and thus creates the level of natural cultural hybrids. This involves the modification of the material foundations of landscape, which in turn influence individual interpretations, evaluations, and actions, which in turn can have repercussions on society. This dimension is both part of World 1, since it is material, but it is also part of World 3, since many of these material objects are also the result of social placements, formations, or transformations. Since World 3, in turn, cannot directly intervene in World 1, but requires the mediation of the individual person, World 2, just as the production of the representations of World 1 in World 3 requires the mediation of World 2, World 2 also flows into this hybrid dimension.

At this point, the principal functional interdependencies between the levels of the model are shown (see visualization, Fig. 2.3), although the interdependencies between the subsystems and their spatial scales of a level are not shown for reasons of clarity of the model (for example, global atmospheric processes also have feedback with the local hydrosphere; the same applies to society and the mass media). The different social subsystems show a special intensity of observation among themselves, whereby the observation systems of the mass media and science have a prominent position here, since these not only provide society with self-observation findings, but also with a greater (science) or lesser (mass media) observation of the natural initial substrate, the world of nature-culture hybrids, and the (selectively) individual living world. The system of science does not have any clear spatial allocations, as it tends to be organized on the basis of sliding boundaries (the modes of scientific landscape observation have already been presented in Sect. 2.2, without having been classified there in the scheme presented here).

However, the structure of the levels and the intensity of the mutual reference are not stable, but variable in time. Thus, a systemically differentiated society is linked to its modernization, a socially less differentiated society - for example, due to the lack of a systematic (self-)observation system of science – is far less able to deal with changes at the level of the natural initial substrate in a reflective manner. However, it is also less able to intervene in the material structure, and the natural-culture hybrids (generally) turn out to be less in the direction of the pole of the 'cultural'.

After presenting the basic structure and principal functions of our approach to the 'neopragmatic urban landscape biography', we will now classify theoretical accesses to the levels and their connections before weighing up the theoretical strengths and weaknesses of the access.

Society is thought of, on the one hand, in a social constructivist way, especially with regard to the question of its influence by individuals (World 2), and on the other hand, with regard to its systems of observation, being in principle a theoretical autopoietic system (Fig. 2.3). By being based on autopoietic system theory, this appoach enables the observation of the specific logics in dealing with space/landscape and also the mutual observation of the individual subsystems within themselves. The observation of the respective environments by the observation systems is thereby framed by discourse theory; this approach has a constitutive focus on the references of the ascertainable and the unascertainable without loss of social recognition. The investigation of individual societal environments is based on a phenomenological approach (especially regarding the component of experiencing materialities as well as social environments), since this approach focuses on the individual in his or her self-perception in the world. If the connection of the individual to the institutions of society, the incorporation of social patterns of interpretation and evaluation, and also the relation of the individual as mediator between the social and the material world, is thematized, a framing takes place with the

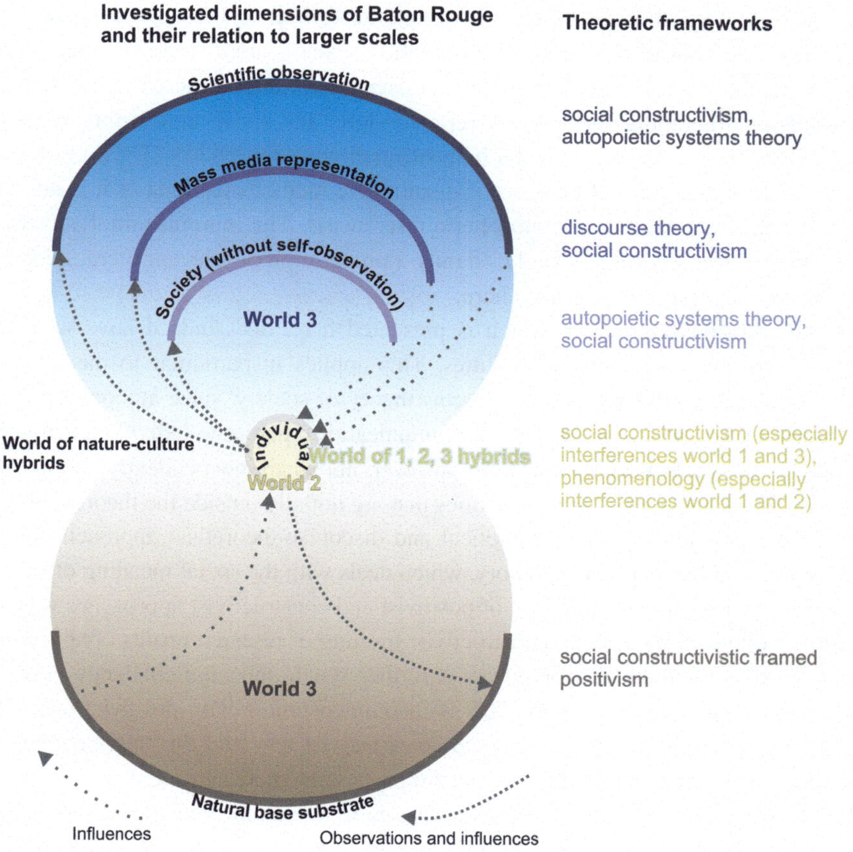

Fig. 2.3 Principle reference structures between the levels and the assignment to the worlds studied according to Karl Popper and the theoretical framework of the study of the individual dimensions

help of social constructivist landscape theory. The level of the 'natural initial substrate' and, in part, the 'natural cultural hybrids' is based on a positivist approach framed by social constructivism, which means that the approach of measure/weight/counting/modelling/forecasting does not achieve 'the objective truth about a landscape'. Rather, a recourse is made in the awareness that, on the one hand, abstractions from measurement results, etc., also represent constructions, and that, on the other hand, a quantifying access to the world is also a result of social conventions, which in turn are subject to a social negotiation process, i.e. are socially constructed. 'Natural cultural hybrids' are then examined on the basis of a 'social constructivist positivist understanding of landscape', if this is not their symbolic meaning, here a social constructivist approach is chosen, or their inclusion in the formation of atmospheres, here a phenomenological approach is taken, with the focus of interest on their materiality.

The methodological accesses are also made according to these theoretical orientations. If the observations of the social observation systems are examined, discourse-analytical or media-content analytical access is used. Individual patterns of interpretation and evaluation, including their social repercussions and those concerning the hybrid natural culture, are recorded with the help of qualitative interviews. These, in turn, are recorded in their materiality by means of quantitative methods (as long as it is not a matter of symbolic occupations or atmospheric references). The 'natural initial substrate' is not empirically recorded in this study, so only a preparation of relevant literature (written from a positivist perspective) is used here.

This neo-positivist approach, which is presented here, does indeed have weaknesses resulting from theoretical incompatibilities. This applies in particular to the integration of the individual into the autopoietic systems theory of society, since autopoietic systems theory does not know individuals (only communication); here, the individual can only be framed as a construct of communication, however, not as an independently acting subject with a will. The same applies to materialities that are initially outside the theoretical framework of the autopoietic-systems-theoretical and discourse-theoretical approach. This gap in turn can reduce the assemblage theory, which deals with the social meaning of matrices. The fundamental incommensurability of positivist and constructivist approaches was, however, further removed by a social-constructivist framing of research results obtained on the basis of positivist theoretical approaches. A further synchronization challenge exists with regard to phenomenological access. This is incommensurable with positivist understandings, but since these are framed in a socio-constructivist way and this approach has large intersections with phenomenology, a connection is possible here as well.

The neopragmatic approach we pursue also aims to ensure that the results achieved not only shed light on the 'Baton Rouge phenomenon' from different perspectives, but also that the results can be used politically and administratively. The transformation of analytical results into administrative and political practice is based on assessments. Assessments are bound by standards and these standards must be presented and justified. In our study, we follow the approach of 'life chance opportunities' of Ralf Dahrendorf (1932–2009), a German sociologist and political philosopher who can be attributed to the tradition of political liberalism. His approach can be placed in the tradition of Max Weber and Karl Popper, which is represented in the United States by John Rawls, Armatya Sens, and Martha Nussbaum (Nussbaum 2009; Popper 2011; Rawls 1971, 2001; Sen 2009; Weber 1976 [1922]). At the center of Ralf Dahrendorf's political sociology and political philosophy are his reflections on life chances. By life chances, Dahrendorf (2007, p. 44) contemplates "choice chances, options". These require two things – firstly, "rights to participate" and, secondly, "an offer of activities and goods to choose from". In his understanding of 'chance' (Dahrendorf 1979a, p. 98; but also 1968b; 1994), he follows Max Weber (1976 [1922]): for him, 'chance' is both a "structurally founded […] probability of behavior" and also understood "as something the individual can have, something as a chance to satisfy interests" (Dahrendorf 1979b, p. 98). Such 'chances of satisfying interests' are linked to "social relations […]. The opportunities themselves are social. Social structures order opportunities" (Lindner 2009, p. 20). Yet life chances are no guarantees:

"They only become concretely lived biographies through individual efforts - or they are forfeited" (Lindner 2009, p. 20). Life chances depend on social contexts "Life chances are possibilities of individual growth, the realization of abilities, wishes and hopes, and these possibilities are provided by social conditions" (Dahrendorf 1979b, p. 50). The life chances of people are not only determined by options, but also by ligatures (Dahrendorf 1979b, p. 50): "Options are choices given in social structures, alternatives of action", ligatures create references and thus constitute "foundations of action" (Dahrendorf 1979b, p. 51). For Dahrendorf (2007, p. 45), ligatures are value concepts, i.e. "deep bonds whose existence [gives] meaning to the chances of choice". These strong affiliations cannot be shaken off without the danger of anomie (Dahrendorf 1983), after all they are "structurally predetermined fields of human action. The individual is placed in bonds or ligatures by virtue of his social positions and roles" (Dahrendorf 1979b, p. 51). Ligatures produce commitment as well as affiliation. There is a fundamental difference between ligatures and options: "Ligatures are given, options are wanted" (Dahrendorf 1979b, p. 108). This also reveals the challenge concerning ligatures: they integrate people into a common community and society. This involvement can give meaning to actions and options, but it can also limit options because certain options are not implemented as ligatures. For Dahrendorf (1987), for example, overcoming overly rigid ligatures and creating options is considered an essential element of social modernization.

The task of politics is to "increase the diversity of life chances of the individual. Diversity is a matter of the nature and characteristics of life chances, as well as their rank and scope. Liberalism wants difference, because difference means freedom" (Dahrendorf 1972, p. 222). Thus, Dahrendorf (1983, p. 183) evaluates economic as well as social inequality positively in principle, since "inequality is a productive force of the social process, because it stimulates initiative and thus change". But according to Dahrendorf (2007, p. 86), inequalities are only productive "if and as long as they do not enable the winners to prevent others from fully participating in society or, in the case of poverty, to prevent people from exercising their civil rights". People who are prevented from exercising their civil rights often suffer "from the lack of 'incompleteness'. They live in 'incomplete families, mostly single-parent families without a father. They do not have full time jobs; temporary work has returned under a variety of names. They are incomplete citizens, perhaps immigrants or asylum seekers" (Dahrendorf 1991, 251). If these 'incompletions' (although this term certainly needs to be discussed) were combined, civil rights and thus life chances would be considerably reduced. In the terminology of John Rawls (2001), he concludes from this that it is a political task to implement not only formal, but also substantial, equal opportunities (Dahrendorf 1983, 1987). He sees education as the key to substantive equality of opportunity, which should be given to the especially underprivileged sections of the population so that they can seize their opportunities in life; accordingly, he formulated a 'civil right to education' in the 1960s (Dahrendorf 1968a).

The advantage of the multi-perspective approach is that it gives a more comprehensive impression of the complex and feedback-driven development of Baton Rouge than would be possible with a mono-theoretical view.

2.5 Neopragmatic Methodological Approach

As has been made clear in the previous section, the neopragmatic approach we have chosen is not limited to the combination of different perspectives in the theory of science but is also reflected in a mixed-methods approach. The different methods applied will be briefly named and contextualized in the following: a more precise specification of the methodological procedure will then be made in the context of the presentation of the results.

With the aim of making the explanations as clear as possible for the reader, we proceed in the order of appearance in the study. Chapters 3, 4, and 5 are based on a classical evaluation of existing scientific literature, sometimes combined with the evaluation of existing statistics and their graphical or cartographic presentation (as presented in detail in Kindler et al. 2019). In Chapter 6, the first Sects. (6.1 to 6.6) add methods of classical geographic spatial analysis (positivist basic attitude): in comparison to the 1959 surveys by Brill (1963), we have mapped and cartographically represented both building heights and dominant land uses of the downtown area in 2019. In preparation for this, Google Maps and Google Street View were used, and on-site mapping was used for comparison with the prepared maps. The aim of this study was to record the structural change of the downtown area. In addition, figure-ground diagrams have been prepared for this chapter to illustrate the layout of certain parts of Baton Rouge (streets, shopping structures, the eastern area of the 'Exxon Grounds'). To document our experience within parts of Baton Rouge, Sect. 6.7 is based on the 'Phenomenological Walk' method, in which perceptions are presented without analytical vocabulary, whenever possible.

In chapter 7, we follow the qualitative and quantitative media content analysis methodology to capture parts of the social landscape of Baton Rouge and is consequently derived from a constructivist theoretical approach. The focus of this part of the investigation - the increasing importance of Internet communication for the social construction of the world in general and landscape in particular - is initially on the presentation available on the Internet: first on (mostly textual) homepages, then on pictures and videos, and finally the tourist view as conveyed via TripAdvisor ratings, on the subject of Baton Rouge. Subsequently, the media view of Baton Rouge is broadened by treating its mass media construction in musical contexts such as the representation of Baton Rouge in song, but also in national newspapers such as the New York Times, which has the highest circulation in the United States.

In chapter 8, qualitative interviews are the focus of the Baton Rouge study. We have used these to gather further expert special knowledge, as well as elements of the stereotypical landscape and the native normal landscape of Baton Rouge, but also possible relationships between these levels of access. The evaluation of the interviews was carried out with the help of qualitative content analysis.

3 Complex Impacts on Baton Rouge: Supraregional Development Contexts and Impulses

In the previous chapter, the landscape theoretical framework as well as the biographical aspects and essential specifics of the 'neopragmatic approach' were discussed. In this chapter, essential aspects of settlement development will be presented that are neither specific to Louisiana nor to Baton Rouge but do follow global or, at least, national logics. Since these connections have been discussed in detail (e.g. in Beauregard 2006; Dear 2004; Gerhard and Gamerith 2017; Hayden 2004b, 2009; Holzner 1996; Sorkin 1992; Yeates 1990; Zukin 1995), this chapter will only give a brief account to contextualize developments in Baton Rouge. These are initially 'modern' developments of suburbanization and de-urbanization, then 'postmodern' developments of urban–rural hybrids, spatial pastiches, and Urbanizing Former Suburbs (URFSURBS).

3.1 Modern Urban Development: Suburbanization and De-Urbanization

The modernization of society, briefly summarized in the dimensions of industrialization, rationalization, secularization, and enlightenment, represented an emancipation from pre-modern society (although this modernization could take place unevenly, both spatially and in its dimensions). The majority of people before modernization had existed "in a constantly repeating cycle of poverty in predominantly rural environments" (Dahrendorf 1987, p. 192). The privileges of the few are based on the work of the many, without the possibility of a way out of this situation. Thus, pre-modern life was an "endless repetition of birth and decay, summer and winter, hard work and simple pleasure" (Dahrendorf 1987, p. 193). Housing space was scarce, and the settlements were small except for a few towns, with generally unpaved roads. The village formed an 'emergency and terror community' (Fertig 2005). With the modernization of society, these conditions

O. Kühne and C. Jenal, *The Multivillage-Metropolis Baton Rouge*, RaumFragen: Stadt – Region – Landschaft, https://doi.org/10.1007/978-3-658-30716-5_3

changed fundamentally: "(1) in the transformation of relatively closed class or caste societies into open class societies, (2) in the breaking down of traditional status hierarchies and the equalization of all wage earners, (3) in the creation of a situation of acute maladjustment and alienation for the workers who are initially alienated from industry, (4) in the creation of growing social emergencies, especially among industrial workers, (5) in the class antagonisms between entrepreneurs and workers that are constantly intensified by these circumstances" (Dahrendorf 1965, p. 65). The production of that time was extensively organized, it was only increased if the input of labor, land, and capital was increased roughly proportionally to the increase in production. Low productivity also kept domestic purchasing power within narrow limits, and the state intervened in the economy only to a limited extent in regulating it (Aglietta 1976; Hirsch and Roth 1986; Ipsen 2000, 2006; Moulaert and Swyngedouw 1989).

But after this early phase of modernization, productivity increases, the expansion of civil rights, the establishment of an effective administration, the increase in education, and the associated technical innovations, among other things, and much more, also brought 'the many' a reduction in pain, a lengthening of life expectancy, and with the increase in personal living space also privacy, a more extensive access to art "through modern reproduction methods" (Dahrendorf 1987, p. 194). However - as we mentioned above - the assertion of the achievements of modernity, especially those of freedom, was not equally distributed spatially and socially: "In the nineteenth century, lack of freedom arose above all from the enormous imbalance between the workers who had to offer their labor and the factory and landowners who owned the capital" (Herzog 2013, p. 72). Here is the essential starting point of Marxism's critique of the beginning industrialization. Despite the problems of early modernization, Dahrendorf (e.g. 1987) connects the process of modernization with the increase of life chances. Connected with this is also an increasingly important "individualization of social conflict in open societies" (Dahrendorf 1994, p. 236), connected with the consequence: "Individual mobility takes the place of class struggle" (Dahrendorf 1994, p. 236). The consequence of this is that "individual competition is stronger than collective solidarity" (Dahrendorf 1963, p. 64), a tendency that is found more in the United States than in the welfare states of Europe.

Modernism, now strengthened by effective administration, technical innovations, increasing economic output, and the will to spread the rationality of the Enlightenment, also manifested itself in architecture and urban planning. Here, the principle of 'form follows function' meets the economic striving for economies of scale and the will to physically manifest these principles, elevated to an urban utopia, in the largest possible units (in the sense of inscribing World 3 into World 1). The normative reference to function thus characterized a globally relatively uniform architectural style (nevertheless, some adaptations, especially to climatic conditions, were still being made), local or regional traditions were ignored (Imbert 2007; Kühne and Franke 2010; Löw 2010; Welsch 1993), which also meant a radical reduction of contingency (Burckhardt 2004b, p. 147): "Functionalism demands the sharp program, the rigorous decision for one building purpose and against the other". This approach to architectural functionalism is characterized by the urban sociologist Walter Siebel (2004, p. 19) as "an engineering utopia that relied on the principles of nature (light, air, sun) and the rationalization of industrial labor to build a good city". The "rationalist concepts of Fordist planning [...] of the European city [...] are intended to exorcise the jungle, the labyrinthine, the mythical and the threatening" (Siebel 2004, p. 20; cf. also Vicenzotti 2008), since winding and narrow neighborhoods were regarded as "breeding grounds for the unholy trinity of illness, crime and revolution" (Sennett 1991, p. 88), i.e. of social deviance. Modernist functionalism in architecture, urban planning and regional development symbolizes "the imaginary relations of individuals to real living conditions, which are inextricably linked to one another by the division of labor, because they are apparently naturally marked by normative divisions" (Prigge 1991, p. 105).

Compared to the welfare states of (Western) Europe, and in particular the efforts to establish settlements that follow a socialist concept of society, which enforced functionalist urban development preferably by building large housing estates with large apartment blocks and often at least state-subsidized, the spatial development of the United States shows a clear dominance of private or private-sector user interests over political or planning and administrative ideas of use. To put it bluntly: "From the beginning American cities have essentially been machines for minting money" (Conzen 2017, p. 18). A major consequence is suburban and de-urban settlement development, which is manifested here with particular intensity. For example, the decentralized settlement structure of the United States has "more thoroughly shaped the culture, the social and economic structure of the country as well as the way of life of its citizens and their ideas of a good life than in other countries" (Müller and Rohr-Zänker 2001, p. 27). Accordingly, more people in the United States live in suburban and de-urban areas, more than in inner cities and rural areas combined (Hayden 2004b). Mass mobilization increasingly made it possible for people with less symbolic capital (see Textbox 2) to turn their backs on the "dirty, crowded and epidemic-ridden American cities" (Karson and Betsch 2007, xv; similar to Hanlon 2010; Palen 1995). The form of mobilization can be traced back to the beginning of the nineteenth century and was successively intensified by the construction of railway lines, tram lines and, since the second decade of the twentieth century, the rapid spread of the automobile (Culver 2017; Holzner 1996). According to J.B. Jackson (2005), cars symbolize "liberation from the narrow confines of the home [...]: they embody the possibility of making contact with the rest of the world easily and quickly, of showing off and - the most important possibility of all - of retreating into one's private life".

The way of settlement in detached single-family houses came very close to the cultural pattern of individual life in a self-chosen community (Dahrendorf 1963). The socio-spatial consequence of the suburban and de-urban migration of people and economic functions was a strong segregation, the residual segregation left behind those who were not able to migrate to suburban communities due to the lack of symbolic capital (Palen 1995). Developments that can be described with the metaphor of the "metropolitan doughnut" (Breckenfeld 1977, p. 231), while in the Suburbs people with a high endowment of symbolic chapter gather, the settlement core is shaped by people with a low endowment of symbolic chapter, which in turn updates the widespread stereotyping of urban settlements, to which a fundamental lack of order has been attributed (Chaney 1997). Häußermann (1998, p. 80) sees here a fundamental difference between US-American and European urban symbolism: while in the United States the freedom of individual property "on the free land represents the highest value of the American way of life", in Europe "the development of a democratic society and forms of individual self-determination are closely linked to urban civilization". A life in a suburban, and especially a de-urban, context makes it possible to aesthetically stylize the ideal of rural life already formulated by Jefferson, albeit beyond the troubles and uncertainties of such a life (Hardinghaus 2004; Palen 1995). The history of suburbanization can also be interpreted as the history of two diverging logics: On the one hand, there is the lifeworld wish of the residents to find happiness and security in their own homes, on the other hand, there is the wish of the investors to profit from this, following the economic system logic (Hayden 2009). The transfer of Fordist principles of mass production to housing construction made the suburban single-family house affordable for large sections of the workforce from the 1940s onwards (Hanlon 2010; Hayden 2004a, 2004b; Krisch 2009; Waldie 2005). These 'sitcom suburbs' resembled the forms of living and working depicted in the popular sitcoms of the 1950s and 1960s evening series, in which the suburban home, rather than the urban apartment, was at the center of family life (Duany et al. 2000; Harlander 2009; Hayden 2004a, 2009; Muzzio and Halper 2002). The aspiration expressed in the Suburbs for a life in a voluntary community of like-minded people (i.e. not a 'pre-modern' traditional European village community into which one is born) can be seen as an expression of the aspiration for a "kind of insurance policy against the risks inherent in daily life in a many-voiced world" (Bauman 2008, p. 131), even if this "immersion in 'sameness' […] cannot reduce or even avert the risks that one wants to avoid" (Bauman 2008, p. 131). Another - related - 'insurance policy' is the 'own house': Unlike the European countries with their extensive state-supported education and social security systems, the home serves as a mortgage 'insurance' against major expenses, from financing a car to sending children to college. In this respect, the inhabitants of a suburban community are concerned to maintain the standard of symbolic capital in it. Developments that could lead to a reduction in the value of the properties are rejected accordingly (such as the construction of trunk roads through one's own neighborhood, waste incineration plants, or the immigration of people with a smaller amount of symbolic capital; cf. Davis 2004). This phenomenon has become known under the acronym NIMBY (Not in My Back Yard) and - in its extreme form - the BANANA (Build Absolutely Nothing Anywhere Near Anything; Bourne 1996; Cuff 2000; Kaplan et al. 2009; Pietila 2010; Wolch et al. 2005).

Textbox 2: Symbolic capital

According to Bourdieu (1979 [French original 1972]), symbolic capital can be understood as the opportunities that are suitable for gaining and maintaining social recognition and social prestige. Symbolic capital appears "as a perceived and legitimately recognized form" (Bourdieu 1985a, p. 11) of economic, social, and cultural capital (Bourdieu 1979 [French original 1972], 1985a):

- *Economic capital is material property that can be exchanged for money.*
- *Social capital refers to the formation of social networks that transform into social recognition.*
- *Cultural capital appears in three forms: firstly, in an objectified form (books, technical equipment, works of art); secondly, in an incorporated form (education, cultural skills); thirdly, in an institutionalized form (diplomas, educational titles).*

The degree of power that is granted to the bearers of the individual types of capital within the fields and the social world as a whole depends: "a) on the volume of capital, b) on the capital structure, i.e. on the relationship of the various types of capital to each other, c) on the ratio of seed capital to current capital" (Wayand 1998, p. 223). From this relationship it can also be concluded whether the capital providers are on an ascending or descending social career path (Bourdieu 1979 [French original 1972]). In addition, the varying availability of symbolic capital results in a vertical differentiation of society, which is manifested particularly in the expression of a class-specific taste. As a result of the different availability of the three capitals, which are in principle exchangeable among themselves, Bourdieu (1979 [original 1972]) sees a class structure of society. It distinguishes between three basic classes:

a) *The 'ruling class' is composed of entrepreneurs (with much economic but little cultural capital) and intellectuals (with much cultural but little economic capital). The ruling class is the bearer of the legitimate taste, which is characterized by 'the sense of distinction' over the 'middle class'.*
b) *The 'middle class' (also petty bourgeoisie) is composed of the descending middle class (mainly older small traders and craftsmen), the new middle class (new professions that do not require educational qualifications, such as salesmen, animators,* etc.*) and the executive petty bourgeoisie (especially employees who invest a lot of time in education and the autodidactic acquisition of cultural knowledge). The middle class is the bearer of the middle taste, which is characterized by 'educational zeal'. They try to emulate the 'ruling class'.*

c) *The 'controlled class' is composed of the rest of the society. They are the bearer of the popular taste, which is based on 'the decision for what is necessary', not on the taste of the other classes.*

Belonging to a certain 'class' of a person or even a social group or part of society is not stable. Since interpretations, habits, patterns of evaluation or even the appearance, i.e. the habitus, prove to be quite stable, the habitus of people proves to be less adequate to their new position. Classical examples are the impoverished aristocrat who clings to the aristocratic appearance, as well as the nouveau riche who lacks the cultural capital to find his way in a class communicating through fine aesthetic differences. Bourdieu (1979 [French original 1972]) calls this persistence in an outdated habitus the 'hysteresis effect'.

Even if today we can assume that society is more differentiated, the approach of symbolic capital and the relationship between classes still seems suitable in principle for examining the basic features of social differentiation and references (Illing 2006; Kühne 2018b) and integrating them into a 'neopragmatic' theoretical framework.

Although the expansion of suburban development was not centrally planned by the federal government, it was significantly accelerated by programs to support housing and highway construction (Culver 2010; Hayden 2004a, 2004b; Kaplan et al. 2009; Klohn 1998; Kühne 2012a; Pietila 2010; Zelinsky 2010; Zsilincsar 1972): the Federal Housing Administration (FHA) was founded as a subordinate agency of the U.S. Department of Housing and Urban Development in the course of the New Deal of 1934 with the aim of helping lower-income sections of the population to buy their own homes. A federal fund for financing highways has existed since 1916, and the Highway Act of 1921 systematized highway construction with the aim of connecting all settlements with a population of over 50,000 to the highway network. The 'Interstate Highway Act' of 1956 further accelerated the expansion of the highest-ranking road network. Thus, those who were willing and able to do so, and who had the availability of symbolic capital, were able to distance living and working from each other as a result of the use of the private automobile.

3.2 Postmodern Housing Developments: From Grids to Urban–Rural Hybrids, Spatial Pastiches, and Urbanizing Former Suburbs (URFSURBS)

In the context of the change in social values in the affluent societies subsumed as 'Western', Fordism has been in crisis since the mid-1960s, so that Fordist companies were increasingly unable to meet increasingly individualized demand. In line with the shift in demand towards more individualized products, there was a shift towards more flexible production, which meant that knowledge (whether about products, organization, design, as well as distribution) became increasingly important in economic activity (Rifkin 2007). In terms of spatial structure, this meant turning away from central large-scale production facilities and moving towards a decentralized industrial economic structure (Dear and Flusty 1998; Hirsch and Roth 1986; Ipsen 2006; Kühne 2012a; Lipietz 1991; Moulaert and Swyngedouw 1989; Soy 1989).

On a global level, a new type of settlement is emerging: the global cities. Its emergence can be traced back to "the fact that advanced internationalization and worldwide organization of economic activities requires 'hubs' of coordination and control of these globally extended economic processes, and that global cities with their concentration of international financial and corporate services function as places of production of global control capacity" (Krätke 2002, p. 49; Gladstone and Fainstein 2003; Massey 1999; Sassen 2000; Soy 2000). Especially in global cities, great wealth (from control activities) meets great poverty, especially of (illegal) immigrants or those whose participation in education has been denied (Amin and Thrift 2002; Sassen 2001). In terms of social space, the quarters of the wealthy often meet the living environment of the poor directly, an example of the loss of habitual coherence of dense spaces (cf. Hall 2006; Jorgensen and Tylecote 2007). On the one hand, historical urban centers are facing competition from new production sites of information and communication technologies (Edge Cities; Garreau 1992), on the other hand, they are staged, are given an identity-giving and often also marketable function, and by means of the commitment of well-known architects or increasing media presence, the interest of investors and tourists is to be increased. In addition to the 'architectural beacon projects', the restoration and staging of historical community images as an expression of the post-modern appreciation of the historical will receive special attention: "While modernity sought to free itself from all history" (Klotz 1985, p. 423), regional, ethnic, and historical aspects are now to be taken up in the sense of creating a 'special' place. However, even with this turn towards the historical, central parts of the area are reshaped according to aesthetic and economic logics, which places them in conceptual proximity to a theme park (Crilley 1993; Knox and Pinch 2010; Sudjic 1992).

In the context of the increase in attractiveness of central parts of the community, gentrification processes arise in which people with a low symbolic capital are displaced by those with a high one. On the whole, different subspaces of larger urban areas are not developing a universal pattern of development; on the contrary, individual impulses, locations, connections, etc. will become increasingly important (cf. Degen 2008; Kühn 2001; Ungers 1990), which is a strikingly modernist idea of "shaping a uniform cityscape" (Löw 2010, p. 154; cf. also Jorgensen and Tylecote 2007). The striving for modern unambiguity in communities was reflected in the attempt to make land uses monovalent (one surface, one use), whereas postmodern spaces are formed (temporarily) from a patchwork of surfaces of different degrees of valency. Nonvalent surfaces (such as brownfield sites) are found next to modernist monovalent (such as purely residential areas) and polyvalent spaces (spaces that have several functions, for example as a function of a symbolic surplus of meaning beyond the actual use, for example in the case of iconographic meanings; Kühne 2006). This patchwork is determined by a high degree of contingency: "Urban development takes place at quasi-random locations. In principle, all locations can be chosen, since they are connected by data highways" (Dear 2005, p. 34). Corresponding to the diversity, divergence, and polarizing developments in the populated area, a metaphor is used in a special way in the description (Basten 2005, p. 57): "Almost all attempts to characterize the new, postmodern structure of the city fall back on the image of a grid or net, whereby, however, not the figure of a spider's web, but that of a goal net or even that of a safety fence is meant". The different compartments of this structure show a varying degree of hybridization from public to private, from public sidewalks to public roads, the use of which requires (mostly) the availability of a motorized vehicle, from shopping malls to gated communities (Wiegandt 2017).

In the course of advancing postmodernization, 'hybridizations' are gaining in importance. Here the word 'hybridization' can first of all be understood as a "metaphor for cultural mixing" (Hein 2006, p. 59; cf. also Ha 2006; Schönwald 2017), whereby culture is not to be understood in an essentialist, and static way, but rather as "work in progress" (Ackermann 2004, p. 144), something that is subject to a constant process of negotiation. Accordingly, hybridity does not offer a "perspective of depth or truth [...]: it [hybridity] is not a third concept that resolves the tension between two cultures or the two scenes of the book in a dialectical game of 'knowledge'" (Bhabha 2000 [English original 1994], p. 168). 'Hybridity' and 'hybridization' are not limited to the cultural world, they are also to be found between the dichotomies of nature and culture, good and evil, male and female, etc., so contrived in modernity (among others: Zierhofer 2003, 2008). These hybridization phenomena are also found in the settlement context: The expression of 'urban–rural hybrids' (Kühne 2012a, 2016b; also Kühne et al. 2016; Kühne and Schönwald 2015b, 2018) expresses an (increasing) differentiation of city and country. This can be found as

- structural (e.g. in terms of building development),
- functional (e.g. more central-local),
- life-worldly (the endeavor to take care of oneself personally, the elements of life such as living, working, providing for oneself, recreation, etc., in view of the economic, administrative, family, etc., situation),
- emotional (i.e. in relation to 'concealment') but also
- cognitively (e.g. in the form of spatial description).

The processes of urban–rural hybridization do not only mean the spread of urban lifestyles into the surrounding area, but also the spread of 'suburban' lifestyles into urban contexts (domestic, family, and car-centered; Kropp 2015). Urban–rural hybridization does not occur along spatial gradients, but rather pastiches of varying degrees of hybridity are created (Hofmeister and Kühne 2016). The term 'pastiche' means "not simply de-differentiation, but *presupposes the formation of difference*, in order to then lead to hybrid crossings, recombinations, reintegrations" (Vester 1993, p. 29, emphasis in original); cf. also Hoesterey 2001). Here difference means "not only a relative, i.e. related to a common ground" (Scherle 2016, p. 61). It means a difference "which is (no longer) held together by a uniform foundation and which breaks up the classical question of the relation of the One and the Many, of the general and the specific". If the emergence of edge cities signified a functional and structural differentiation process, this is continued with the emergence of the edge-less cities (Lang 2003), whereby these edge-less cities come into being in "many different shapes, sizes and densities, they are found in correspondingly diverse arrangements" (Lang et al. 2009, p. 727). Since they lack a clear 'outside', the 'inside' is also difficult to define. Due to their lack of spatial specificity, they are not attributed an 'identity', so that "they are not perceived as a place" (Lang et al. 2009, p. 732). They form correspondingly novel compartments in the spatial pastiche of the urban–rural hybrids. As a result of the openness of the concept of 'spatial pastiches', the persistent developments of the 'grid' postmodernism can be integrated without any problems. Constitutively, however, the 'spatial pastiche' refers to those emergent developments that did not emerge through demarcation and exclusion, but through connection and graduality, such as the Edgeless Cities and the Urbanizing Former Suburbs to be dealt with here (URFSURBs; Kühne et al. 2017; Kühne and Weber 2019b; Schönwald 2017).

The term Urbanizing Former Suburbs (URFSURBS) describes the extension of 'urban' lifestyles, but often also functions (e.g. 'upscale' services) and structures (e.g. in buildings typical of the city center, such as office (high) buildings or apartment blocks) into suburban spaces, especially near the city center. It can thus be understood as a continuation of reurbanization processes, which in Anglo-Saxon discourse are also - often as a result of the connotation of gentrification, i.e. the displacement of persons with a lower endowment of symbolic capital by persons with a higher endowment of symbolic chapters - critically occupied by the terms revitalization, restructuring, and urban renaissance (Gerhard 2017; Lees 2008; Peck 2005; Smith and DeFilippis 1999). The emergence of the URFSURBS is an outcome of the loss of attractiveness of suburban to de-urban living and working, as well as to the increase in attractiveness of inner-city locations. In addition to the diversification of living together and the increasing importance of diverse face-to-face contacts (Gallagher 2013; Grant et al. 2013; Hanlon 2010; Hesse 2010; Zukin 1992, 1998), which make living in or near the city center more attractive, the increase in energy prices also speaks against a suburban lifestyle (daily commuting to work and the temperature control of a mostly detached single-family house; see also Häußermann 2009; Kühne and Schönwald 2015a). As a result of the limited availability of land in the classic inner-city locations, urbanization is spreading to formerly suburban community areas. The degree of intensity of 'URFSURBanization' varies:

1. The URFSURBanization becomes clear in the creation of new material structures that differ significantly from the previously existing structures, such as apartment and

office high-rises with upscale gastronomy instead of suburban single-family houses or industrial buildings.
2. The change becomes less obvious when the physical structures are retained but the use changes (e.g. the transformation of cheap shops into cafés etc.).
3. The change is even less evident when existing material structures and uses are preserved (mostly in residential buildings), but under their renovation.

Common to these developments is the increase in symbolic capital among the inhabitants. This (mostly) takes place through the replacement of the residents, but it can also be achieved through the appropriation of the different capitals (especially economic ones) by portions of the resident population (Kühne and Weber 2019a). At the same time, processes of segregation according to the endowment of symbolic capital, according to lifestyles, political preferences, etc., are also taking place (Bishop 2009), resulting in different overlaps and differentiations of architectural structure and social structure. Accordingly, the spatial pastiche becomes more delicate in terms of area and social structure.

These findings of postmodern social and residential development - in addition to the theoretical considerations on a neopragmatic research design presented in Chapter 2 - form the framework for the investigation of the developments in Baton Rouge.

4 The Regional Contextualization: Main Features of Louisiana's Development

Within the framework of the regional contextualization, in which the basic features of Louisiana's development are outlined, the following section will first look at the World 3 of Louisiana's natural basic substrates which are the physical-spatial foundations of Louisiana in accordance with the chosen approach (Sect. 4.1). Subsequently, the historical contextualization focuses on essential lines of development in Louisiana (Sect. 4.2), in order to then focus on the regional procedural and structural integration of Baton Rouge (Sect. 4.3).

4.1 The Context of the Natural Initial Substrates: Physical and Spatial Foundations

In terms of Louisiana's physical and spatial foundations, the Mississippi River forms a central element of its current shape: thus, the Mississippi River, whose Alongquin Indian name Misi Sipi means 'Great River' (Yodis et al. 2016, p. 3), crosses eight US states from its source in Lake Itasca, Minnesota, flowing some 2,400 miles from north to south into the Gulf of Mexico along the coast of Louisiana. It collects water and sediments from a drainage basin covering 1.2 million mi^2, so that about 40 per cent of the water of the United States flows through it (Yodis et al. 2016, p. 17), whereby "[l]ooking at Louisiana on the continental scale, one can easily envision it as the nation's drain" (Mathewson 2014, p. 2). In the process, the dynamics of the river and its sedimentary deposits, which have persisted for about two million years, have formed about two-thirds of Louisiana's subterranean land mass and have produced an extensive accumulation of sediment in the Gulf of Mexico (Penland and Suter 1989; Yodis et al. 2016). Louisiana is accordingly dominated by the alluvial plains produced by the Mississippi River, which runs along a broad north-south axis. These lowland plains, which have been

O. Kühne and C. Jenal, *The Multivillage-Metropolis Baton Rouge*, RaumFragen: Stadt – Region – Landschaft, https://doi.org/10.1007/978-3-658-30716-5_4

formed primarily in the last 7000 years, are traversed by meanders of present and past river courses (Yodis et al. 2016). So-called bayous form 'natural dams' to these meanders as backwater areas, which fill the deeper lying areas between present and former riverbeds (see in more detail Yodis et al. 2016). The Mississippi River is not only a central element in terms of its physical and spatial foundations and its shape but is also of decisive importance in the context of Louisiana's settlement, agricultural cultivation, and economic development. Even though the Mississippi River is "Louisiana's most prominent physical feature", "Louisiana and its physical landscapes are much more than just a mighty river and its immense contents" (Mathewson 2014, p. 2), as will be shown below.

In the global context, the state of Louisiana is assigned to the so-called 'Sun Belt' and as a state lies on the same latitude as, for example, northern Mexico, the Sahara, and the Middle East (Yodis et al. 2016). Louisiana's humid subtropical climate is, however, the result of its location on the Gulf of Mexico, which supplies warm and humid air to the states located in the southeast. Accordingly, the climate is characterized by hot summers with high humidity and heavy rainfall events, as well as mild winters with midlatitude cyclones, which can, however, also reach temperatures below the freezing point due to polar air invasions (Keim and Muller 1993; Yodis et al. 2016). Temperatures in the north of the country are on average about five degrees Fahrenheit warmer in summer and the corresponding number of degrees colder in winter, and on average about 10 to 15 inches more rain falls in the north than in the south (Mathewson 2014). In terms of its regional-geographical classification, however, Louisiana is a south-central state with access to the Gulf of Mexico, the entire state-wide territory of which lies within the physiographic region of the Gulf Coast Plain. Accordingly, the earth's surface is characterized by a low relief between six feet below sea level in southern New Orleans on the coast and elevations just over five hundred feet high above sea level in northern Louisiana, although more than half of the country is only slightly below one hundred feet above sea level, especially in the areas formed by the Mississippi River (Yodis et al. 2016). In geological terms, these formations are comparatively young and taper off, especially in the course of the north-south direction, so that "[a]long its southern margin, today Louisiana is more a work in regress than progress due to sea level rise, subsidence, and silt starvation" (Mathewson 2014, p. 3). As a result of climate change, the majority of the calculations indicate that the Wall Islands and a large part of the southern Parishes – the Louisiana equivalent of 'counties' – will become isolated government units (see also Sect. 4.2) and will disappear under the rising waters of the Gulf of Mexico. As well, the intensity and frequency of extreme weather events such as tornadoes, hurricanes and rainstorms will also increase with a warming of the climate in Louisiana of three to seven degrees Fahrenheit by the year 2100 according to the scenarios, thus further intensifying the extreme weather events already common in the region due to its geographical location and climatic conditions (Couvillion et al. 2011; Couvillion et al. 2017; Mathewson 2014).

The two characteristic landforms of Louisiana are the Lowlands and the Uplands. The Lowlands were formed after the melting of the last ice age during the Holocene, when the accumulation of sediments from the Mississippi and other rivers in conjunction with the simultaneous flooding of their estuaries by rising sea levels formed a huge plain that slopes down to the sea (Yodis et al. 2016, p. 16). This level is characterized by three larger sections: the alluvial plain of the Mississippi as well as the Red River, the delta plain, and the Chenier Plain (Yodis et al. 2016). In the alluvial plains, especially along the meandering Mississippi which flows in an S-shape through low gradients and carries a high sediment load, regular floods enrich the soils on the left and right side of the river with sediments. These occur mainly from January to June (see in detail Yodis et al. 2016, pp. 16ff.). The delta plain located in southeastern Louisiana forms a rugged coastline and is composed of six different active and dormant Mississippi estuary deltas that have been deposited over the past 7,000 years (see Yodis et al. 2016, pp. 40ff.). The Chenier Plain on the southwest coast – derived from the French chêne for 'oak' and roughly meaning 'place where the oaks stand' – is, here as well e.g. however, an approximately twenty-mile-wide swamp area characterized by long crests composed of sand and shells running parallel to the coast and tidal flats between the crests with swamp and bog vegetation – the so-called Cheniers (see in detail McBride et al. 2007; Otvos and Price 1979). The geology of the Uplands, on the other hand, is determined by several layers of sedimentary rock (Yodis et al. 2016). These strata, deposited in the Mesozoic and Cenozoic, slowly tilted towards the Gulf of Mexico as a result of coastal subsidence and isostatic uplift in the interior and tapered towards the coast in terms of their geological age (Yodis et al. 2016, p. 61). Thus, the Uplands are composed of pine-covered hills in the north (van Kley 1999), whose surfaces consist of Tertiary sediments, but which are covered by Mississippi and other Quaternary rivers with sediment depths of several thousand feet towards the coast (see in detail Yodis et al. 2016, pp. 61ff.).

Thus, based on the physical, spatial, and climatic conditions, there is a great potential for intensive agricultural cultivation, especially of the loamy soils in the alluvial plains of the Mississippi – here in particular sugar cane, rice, cotton, corn, and soybeans with the corresponding secondary consequences for the ecology of the soils and waters as a consequence of the high fertilization inputs into the soil (Bedient et al. 2005; Larson et al. 1995) – whereas heavily weathered or acidic soils are usually used as pasture land or for timber production (Rehder 1971; Yodis et al. 2016). The numerous waters are especially important for fishing and fish farming, as well as for hunting and trapping.

In the north, south, and offshore, Louisiana has significant oil and gas deposits which, after their discovery in 1901, were also commercially produced (LMOGA – Louisiana Mid-Continent Oil and Gas Association 2019a) and whose production peaked in the late 1970s at around 550 million barrels per year (Yodis et al. 2016, p. 188). Currently, annual production is around 50 million, making it one of the top ten oil and gas producing states in the United States (LMOGA – Louisiana Mid-Continent Oil and Gas Association 2019b). However, the extensive extraction of oil and gas also leads to a number of undesirable side effects, which continue to cause serious damage to the environment and health to this day. Until the late 1960s, the Mississippi and the bayous served as disposal sites for waste products and wastewater from the chemical industry (cf. Hochschild 2016), so that for decades highly toxic substances were introduced into Louisiana's environmental systems without being purified, even in the course of accidents in the production process, which to this day continue to pollute large parts of the country and (drinking) water with toxic substances as well as the emissions from the production sites generating air pollution that is hazardous to health (cf. and many others 2003; Colten 2000, 2006b, 2014a; Diaz 2011; Gottlieb 1980; Lincoln et al. 2011; Luckett et al. 2012; Page et al. 1976; Priest and Theriot 2013; Tsai et al. 2004). It was not until a series of serious fish die-offs in the early 1970s that the first political attempts were made to regulate this practice (Yodis et al. 2016), although environmental requirements for companies are still considered to be low today, and those sections of the population, those who have the least amount of symbolic capital at their disposal in particular, are exposed to these substances, which are hazardous to the environment and health (see, among others, Blodgett 2006; Bullard 1997; see Lee and Black 2017; Wright 2005).

After the examination of the physical and spatial foundations of Louisiana, the following section will provide a historical contextualization, outlining the history of settlement and the main lines of development in Louisiana.

4.2 Historical Contextualization: Major Lines of Development in Louisiana

According to the current state of research, the first settlement of the area within modern day Louisiana is dated between 10,000 to 8,000 years before our time (Cummins 2014). The so-called Paleo-Indians lived mainly in today's north and may have been nomadic hunter-gatherers who moved around in small communities, about whose dynasty, however, little is known due to only isolated indications of their presence (Cummins 2014; Yodis et al. 2016). In the Meso-Indian era, which is dated between 8,000 to 4,000 B.C., the transition from a hunter-gatherer society to a sedentary way of life took place, in the course of which permanent settlements were established in all parts of the country and also a number of cultural skills (such as making simple ceramics, baskets, fishing nets etc.) and rites (such as burying the dead) developed (Cummins 2014; Yodis et al. 2016). In the Neo-Indian period, about 4,000 to 2,600 years before our time, complex cultural centers were formed, the finds of which point to trade connections as far as Missouri and Texas, as well as to an early center of activity in northern Louisiana (Yodis et al. 2016). More complex cultural techniques such as ornamental ceramics or the agricultural cultivation of farmland were also developed here (Yodis et al. 2016). Thus, more and more tribes differentiated among the Native Americans, creating ever more complex societies and forms of settlement, developing new hunting weapons and methods – such as bows and arrows – as well as new cultural techniques and refining existing ones, so that in the course of the following centuries up to the European Middle Ages, various highly organized regional indigenous tribes emerged (see Cummins 2014 for details).

However, with the arrival of the first Spanish explorers around Hernando de Soto in the sixteenth century and later the French immigrants from French Canada in the eighteenth century, many members of the indigenous tribes died from diseases introduced by the Europeans. During the early phase of colonization many explorers and settlers first joined the Indian tribes in order to be able to survive in this environment which was very unfamiliar to them, this close constant contact allowed the diseases to spread easily (Jackson 2014). With the increasing colonial territorial claims, the Native Americans fell victim to the military attacks and enslavements of the colonial powers (Cummins 2014; Webre 1984; Yodis et al. 2016). Moreover, the Native Americans found themselves caught up in the rivalries between the various colonial powers, with whom they in turn formed changing alliances – some tribes allied themselves with the Spanish, others with the French or the British (Cummins 2014). At the beginning of the nineteenth century the population of the Native Americans comprised only a fraction of their population as at the beginning of the eighteenth century (Yodis et al. 2016, pp. 115ff.), so that they "had become an 'invisible people'" (Cummins 2014, p. 23). Still today, numerous place names can be traced back to the translations of the original names into Native American languages (see also Jackson 2014).

The present-day Louisiana formed from a part of the Spanish kingdom covering wide areas during the sixteenth and seventeenth century, but the interest of the Spanish crown in the settlement and further development of the area was extremely low, so that accordingly, only a few Spaniards came or even stayed in the region. When, at the end of the seventeenth century, the French nobleman René-Robert Cavelier Sieur de La Salle became the first European to reach the mouth of the Mississippi River from Canada, following the route of earlier expeditions, he claimed all the territories affected by the Mississippi River for the French crown in 1682 and named the territory 'La Louisiane' in honor of King Louis XIV (Cummins 2014). The settlement of the area was rather slow at first, so that consequently several years passed without any significant progress. The first city foundations were made with the settlements of Natchitoches and New Orleans in 1714 and 1718 respectively (Yodis et al. 2016), which served as control and trading posts on the Spanish settlement border with Texas and to control the access to the Mississippi from the Gulf of Mexico (Mathewson 2014). Although a number of further settlements were founded over the following decades, the French were only able to attract a few people to the remote colony despite their great efforts – for example, ten years after its foundation, New Orleans had only about 1,000 inhabitants – and there was also a lack of skilled workers in agriculture and handicrafts (cf. Yodis et al. 2016). In order to supply the young colony with labor, the French colonial power brought several thousand people from Africa, mostly from Senegambia/West Africa, to Louisiana in the 1720s and 1730s, so that in the early 1740s more than three times as many people came from Africa than from Europe (Yodis et al. 2016). The group of Africans from the French Antilles, who came to Louisiana during and after the Haitian revolution (1791–1804), joined the group on a voluntary basis (also known as the 'gens de couleur libres'), as well as those slaves who were forced to move by their owners, the plantation owners (Jackson 2014). While the slaves deported from West Africa introduced rice cultivation to Louisiana, the Caribbean immigrants cultivated sugar cane in Louisiana (Yodis et al. 2016). For many slave owners "slavery was no longer considered a necessary evil, but a paternalistic institution that brought the supposedly barbaric Africans the blessings of civilization" (Berg 2017, p. 74). In addition, in the early 1720s, ethnic Germans were recruited to cultivate the land and develop agriculture. They successfully set up farms to the west and east of New Orleans along the Mississippi – the 'German Coast' (Mathewson 2014), especially for the cultivation of vegetables and fruits, but also for livestock (Yodis et al. 2016).

With the end of the French and Indian War (1754–1763; Treaty of Paris 1763), in which the colonial powers France and Great Britain fought with their respective Indian allies for supremacy in North America, the French were defeated by Great Britain with large losses of territory (thus, France and French Canada lost a large part of their territory to Great Britain) and transferred Louisiana to the royal houses linked by the Bourbon nobility to Spain, since Great Britain had no interest in the colony, and so Louisiana was not to remain under French rule (Cummins 2014). At that time New Orleans became an economically prosperous trading city, which not only had a firm place in the French economic and trading system, but also developed into a "lively, vibrant, and teeming port city" (Cummins 2014, p. 66). Although Spain officially took over Louisiana in 1763, it took several years until the Spanish crown also established a colonial government (Cummins 2014). Although Spain was able to attract few immigrants from mainland Europe, it did succeed in recruiting immigrants from the Canary Islands, the Isleños, who settled in the southeastern coastal marshlands of New Orleans on the border with British West Florida (Mathewson 2014). In addition, it was the French-speaking exiles, the so-called Acadians or Cajuns, who, as a result of the French defeat in the French Indian War, were expelled from their original settlement area in French-Canadian Acadia and immigrated to Louisiana (Trepanier 1991; Yodis et al. 2016). They first settled along the Bayous Lafourche and Teche, west of the Mississippi (Mathewson 2014). As part of a worldwide colonial power, people from other Spanish colonies such as the Philippines and Mexico now also came to Louisiana, so that in the French as well as in the Spanish colonial period of Louisiana and in the early years of its affiliation to the United States, the largest ethnic groups were of non-European origin (Mathewson 2014; see for details on the various cultural influences in Jackson 2014).

The French Revolution, which began in Europe at the end of the eighteenth century, changed the balance of political power and France, which had intended since the Treaty of 1783 to put Louisiana back under French leadership one day in order to implement the vision of a French empire in the New World, successfully demanded under Napoleon Bonaparte that the Spanish reattach Louisiana to France, which was sealed with the Treaty of San Ildefonso in October 1800 (Cummins 2014). This transfer of the Louisiana colony to France, initially kept under wraps, became known, however, and led to irritation for the leadership of the United States, as a new rivalry for control of the Mississippi was feared (Cummins 2014). To avoid this, the strategy pursued by the United States was to acquire the city of New Orleans (Cummins 2014). Although an initial purchase attempt failed, the two American negotiators James Monroe and Robert R. Livingston were made a surprising offer after further negotiations: France offered not only New Orleans but Louisiana as a whole, with a complete waiver of all French rights to the entire area for 15 million dollars (Cummins 2014). Although the negotiators were not officially authorized to make a deal of this magnitude, they seized the historic opportunity and signed the Louisiana Purchase Agreement in Paris in April 1803 (Cummins 2014). It was "not clear what they had in fact purchased, since 'Louisiana's' boundaries had never been well defined" (Cummins 2014, p. 98; see also Stenberg 1934). This surprising turn of events is attributed by historians to the fact that Napoleon both urgently needed money for his military operations in Europe, and simultaneously, he probably realized that Louisiana without New Orleans is of no value to him as it would leave him unable to defend the vast Louisiana territory against the Anglo-American expansion efforts in the long run, especially against the background of large troop losses from the uprisings and guerrilla wars in the Caribbean colonies (Cummins 2014; see in detail at Cerami 2004).

With the sale of Louisiana to the United States, more and more Anglo-Saxon settlers moved into the area. Some of them were large-scale farmers who pushed into the fertile alluvial plains of the Mississippi and the Red River and bought up large amounts of land – from the Acadians among others, who had little financial capital and were often in debt, and who subsequently moved westward into the Swamps or into new settlements around St. Martinsville (Yodis et al. 2016). The large-scale farmers, subsequently, built up huge agricultural enterprises with the cultivation of cotton and sugar cane, which required an even greater demand for labor. Consequently, the number of African-Americans in the pre-war period rose from 34,000 in 1810 to over 360,000 in 1860 (Yodis et al. 2016). But also, other immigrant groups such as Irish, Germans, and Italians arrived after having left their home countries because of famine, for political reasons, or in search of a better life in the New World (Yodis et al. 2016, pp. 124ff.). As a result, numerous ethnic and cultural influences have come together in today's Louisiana, which are reflected in cultural customs (such as Mardi Gras), cuisine (e.g. Jambalaya, Gumbo), music (e.g. Zydeco, Dixie), and languages (such as Cajun, Creole etc.; Gaudet and McDonald 2003).

The American Civil War (1861–1865) with the defeat of the southern states led to great poverty among the majority of both white and black population groups, which was to consolidate for several decades through widespread sharecropping and the crop system (cf. Clark 1946). During the reconstruction of the state, despite the abolition of slavery and the constitutional amendment guaranteeing equality before the law and the right to vote, this period was marked politically by electoral fraud and corruption (see in detail Kelleher Schafer 2014): "Finally, during Reconstruction, Louisianans grew even more accustomed to fraudulent elections and corruption in politics, and this, too would have its degrading effects on the future" (Kelleher Schafer 2014, p. 226). For the Civil War did not lead to a renewal of the political-economic elites, but rather the pre-war battles amongst plantation owners, merchants, bankers, etc. continued to control the state long after the end of the war and reconstruction, whereby “that situation would not change until well into the twentieth century” (Kelleher Schafer 2014, p. 226). Until the 1930s, about sixty percent of Louisiana's population lived in rural areas, and the average income of about $286 was about $200 lower than the national average of $472 (Haas 2014, p. 321). Accordingly, despite the developing petroleum industry, which was to establish itself as Louisiana's dominant economic sector, the majority of Louisiana's inhabitants lived at or below the poverty line until the middle of the twentieth century (Haas 2014).

4.3 Current Spatial Contextualizations: The Regional Procedural and Structural Integration of Baton Rouge

Louisiana is thus facing enormous challenges, both physically-spatially and economically-socially, which – as will be shown in the following – culminate in a special way in Baton Rouge, its capital, or have a retroactive effect on Baton Rouge. One effect derives from the physical and spatial coastal erosion and the associated loss of land, which will be exacerbated by the consequences of climate change, such as rising sea levels and an increase in the intensity and frequency of extreme weather conditions, and which will change the living and housing situation of the inhabitants through extensive resettlement, particularly in the southern coastal areas, but also in the uplands and plateaus. In addition, as a result of decades of low political regulation of the oil and gas markets, oil and gas prices have fallen; gas production and processing, the production process and the disposal of its waste have to date caused severe environmental damage. This has not only contributed to high levels of pollutants in the soil, water, and air, but also threatened Louisiana's drinking water reservoirs through the ingress of salty seawater (Colten 2012), with the result that, against this background, the US American South “has undergone the transition from a region of surplus to one of emerging shortages” (Colten 2014b, p. 3).

Economically, Louisiana, and especially Baton Rouge, is dominated by the mineral oil and petrochemical industries, whose long-term future is questionable in view of the finite nature and foreseeable replacement of fossil fuels by new forms of energy production and requires new plans for the future. Socially, as will be explained later, a strong ethnic and social segmentation becomes apparent, which is reflected in the almost immediate coexistence of the highest and lowest human development indices. This is particularly evident in Baton Rouge (see in particular Chaps. 5 and 6).

5 Historical Aspects of the Urban-Rural Hybrid Landscape Biography of Baton Rouge

In his dissertation, published in 1963, Dieter Brill writes in the introduction: "At first glance, Baton Rouge does not appear to be a very big city, despite its 187,689 inhabitants. The extensive urban area is rather loosely built up, the central business district, the City [the term for the Central Business District used in Germany; notes O.K. and C.J.], shows only modest dimensions both vertically and horizontally compared to other North American cities of similar size" (Brill 1963, p. 15). Even today, in 2019, Baton Rouge – which has now grown to about 225,000 inhabitants, was rather loosely built up, while the revitalization efforts of the downtown area, which has been going on for about 20 decades and includes not only the Central Business District (CBD) but also the government district to the north of it (cf. also Brian 2003), are clearly tending to expand vertically as well as horizontally, which is also reflected in a modest URFSURBanization. Although the first documented human presence in the present-day Baton Rouge urban area dates back to about five thousand years ago, when Indians built two mounds on what is now the LSU campus, which served ceremonial or social purposes rather than housing (Draughon Jr. 1998; Rodrigue and Phillips 2011; see Fig. 5.1), it was not until the eighteenth century that the first signs of permanent settlement in the area were found.

In this chapter, the development of Baton Rouge will now be traced in essential features.

5.1 Moderate Settlement Formation: Baton Rouge Prior to The Annexation with the United States

Baton Rouge is located on the edge of the youngest Pleistocene terrace, the Prairie Terrace, which is cut by the Mississippi River and gently slopes southeast at about 0.5 m per kilometer. In the north of Baton Rouge, it reaches a height of about 21 m above sea

O. Kühne and C. Jenal, *The Multivillage-Metropolis Baton Rouge*, RaumFragen: Stadt – Region – Landschaft, https://doi.org/10.1007/978-3-658-30716-5_5

Fig. 5.1 Indian Mounds on the LSU site; own photo taken in 2019

level (Brill 1963). Independent of a certain agitation of the surface of the terrace – quasi following the blueprint of American urban development (see among others: Hofmeister 1971; Kaufmann 2005; Schneider-Sliwa 2005; Yeates 1990) – the city's ground plan is oriented towards the points of the compass, with the consequence that until the nineteenth century, when the city did not yet have an effective drainage system (let alone paved streets), lower-lying districts were often difficult to reach after rainfall and only accessible via temporary bridges and footbridges. While the central urban area was spared the floods of the Mississippi River because of its terraced location, the lower-lying districts were repeatedly flooded, especially as a result of heavy rainfall (Brill 1963). The vegetation in Baton Rouge and its surroundings before the settlement by European immigrants (and their descendants) was (and still is today in large parts) characterized by dense forestation (Brill 1963) due to the humid subtropical climate (Cfa, according to Köppen's climate classification).

'Baton Rouge' was already a topographical term before the settlement by Europeans. This goes back to the co-founders of the colony of French Louisiana, Pierre le Moyne, Sieur D'Iberville, and his brother Bienville, who, together with La Salle, co-founded the colony of French Louisiana. On an expedition in 1699, the place was so named, either after a red stake placed there by the natives or after a huge cypress tree standing there, both interpretations have been presented (Carleton 1981; Draughon Jr. 1998; Gleason and Brockway 1991). The name first referred to the terraced area with its steep bank facing the Mississippi and was later transferred to the settlement, which was influenced by Europeans (Brill 1963). The history of the settlement of and around Baton Rouge by European immigrants dates back to 1721, when the D'Artaguette brothers started farming there. At that time, about 700 Indians settled in the Baton Rouge area, subsisting through agriculture, fishing, and hunting (Rodrigue and Phillips 2011). However, the settlement efforts on the part of the French remained restrained, and even the assumption of power by the English as a result of the Seven Years' War in 1763 (after a brief Spanish intermezzo) changed little in the intensity of the settlement by Europeans, at first. A settlement with colonists of English origin did not take place until the 1770s. Under English rule, not only was slavery maintained, but the power of disposition of the slave owners was further extended, and the law of Roman tradition guaranteed only minimal slave rights, such as the right to buy one's own freedom: "Their mastery was absolute" (Rodrigue and Phillips 2011, p. 8). In 1779, as a result of the American War of Independence and the presence of the Spanish allied with the Americans on the west side of the Mississippi, 'Fort New Richmond' was hastily built. This fort was taken over the same year by Spanish troops after three hours of cannon fire, an element of the Spanish take-over of West Florida (Brill 1963; Carleton 1981; Rodrigue and Phillips 2011), the first of three battles of Baton Rouge, the other two followed in 1810 and 1862, which also completed the direct involvement of Baton Rouge in military conflicts (Carleton 1981).

The change of colonial rule meant considerable political and economic changes for the inhabitants: after the Spanish took over power, all inhabitants of the settlement were forced either to swear allegiance to Spain or to leave the settlement within six days – and with it most of their possessions. As a result of this ultimatum, most of the British-born population remained in Baton Rouge. The mode of local government also changed: the Spaniards did not allow elements of democratic self-government; the officials were appointed by authorities of the Crown. Trade was also restricted: it was not – as in the time under English rule – a de facto possibility with anyone, but was limited to Spanish ports and Spanish citizens, which limited the economic development capacity of West Florida and thus also of Baton Rouge. On the other hand, under Spanish rule there was no property tax, which in turn meant that landowners themselves were responsible for maintaining the local infrastructure (Carleton 1981).

The fort, which was renamed 'Fort San Carlos' by the Spanish, became the nucleus for the development of the urban settlement of Baton Rouge. From the Baton Rouge garrison, the West Florida District was administered by a governor and it was here that the trade between the northern and southern Mississippi ports was controlled. In addition, the flood protection of the terraced site made the area of today's Baton Rouge attractive for the development of civil settlement. However, from the 1790s onwards, this initially took place rather unsystematically, driven by different landowners and in differently localized settlement cells, from which a coherent settlement only gradually developed. The most important of these cells was the Beauregard Village, founded on the plantation of Elias Beauregard in 1806, which – influenced by the Spanish influence – had a radial ground plan (Brill 1963). The first settlement activities, which suggest a certain central-local importance, took place in the first decade of the nineteenth century, along today's First Street, but in the flood-prone portion; the proximity to the shore with the possibilities of trade seems to have been overwhelmed by the danger of flooding, as the further settlement development then took place eastwards on the terrace (Brill 1963). These developments are in the context of the trade of products of an intensifying agriculture, especially between Baton Rouge and New Orleans, where melons, pumpkins, beans, and tomatoes were grown whereas the 'German Coast' area in the west of New Orleans supplied corn. Towards the end of the Spanish era, cotton and sugar cultivation also intensified (Rodrigue and Phillips 2011) establishing an agricultural tradition that continues to this day, which in turn - until the American Civil War – was associated with an expansion of slave-based farming (Rodrigue and Phillips 2011).

Especially among the British-born population, Spanish rule was not widely accepted, the Spaniards were regarded as strangers and enemies, and Spanish law was considered a restriction on their own way of life. In addition, during the period of change from the 18th to the nineteenth century, tolerated by the Spanish governance, which expected an increase in prosperity from it, numerous immigrants from the United States settled in the Mississippi Valley. These enforced extensive trading rights (including the use of the port of New Orleans for US trade). West Florida came under increasing American influence (Carleton 1981). By the end of the first decade, a large part of the population, especially those of American and English descent, had grown weary of the corrupt Spanish rule and were striving for independence (Carleton 1981). But the end of the Spanish rule came sooner than circumstances would have suggested. On September 23, 1810, rebels stormed the 'Fort San Carlos' in Baton Rouge and overpowered it – without much resistance from the Spanish garrison soldiers. This 'second battle of Baton Rouge' cost the lives of two Spanish soldiers and injured four others, while there were no victims on the rebel side (Carleton 1981). Although the rebels sought to annex the West Florida district to the United States, they raised in place of the Spanish flag a blue banner with a white star in the middle, the flag of the Republic of West Florida, with 'capital' Baton Rouge, an entity which, however, three months later merged with Louisiana territory of the United States, making those of Baton Rouge American citizens, and in 1812 became part of the newly created state of Louisiana (Arthur 1935; McMichael 2008; Rodrigue and Phillips 2011).

5.2 The Gradually Developing Southern State City: Baton Rouge from its Integration into the United States to the Establishment of the Petrochemical Industry

After an unsteady beginning, the development of Baton Rouge consolidated with the takeover of power by the United States. Prior to the American Civil War (1861–1865), Baton Rouge had developed from a disorderly settlement with only a few hundred inhabitants, largely dependent on the Mississippi steamboat service which operated since 1814, into a major Southern city, with a designation as a state capital, and possessing a population differentiated by profession (Douglas 1955). With the annexation, the ban on the import of slaves from Africa, issued by the U.S. Congress in 1808, also took effect. Independently of this, the slave trade in the United States flourished and of the 1,463 people living in Baton Rouge in 1810, 675 were slaves. Female slaves worked in the small country town at that time as cooks, maids, nurses, or laundrywomen; male slaves worked as carpenters, vendors, painters, shoemakers, etc., while they cultivated cotton and cane sugar plants in the agricultural surrounding area, thus contributing significantly to the economic growth of the region (Rodrigue and Phillips 2008, 2011). Before the Civil War, Baton Rouge then had a population of about 5,500 (of which almost a quarter were slaves), only to grow to just over 11,000 at the turn of the century from the 19th to the twentieth century (Brill 1963; Rodrigue and Phillips 2011; Fig. 5.2). The growing population was reflected in the number of churches as well as their different cultural origins: in 1852 there were four churches in Baton Rouge, one Roman Catholic, one Episcopalian, one Presbyterian, and one Methodist. On the basis of the predominant language spoken during the church services, the increasing Americanization of Baton Rouge is also evident: while in the early 1830s the church services were held mostly in French, the English language dominated in the 1850s (Carleton 1981).

After the American seizure of power, the fort was formally upgraded and expanded between 1819 and 1836. The 'Pentagon Barracks', which have been preserved to this day and were built for auxiliary personnel, are a physical manifesto of these developments. The troop presence in Baton Rouge remained variable, however, since New Orleans, as the most important port in the region, was deemed to be more vulnerable to external enemies, and Baton Rouge – being located in the plantation area – was a more likely location for the suppression of possible slave uprisings (perceived in the sense of an "internal threat"; Brill 1963).

After the election of Abraham Lincoln as President of the United States in 1860, Louisiana's leaders decided on January 26, 1861 – less than fifty years after joining the United States – to leave it and join the Confederacy, a step that was taken the following March, one month before the beginning of the Civil War (Rodrigue and Phillips 2011). According to Carleton (1981), large parts of the population of Baton Rouge were critical concerning the threat of war with regard to withdrawal from the Union but attempts at

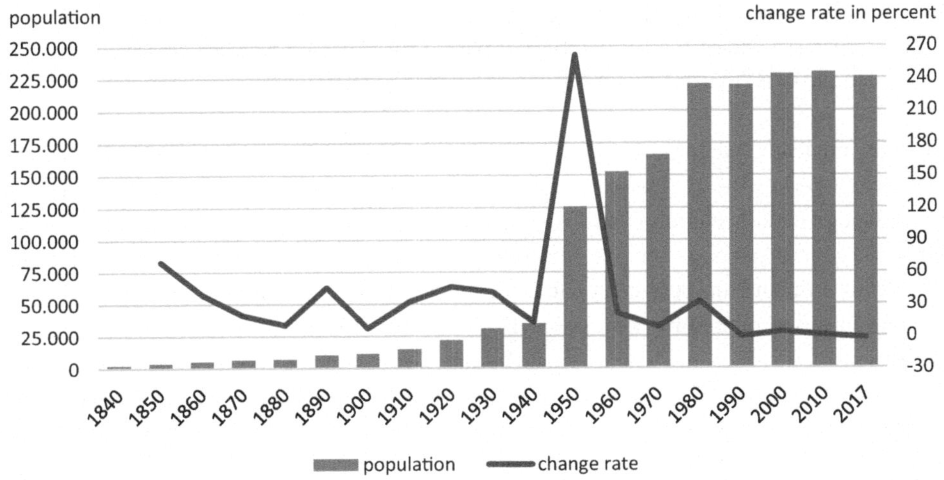

Fig. 5.2 The population development of Baton Rouge between 1840 and 2017, in absolute terms and as a rate of change between the reference years (own presentation and calculation, data basis: World Population Review 2019a)

mediation failed. In the following civil war, shortly after the conquest of New Orleans, Baton Rouge was taken by Union troops in 1862 without any significant resistance. After the failed attempt of reconquest by the Confederates in the same year, Baton Rouge lost its military importance after the war ended – at about the same time as slavery was abolished – and the garrison was disbanded, the fortifications were razed to the ground. In 1886, the area of the former fort was given over to the Louisiana State University (LSU), which was then under construction (Brill 1963).

The establishment of LSU proved to be one of those components of the settlement biography of Baton Rouge that more than compensated for the loss of its military function: as early as 1834, Baton Rouge became the location of the Louisiana State Prison (demolished in 1918), which – as a product of the work commitments of its inmates – quickly developed into a flourishing industrial enterprise. As a result of political rivalries, especially the mistrust of the rural areas towards the economic as well as political importance of New Orleans, it was decided in 1845 to move the Louisiana legislature from there (this separation of economic importance and political power is not a special case in Louisiana, but was systematically pursued throughout the United States; see Schneider-Sliwa 2005). Baton Rouge prevailed over numerous competitors, not least because, in addition to a prestigious building plot on North Boulevard with a view of the Mississippi, it was able to provide funds of $130,000 for the construction of the Capitol, which was then built between 1847 and 1849 in the 'castellated Tudor Gothic style' and cost a total of $396,000 to build (Gleason and Brockway 1991). In 1850 the parliament of Louisiana met here in its first legislative period. Today, a somewhat contemporary historical counterpoint to modernist high-rise construction, it dominated the view of Baton

Rouge from the Mississippi and its opposite bank in the second half of the nineteenth century (Gleason and Brockway 1991). Although New Orleans' desire for a permanent relocation grew after the end of the Civil War, including the Capitol having been partially destroyed by the war and by 1864 with the parliament again meeting in New Orleans, yet the 1879 state constitution established Baton Rouge as the capital of Louisiana. After the restoration of the Capitol, parliamentary and governmental activity were resumed in Baton Rouge in 1882 (Brill 1963; Rodrigue and Phillips 2011; Fig. 5.3).

The end of the civil war was linked to several challenges: firstly, politically integrating the formerly Confederate States into the Union; secondly, rebuilding the cities and economic enterprises damaged by the civil war; thirdly, providing education and support for the freed slaves. This reconstruction era was highly conflict-laden and is still linked to the north in large parts of the collective memory of white southerners as the genesis of their economic, political, and moral devaluation (Berg 2017; Hochschild 2016). The plan of President Andrew Johnson (1865–1869) included a recognition of the abolition of slavery by the ex-Confederates. However, starting in 1868, the Congress, dominated by radical Republicans, demanded the ratification of the 14th and 15th Amendment, which made the freedmen equal citizens of the USA and gave black men the right to

Fig. 5.3 The Old State Capitol as it presents itself today, after renovation, as seen from the Mississippi; own photo 2019

vote. However, a land reform in favor of the black population was not implemented. With the end of the occupation of the Southern States in 1877, the Reconstruction era ended and numerous achievements in favor of the black population were undermined by the old elites and the regime of white supremacy, a paradigm known as the Jim Crow system was installed, rooted in the popular stereotype of the stupid and lazy black. This was based on omnipresent racial segregation, political disenfranchisement, and economic exploitation of the black population, a system that even the Supreme Court supported with its separate but equal ruling of 1896 (Berg 2017; Foner 1995; Kolchin 2003). Even today, one and a half centuries after the civil war, cultural and social attributions are still being made that assign a special role to the Southern States, which contrast with the "'real' America with its Anglo-Saxon values, norms and rules to the ideal" (Gamerith 2017, p. 116) and are stereotyped by the 'real America' as "conservative, detached from the world, backward, agrarian, obsessed with the past, in love with guns and unteachably racist" (Gamerith 2017, p. 116).

In Baton Rouge, too, postwar years were not only marked by the loss of military and political significance, the war had also interrupted traditional trade relations, economic development remained subdued, numerous businesses were abandoned, and "crime and violence were common during the late 1860s, with fights, robberies, murders, drunkenness, and prostitution heading the list of convictions" (Carleton 1981, p. 100). The situation of the black population after the civil war also remained precarious, despite the efforts of the federal government to enforce the human and civil rights of the former slaves. Despite the opening of the first school for blacks in 1864 – the presence of Union troops having made that possible – opportunities for blacks remained limited because the white population refused to pay taxes for public schools, which would have benefited black children in particular, while the children of (mostly wealthier) white parents attended private schools (Rodrigue and Phillips 2011). Accordingly, the opportunities for professional advancement remained limited. Withholding education also made it more difficult for black people to organize themselves politically, with the result that they remained significantly underrepresented in political office, measured by their share of the total population. Threats and intimidation by the Ku Klux Klan also hindered the enforcement of the constitutional rights of the black population (Rodrigue and Phillips 2011). The liberation of slaves also had unintended side-effects in the social sphere: as slaves, blacks generally lived in the buildings of their respective slave owners; with their liberation, black and white sections of the population largely segregated themselves (Rodrigue and Phillips 2011). At the same time, the proportion of black population in Baton Rouge (and other Southern cities) increased. Liberated slaves from rural areas moved to the cities in the hope of finding work, so the proportion of black population in Baton Rouge rose from 32 percent in 1860 to 52 percent in 1870, and in 1880 59 percent of the population was black, with this shift also occurring concurrently with the departure of the white population from Baton Rouge (Carleton 1981; see also Rodrigue and Phillips 2008). The tide washing through the cities during the Reconstruction era, which shook with racial unrest throughout Louisiana, reached Baton Rouge on

November 7, 1870, an election day when Republicans, including several black candidates, won both houses of the legislature. During the riots, the exact cause of which is still unclear, armed groups engaged in shootings on the streets of Baton Rouge, killing four black people and injuring 20 others (black and white). The riots could only be ended with the use of the army and the imposition of martial law, which led to the arrest of 60 men, including only one black man, whose release was, however, only a short time in coming, as the investigation of the incidents came to a standstill. (Rodrigue and Phillips 2011). In 1876 white democrats regained power in the state and with the withdrawal of federal troops in the context of Louisiana's readmission to the Union on July 9, 1868, the Reconstruction era ended and "effectively silenced the black political voice for generations" (Rodrigue and Phillips 2011, p. 23). This development went hand in hand with increasing socio-spatial segregation, a widely tolerated lynch law against blacks, and thus an extremely divergent development of life chances (Rodrigue and Phillips 2011, among others).

Compared to the importance of state institutions, the economic importance of Baton Rouge remained rather modest in the last decades of the nineteenth century. No major companies in the secondary economic sector settled in the area, nor did the importance of central location in the areas of trade and services come close to that of the state as a whole in terms of politics, administration, and university education. A contribution to this restrained development was made, not least, by the late and rather peripheral integration into the emerging nationwide railway network. For a long time, the emerging supra-local traffic was handled via the Mississippi, at a time when steamship traffic on the river had clearly lost importance. It was only with the railway connections, which were built late between 1883 and 1903, that trade in the city began to grow (Brill 1963; Rodrigue and Phillips 2011). Industrial production in the second half of the nineteenth century was based on the processing of raw materials (wood, cotton, sugar) from the surrounding countryside or to meet the needs of the latter and of Baton Rouge itself (e.g. the manufacture of drainage systems, sugar mills, carts, agricultural equipment, and bricks). In the last decade, only the production of bricks and sawn timber as well as the production of oil, oilcake, and fertilizer from cottonseed reached supra-regional importance. These goods were transported mainly via the Mississippi River (Brill 1963). The state of politics, administration, and public infrastructure in Baton Rouge (as elsewhere in Louisiana and across the southern states) was worrisome at the end of the nineteenth century: corruption was rampant, the public school system was inadequate, the condition of the roads were poor, and poverty was widespread, especially among the black population (Carleton 1981).

Despite these political and administrative conditions, the downtown area developed into a regional business center and with the growth of the community in the last decades of the nineteenth century, the number of non-wooden buildings increased, especially in the central downtown area. At this time, Third Street developed into the central business street of Baton Rouge (Brill 1963). An expression of the (modest) prosperity of Baton Rouge was the opening of a tram line that ran west along Main Street, turned south into

Lafayette Street, east on Government Street, and then – following Dufrocq Street (now part of 19th Street) to the north – turned back onto Main Street, although the streets on which the tram ran were largely unpaved (Carleton 1981). However, these ceased operation as early as 1936 (Rodrigue and Phillips 2008), were considered too inflexible in view of the emerging automobile traffic, and – in view of the road space shared with automobiles – they also had no speed advantage (see in general: Culver 2017; Sachs 1989) (Fig. 5.4).

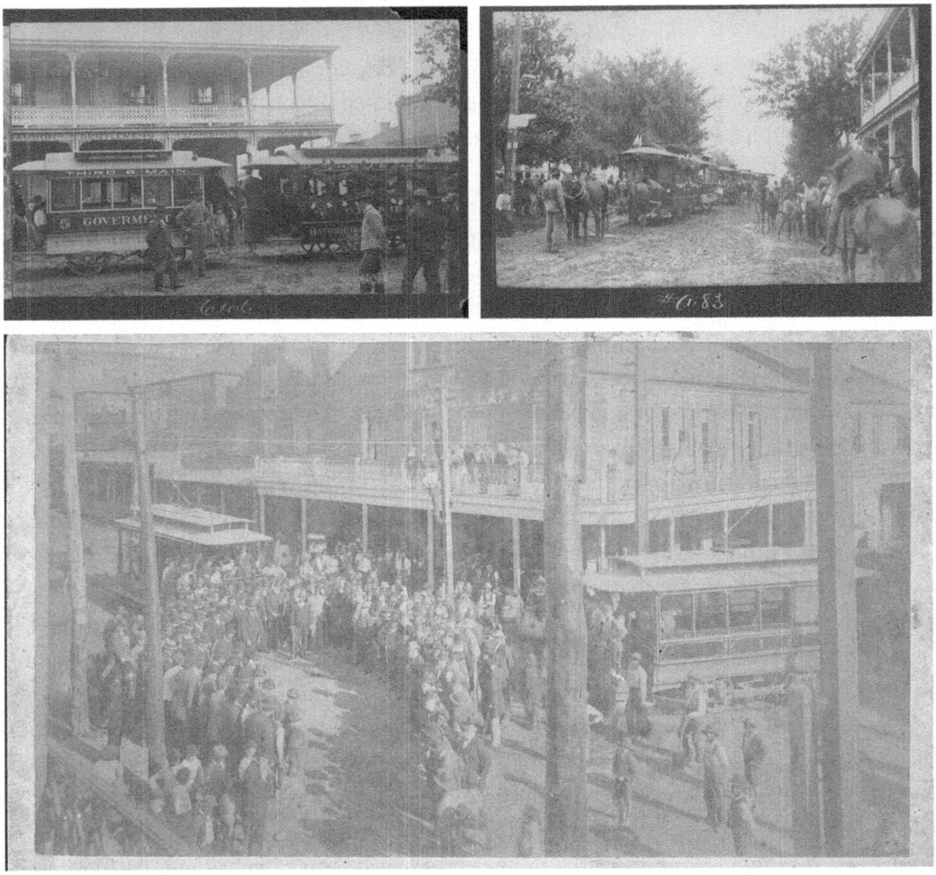

Fig. 5.4 top left: Celebrating inauguration of the mule-drawn trolley-system, 1890 [questionable] (image: Andrew D. Lytle Collection, Mss. 893, 1254, Louisiana and Lower Mississippi Valley Collections, LSU Libraries, Baton Rouge, La. Item Number 1254_C-106); top right: celebrating inauguration of the horse-drawn trolley line, 1890 questionable (Andrew D. Lytle Collection, Mss. 893, 1254, Louisiana and Lower Mississippi Valley Collections, LSU Libraries, Baton Rouge, La. Item Number 1254_A-083); below: first electric cars in Baton Rouge, 1890 [approximately] (image: Joseph St. Clair Favrot Family Papers (Mss. 3317), Louisiana and Lower Mississippi Valley Collections, LSU Libraries, Baton Rouge, Louisiana, USA. Item Number 33170005r)

The majority of the population was black. The white population of that time was mostly made up of the descendants of French, Spanish, and English immigrants, although the influx of American (white) population also intensified after the end of the Civil War. During this period, the cultural homogenization that – as shown – had begun before the Civil War continued. Of the cultural – also linguistic – diversity as a result of 'Americanization', it was primarily the name of the city (and some of its districts) and some of the names of its inhabitants that remained to remind us of the eventful settlement and territorial history. During this period – before mass motorization – Baton Rouge had a social gradient from downtown to the outside: the neighborhood expansions near the business center were inhabited by people with higher symbolic capital; short distances to the shops and offices located in the downtown area formed a basis for this preference for living. Beyond these residential locations close to the downtown area, which were almost exclusively inhabited by whites, a mosaic of neighborhoods was created, each with a dominant white or colored population; a strict separation according to neighborhood had not yet taken place (Brill 1963).

On the eve of the massive industrialization that began with the establishment of the petrochemical industry in 1909, Baton Rouge can be described as a tranquil riverside country town of regional economic importance, at best, and with a state-wide political and administrative as well as educational function. The population, which before the Civil War was still highly culturally differentiated, had largely 'Americanized' itself. There was no strict ethnic differentiation according to residential areas, except for the white, affluent neighborhoods around the economic and political-administrative center of the city.

5.3 The Phase of Forced Industrialization: Baton Rouge and the Oil and Petrochemical Industries

With the forced industrialization of Baton Rouge, a complete transformation began to unfold that "only a few elements of the city of the nineteenth century survived" (Brill 1963, p. 73). The development of Baton Rouge as an industrial city was and is strongly influenced by the regional oil and gas deposits. The city is located on the northern edge of the Gulf Field, a field that extends from the mouth of the Mississippi River to Texas and the Gulf of Mexico. The Caddo and DeSoto Oil Fields are located in the hills of north-west Louisiana and includes not only oil but also significant natural gas deposits (Jones and Bryan 1957). But it was not only the availability of regional raw material deposits that drove the industrialization of Baton Rouge, the city was also involved in the general industrialization of the southern states, especially in the first half of the twentieth century. This process integrated the southern states, which were characterized by monoculturally driven agriculture with an intense high labor input and low yield output – also through the expansion of technical infrastructures – into the industrialized world, which was not least associated with increasing prosperity for large parts of the population

(Brill 1963). This change can be attributed to several factors (Brill 1963; Walton and Rockoff 2014; Woodward 1981):

- The previously unused or only partially used raw material base in the form of forests and agricultural land – as well as mineral resources such as oil and gas, sulphur, coal and iron ore – were increasingly developed.
- The reservoir of labor at low wages was large because of the high birth rate in conjunction with the reduction of labor needed as an outcome of agricultural mechanization.
- The rising level of prosperity generated through industrialization, in turn, produced a stronger demand for industrially manufactured products, which in turn was (also) satisfied by increasing industrial production in the South.

In the first decades of the twentieth century, industrial production in Louisiana concentrated largely on the processing and onward transport, or the first stages of industrial production, of the regional raw materials, both from agriculture and forestry (such as cotton mills, sawmills and sugar factories), and from oil production (refineries). Baton Rouge developed into one of the industrial centers of Louisiana, along with New Orleans and Shreveport – to which Lake Charles was added after the Second World War (Brill 1963; Colten 2012).

The paramount year for the accelerated industrialization of Baton Rouge was 1909, when Standard Oil built a refinery in the north of the city (see also Fig. 5.5), which "abruptly created 700 new jobs and thus at least doubled the number of industrial employees" (Brill 1963, p. 81; see also Rodrigue and Phillips 2011), and simultaneously established the refinery as the nucleus for the later development of chemical and petrochemical industries. Two factors were essentially decisive for the Baton Rouge site (Brill 1963; Burby 2000; Granitz and Klein 1996; McMichael 1961): first, the orographic location on the flood-safe terrace, a factor that lost importance as flood protection techniques progressed – resulting in the industrialization of the former flood plains on the banks of the Mississippi River between New Orleans and Baton Rouge. Second, the favorable geographical location, both in terms of the relevant raw material deposits and in terms of transport possibilities; after all, the Mississippi River from the Gulf of Mexico to Baton Rouge is navigable by ocean-going vessels. Third, the presence of sufficient water for manufacturing processes. Fourth, possibilities for the disposal of production waste in the area. Fifth, moderate winter temperatures, which made extensive frost protection of the production processes seem obsolete. Sixth, the political climate in Baton Rouge was very acclimating toward Standard Oil, which was very different from other locations where the company's rigid business practices were criticized.

With the construction of the Standard Oil plant, a self-reinforcing development began, with the establishment or expansion of infrastructures such as port facilities, pipelines to the crude oil production areas, the expansion of the local roadways, the connection to the nationwide road network and the national air transport network, as well as the production

Fig. 5.5 Standard Oil plant in northern Baton Rouge at the turn of the 1950s and 60s. (From: Brill 1963, p. 255; courtesy of the Institute of Geography of the University of Kiel)

of electricity, combined with expansions of the refinery, giving rise to further production facilities in the chemical and petrochemical industry (Brill 1963). At the same time, the oil refinery remained the industrial core of Baton Rouge: on the eve of the entry of the United States of America into the Second World War, in 1940, one third of the population of Baton Rouge (about 34,700 inhabitants) earned their incomes from the refining of oil (Carleton 1981). The establishment of the Standard Oil plant also meant an expansion of the residential area from Baton Rouge to the north. This expansion was carried out according to the settlement pattern that is characteristic of Baton Rouge – in principle since the beginning of settlement – which is less dense and thus takes up large areas (Brian 2003).

From the petrochemical industrial base at the beginning of the twentieth century, the three development paths of Baton Rouge that are still dominant today are now completed, these began in the second half of the nineteenth century with the selection of the site for state policy and administration plus the establishment of the LSU.

With the creation of new jobs, the population of Baton Rouge also increased; in 1910, the city had almost 15,000 inhabitants, and in 1920 it reached almost 22,000 (Fig. 5.2). Accordingly, new residential areas were built, such as the Garden District and Kleinert Terrace, whereby these areas – in comparison to later developed neighborhood extensions – varied in size and architectural style (also on a small scale): shotgun houses were built, as were Craftsman Bungalows, English Cottages, and houses in the Colonial Revival style (Armstrong 2010; Carleton 1981; Rodrigue and Phillips 2011). With the exception of the

Fig. 5.6 Baton Rouge at the turn of the 1950s/60s. (From: Brill 1963, p. 253; by courtesy of the Institute of Geography of the University of Kiel)

black population, the districts were separated along economic lines, because ethnic dividing lines, "for example, Italy, Cajuns, Whites, and East Asians often lived together in a neighborhood" (Rodrigue and Phillips 2011, p. 35). African-Americans, on the other hand, settled primarily in the independent Scotlandville, the Park, and the Old South Baton Rouge, located between downtown (see Fig. 5.6) and the LSU campus (Rodrigue and Phillips 2011).

However, the integration of Baton Rouge into the national railway network remained restrained even during the phase of accelerated industrialization and did not reach the level of far less politically and economically important cities such as Alexandria, Opelousas, or Lake Charles by the middle of the twentieth century (Brill 1963). The opening of Huey P. Long – O.K. Allen Bridge (in Baton Rouge, however, primarily known as the 'old bridge'), a combined railway and road bridge, in the north of Baton Rouge in 1940, did little to change this; before that, the Mississippi River was crossed by ferries (cf. Fig. 5.7), which could be associated with considerable delays (Rodrigue and Phillips 2011). The first bridge crossing the Mississippi north of New Orleans at this time was of greater importance for local rail traffic, between Baton Rouge, on the eastern bank of the Mississippi, and the 'Port of Greater Baton Rouge' on its western bank, as well as for the rapidly growing automobile traffic (Brill 1963).

Rapidly developing automobile traffic has had a significant impact on the settlement structure of Baton Rouge, as in other cities in the United States and beyond.

Fig. 5.7 Ferry traffic in Baton Rouge across the Mississippi in the period around the turn of the 1950s and 60s. (From: Brill 1963, p. 255; courtesy of the Institute of Geography of the University of Kiel)

With the suburbanization of people with a higher availability of symbolic capital, the social gradient was reversed: the residential areas around the city center – no longer in line with the current architectural fashion, the then current state of the art in building technology, and the demands regarding the size of the plot of land – increasingly had a population with a low level of symbolic capital, while people with a higher level of symbolic capital settled in the suburbs which were developing specifically in the east and south of the city. At the same time, the segregation between black and non-black continued to increase: the black population was concentrated between downtown and LSU, east of downtown and east of Southern University. Whites dominated the rest of the city (Brill 1963). Although the bipolar development of LSU ended with the opening of the new 120-hectare campus in the south of the city, a bipolarity of higher education remained in Baton Rouge: Southern University (SU), founded in New Orleans in 1879, was moved to Baton Rouge in 1914. In 1963, Brill (1963, p. 93) described SU as "a small and less perfect counterpart to L.S.U. for the colored population", a characterization that, given the relationship between the condition of the buildings, the number of students, and the color of the students and faculty, has persisted to this day (see Sect. 6.3 for more details).

In comparison with other American cities, Brill (1963, in evaluation of the data obtained by Bartholomew 1945–1948) attested four special developments of Baton Rouge:

1. The area taken up by industry and public administration proved to be above average
2. The Central Business District of Baton Rouge, conversely, was small both in size and functional importance.
3. As a result of the low level of development of the transport infrastructure, a comparatively polycentric settlement structure developed, which was oriented towards the respective major employers (LSU, CBD, administration, and the petrochemical industry).
4. The rapid growth of the city with a strong importance on land speculation combined with a limited influence of spatial planning has led to a fragmented and disparate use of land.

The growing importance of public administration in the first half of the twentieth century is manifested in the construction of the new Capitol under Governor Huey P. Long, known as the 'Kingfish', in 1931–1932, north of downtown, and other government buildings in its vicinity – a development that continues to the present day. The special importance of public administration for Baton Rouge is symbolized to this day by the fact that the Capitol, with its 140 m of height, is still by far the tallest building in Baton Rouge, the next tallest building, 'One American Place' built in 1974, is topped by the Capitol by 46 m (EMPORIS n.y.). The new Capitol was erected as the highest in continental America (Gleason and Brockway 1991; cf. Fig. 5.9, top left), which jointly expressed the increased self-confidence of the state of Louisiana and the ambitions of the governor who was in charge of its construction. This could also be said for the establishment of a new Governor's Mansion, whose antebellum predecessor he had demolished in a cloak-and-dagger operation in 1929 (Rodrigue and Phillips 2008; see also Fig. 5.8 for Governor's Mansions).

The lawyer Huey P. Long, who – at that time thirty-four years old – was elected to the office of governor for the second time in 1928, developed into a figure who decisively influenced the development of Baton Rouge. The charismatic Huey P. Long was certainly one of the most contradictory figures in the history of Louisiana, if not the United States. He was considered self-confident and charismatic, but he was also "aggressive, determined, blunt, and often unscrupulous" (Carleton 1981, p. 164). He centralized political power in the executive branch, while also pushing through important reforms for Louisiana (Hair 1996; Williams 1969). The expansion of state administration and investment in public infrastructure that he had initiated was in the spirit of the New Deal, in its response to the Great Depression following the stock market crash of 1929, with the opening of the new Municipal Airport at the eastern end of Government Street in 1931 and work on the aforementioned Huey P. Long – O.K. Allen Bridge in 1932 over the Mississippi River in the northern part of the city (see Fig. 5.9, below). In addition to the construction having provided work for nearly 1,500 people, this period of his administration also saw the construction of a post office building and a courthouse, he initiated investments in education and healthcare, as well as road construction in Louisiana (Rodrigue and Phillips 2011). In 1935, the Democrat was elected senator and planned to

Fig. 5.8 Governor's Mansions in chronological order; left in the 1920s © Louisiana Collection State Library of Louisiana; The Old Governor's Mansion right in the 1930s © Louisiana Collection State Library of Louisiana. Built in 1930, the building served as the residence of the governors and their families until 1963, when it housed the Louisiana Arts and Science Museum for a time and in 1998 the Foundation for Historical Louisiana moved in, an organization whose mission is the "preservation of our state's cultural and architectural heritage through education, advocacy and stewardship" (Preserve Louisiana 2019, o. S.; above). The current Governor's Mansion follows the Greek Revival style and is located east of the Capitol (see also: Rodrigue and Phillips 2008)

run for president in 1936; his political program of income redistribution included a guaranteed basic income for families, which he wanted to finance through higher taxes on the wealthy (Brinkley 1981; Jeansonne 1980; Lee 2006). He was controversial not least because of his autocratic style of leadership, his patronage of offices, and his rude treatment of political (and other) opponents. Long's ambitions came to an abrupt end on 8 September 1935 when he was the victim of an assassination attempt. The assassin was presumably the ear, nose, and throat doctor Carl Austin Weiss from Baton Rouge, who in turn was shot by Long's bodyguards. On September 12, 1935, Long was buried in the presence of about 175,000 people (Rodrigue and Phillips 2011; see also Fig. 5.9, top right, for the tomb of Huey P. Long). His rigorousness extroverted way of life, and his sudden death made Huey P. Long an almost exemplary model for literature and film (Perry 2004).

Thirteen years later, Huey P's brother Earl was elected Governor of Louisiana and Russell B. Long, Huey P's son, was elected Senator. Earl continued his brother's politics and political style, built the social and educational system, invested in infrastructure. To finance this expenditure, taxes were increased with the tax burden on citizens reaching 50% at the end of his term of office. As a result of the electoral law at the time, he was not allowed to be re-elected after one legislative period, but he ran for election in 1956 and won again. He was elected to the U.S. House of Representatives in 1960 but died before he could take up his mandate. Certainly, one of the greatest achievements of his tenure was the reform of the electoral law, which increased the ability for black people to exercise their right to vote (Rodrigue and Phillips 2011; see also Carleton 1981).

The Second World War had consequences for Baton Rouge, as it did for large parts of the United States, but more so for the city itself: in the first case, the recruitment of

Fig. 5.9 Above: Capitol and Huey P. Long Tomb and Memorial Statue at Capitol's Ground; below: HPL-Bridge; own photographs 2019

soldiers, but also bottlenecks in the supply of daily necessities such as coffee, pepper, meat, etc. and the rationing of gasoline, tires, and sugar. In the second case, the arrival of an aluminum plant and the accelerated expansion of industry (for example, the production of weapons and fuels; Rodrigue and Phillips 2011), which transformed the 'sleepy little river town' into a 'boom town' (Carleton 1981; cf. also Fig. 5.10), thus boosting its population by 261.8 percent to almost 126,000 between 1940 and 1950 (Fig. 5.2). The booming economy has meant a motorized boost for Baton Rouge, coupled with the expansion of roads in terms of quantity and quality (many gravel roads now having a solid surface; Rodrigue and Phillips 2008).

Although recruitment to the army and greater integration into the war economy meant an increase in symbolic capital for the black population (in Baton Rouge, as elsewhere), they remained 'second-class citizens' (Rodrigue and Phillips 2011), whose

Fig. 5.10 Baton Rouge upswing in the 1950s. (From: Brill 1963, p. 259 (top left), p. 277 (top right), p. 259 (bottom); courtesy of the Institute of Geography of Kiel University)

neighborhoods were highly segregated, with concentrations in Scotlandville, north of the Baton Rouge city limits, and north and south of Florida Street, plus east of downtown (Fig. 5.11). Soon, in the 1960s, there was growing resistance from the black population of Baton Rouge as part of the civil rights movement. The anger of large sections of the African-American population erupted in the summer of 1969 after two young black men were shot in the back by police and peaceful protests at the site of the second shooting escalated into a riot that was eventually suppressed by the National Guard. However, these events increased mistrust of the authorities in general and of the police in large

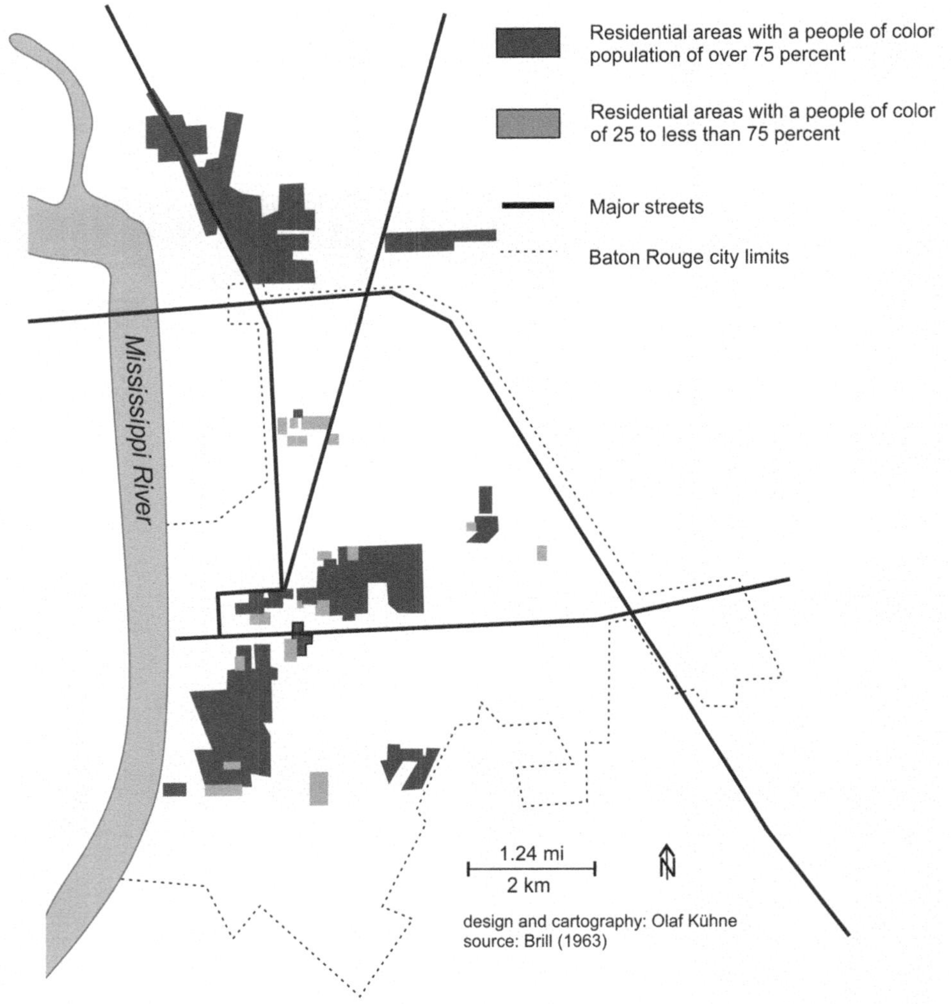

Fig. 5.11 The ethnic grouping of Baton Rouge 1959 (own representation according to: Brill 1963)

parts of the black population (Hebert 1999; Rodrigue and Phillips 2011). By the early 1970s, "The civil rights movement in Baton Rouge ended segregation, led to African-American participation in government, and opened jobs to blacks" (Hebert 1999, p. 359). What has not changed to date, however, is an almost dichotomous separation of residential neighborhoods in Baton Rouge, extreme inequality of life chances between African-American and white segments of the population as a result of high crime rates, lack of educational opportunities, and high poverty in the predominantly black parts of the city (for more details see Sects. 5.4 and 6.3; see also Richard 2006).

The boom of Baton Rouge, which had begun during the Second World War, continued on afterwards without a break and punished all three poles of the 'tripole' metropolis: the combination of suburbanization and mass motorization including the ever-increasing importance of air travel demanded increasing amounts of fuel, the incipient expansion of education (including college places for war returnees) called for the enlargement of higher education institutions, and the increase in governmental responsibilities in a state with a growing population meant the expansion of public services (see also Carleton 1981). The population focus of Baton Rouge was aligned with these three main areas of job development (Fig. 5.12): in 1958, the area of highest population density in Baton

Fig. 5.12 The population density of Baton Rouge and its environs in 1958 (own representation according to: Brill 1963)

Rouge was roughly located between the Exxon Plant and LSU, and east of downtown. Especially along the major roads (particularly Airline Highway, Scenic Highway, and Florida Boulevard, less Plank Road), clear suburbanization tendencies can be seen. By 1970, Baton Rouge had become the second most populous city in Louisiana with nearly 166,000 inhabitants, and its integration into the state's transportation system, particularly via interstate roads, is well advanced. In particular, the Horace Wilkinson Bridge (better known in Baton Rouge as the 'Interstate Bridge' or 'new bridge'), completed in 1968, which crosses Interstate 10 south of downtown over the Mississippi River, has made Baton Rouge a hub of Louisiana's road traffic, with an associated concentration of air pollutants and noise. The technical and social infrastructure (e.g. in the form of new hospitals, schools, the expansion of universities) experienced significant growth. However, it should soon become clear that the economic development path of dependency on industrial development also brings forth considerable problems in terms of economic structure, urban development, and the state of the environment.

5.4 The Moderate Post-Modernization: Baton Rouge's Diversified Developments

In 1972, another dazzling governor was elected to the capital, the democrat Edwin W. Edwards. Edwards, who is often compared to Huey Long for his power orientation and populist policies, served a total of four terms of office (1972–1976, 1976–1980, 1984–1988, and 1992–1996). During his first term of office – with the State of Louisiana benefiting from the oil boom during these years – he lowered taxes and pursued a diversity-sensitive policy, which was also expressed in the appointment of numerous African-Americans and women to his cabinet. Denied a third consecutive term by the Louisiana Constitution, he left office in 1980 but returned in 1984 to beat his 1980 successor, David C. Treen – who was the first Republican in over a century. However, his third term in office was befallen with misfortune: as a result of increasing global oil and gas production combined with Ronald Reagan's deregulation policy, the price of oil fell causing the state's tax revenues to shrink and the subsequent general tax increases diminished his popularity greatly. As a result of the loss of jobs in the petrochemical industry, the population growth rate of Baton Rouge has been in decline since the mid-twentieth century (Rodrigue and Phillips 2011). An indictment against him for fraud, obstruction of justice, and bribery in connection with the award of contracts for state hospitals did not lead to any convictions, but did reduce his chances of re-election to such an extent that he decided not to run for re-election. Although he was perceived to have slim odds, he again won the governorship for the period 1992–1996. He owed his election less to his persuasiveness than to the fact that he was considered the lesser of two evils by the electorate of Louisiana: his opponent was the Republican David Duke, the former leader of the Ku Klux Klan and considered unelectable even by many prominent Republicans. The alternative candidates, Duke and Edwards, were commented upon by numerous drivers in Louisiana who adorned their cars with the bumper sticker: "Vote for the crook, it's important!" (Internet Archive n.y.). During his last term of office, Edwards pushed through a project he had been pursuing for some time: the (strongly regulated by the state) legalization of gambling. As a result, he not only built a casino in New Orleans, he also issued licenses to operate casinos on riverboats, two of them in Baton Rouge – the Belle of Baton Rouge and Casino Rouge (Fig. 5.13). Two years after his last term of office expired, he was convicted of racketeering, extortion, money laundering, and fraud which resulted in a prison sentence (Rodrigue and Phillips 2011). However, his time in prison did not make his political ambitions disappear, in 2014 he ran for election to the U.S. House of Representatives, which he lost (Biographical Directory of the United States Congress n.y.).

Murphy J. Foster III, the Republican who succeeded Edwards, was Governor of Louisiana for two terms. His style of government was more businesslike than that of his predecessor. His term of office, however, changed the physical shape of Baton Rouge downtown intensively: during his term of office, the restructuring of the government district was approved and initiated. In 2003, three old government buildings were

Fig. 5.13 The 'Casino Rouge', one of two casino ships in Baton Rouge, whose shape and design are reminiscent of nineteenth century Mississippi shipping and are firmly connected to the land-based berths. This can be understood as postmodern (unintended) irony, just as design can be understood as a postmodern simulation of the historical; own photograph 2019

demolished, and in 2006 work on the new buildings was completed. The northernmost one (Poydras Building) is located on the northern edge of the Capitol Lake, the others are concentrated around the intersection of North Street and Third Street, south of the Capitol. These buildings have a height of up to twelve stories and thus contributed to a stronger vertical structuring of the downtown area. This publicly funded transformation of the downtown area had already begun south of the CBD in the wake of the oil boom of the 1970s. In 1975, the Louisiana Art and Science Museum moved into the abandoned Yazoo and Mississippi Railroad Company station and across River Road, a twelve-block site was transformed into the Riverside Complex – this included a ten-story administration building, an arena, a library, and parking garages. In 1979, the majority of the city's administrative offices and those of East Baton Rouge moved into this complex (Rodrigue and Phillips 2011; see Sect. 6.1 for the development of the downtown area). In 1982 the decommissioned destroyer, 'USS Kidd' from the Second World War, became part of the Naval War Memorial at the western end of Government Street. Since the turn of the millennium, an intensified development of the downtown area has taken place, primarily by the public sector, but also by private investors: in 2001, the Sheraton Baton Rouge Convention Center Hotel opened, in 2003, the Louisiana Art and Science Museum was expanded to include a planetarium, and in 2004, the Baton Rouge River Center opened – the center included a convention center, an exhibition hall, a theater, an ice skating arena, and a ballroom. In 2005, the Shaw Center for the Arts on Lafayette Street opened its doors. In addition to the LSU Arts and other museums, it included a rooftop restaurant overlooking the Mississippi River (Brian 2003; Rodrigue and Phillips 2011; see also Fig. 5.14 for more intensive downtown development). Thus, since the 1970s, urban development dynamics

Fig. 5.14 Intensified downtown development especially since the 2000s. Top row left: Downtown North Blvd.; center: River Center Branch Library; right: Rooftop Restaurant. Bottom row, left: Raising Cane's River Center, center: 'USS Kidd', right: Planetarium and Louisiana Art and Science Museum; own photographs 2019

had developed bipolarly, north of the CBD, supported by the administration of the state of Louisiana, and in the southwest, supported by the administration of the Parishes, the city, as well as public institutions of the State of Louisiana and private investors.

Louisiana's depression in the 1980s slowed the development of Baton Rouge downtown, but with the initiation of the new Baton Rouge Downtown Development District, efforts to revitalize Baton Rouge downtown intensified. In the context of these efforts is the renovation of the interior of the Old State Capitol in the early 1990s and the façade in the mid-1990s, as well as the renovation of the Old Governor's Mansion in the 1990s (Rodrigue and Phillips 2011). This renovation work can also be understood with the emergence of an appreciation for the historical in the context of social postmodernization processes. However, this historical appreciation is particularly evident where the pressure for change (whether economic or political-administrative) is minimal, such as in functionally replaced public buildings of high symbolic value (the examples cited above) or in suburban historical quarters with a higher level of symbolic capital and residents who follow more postmodern values, such as Spanish Town and Beauregard Town, which were being entered into the National Register of Historic Places as early as the 1970s (see Gleason and Brockway 1991; Rodrigue and Phillips 2011). Far less successful was the conservation of historical buildings in the CBD area. As a result, numerous historical buildings gave way to multi-story car parks, office buildings, and apartments. So the Paramount Theater on Third Street, a landmark of Baton Rouge, was demolished in 1979, after presenting its last show to make way for a parking garage, and again in 2003, the old Baton Rouge Ice House on Main Street, one of the few antebellum buildings in Baton Rouge, was demolished. It was replaced by a condominium building (Rodrigue and Phillips 2011; the effects of these changes on the settlement structure are presented in Sect. 6.1.1).

In 2005, Hurricanes Katrina and Rita hit Louisiana, and neighboring states, permanently altering Baton Rouge, although largely being spared the direct effects. Weather-related catastrophes had already occurred earlier, such as the precipitation-related floods in and around Baton Rouge in April 1983, which flooded lower-lying parts of the urban–rural hybrid, also Hurricane Andrew in August 1992, whose peak wind speeds were lower than those of Hurricane Betsy in 1965, but, however, whirled much longer above Baton Rouge and thus produced higher losses from both the extended fall of precipitation and the constant destructive winds as well as the higher amounts of property damage, yet the consequences of Katrina and Rita were without comparison. When Katrina hit land on August 29, 2005, the storm destroyed numerous towns along the Mississippi coast and flooded downtown Mobile, Alabama. The extensive damage, however, it caused in New Orleans, where it completely flooded the Ninth Ward and other quarters was the most severe. By this time, thousands of New Orleans residents had already followed the National Weather Service's warning and fled the storm, including to Baton Rouge, where they stayed in hotels, with friends, and relatives. There they experienced how the storm in Baton Rouge caused considerable property damage and loss of power, fortunately the people in the state capital survived the storm without serious injuries or any deaths. The situation was different in the New Orleans' Parishes, which at that time were about 80 percent flooded. More and more people from New Orleans sought refuge in Baton Rouge (and other Louisiana communities), where emergency shelters were provided in public and private buildings. The longer the people from New Orleans were unable to return to their city, the more tense the situation became, although initially marked by the helpfulness of the people of Baton Rouge. Food and fuel supply shortages, coupled with price increases, and the additional amount of cars provoking increased traffic congestion, all causing everyday life to become more difficult, and subsequently compounded with the emergence of rumors that the notorious New Orleans gangs were now carrying out their criminal activities in Baton Rouge – fear and rejection of the Katrina victims spread (Rodrigue and Phillips 2008, 2011). Developments that are also reflected in the image of Baton Rouge by New Orleans residents (see Textbox 5.1).

Textbox 3: The view of regional rival New Orleans: '10 things you hate about Baton Rouge'

As a result of Baton Rouge's low public profile, Diana Samuels, managing producer of the online platform NOLA.com and the daily newspaper The Times-Picayune, asked her readers why New Orleans residents had such strong animosities towards Baton Rouge. The results from "500-plus comments, plus a few hundred more on Facebook" (Samuels 2015 n.p.), presented in a 'top ten' format which is popular in the United States, certainly shows the animosity of the topic – typically strongly focused and overstated – yet incomplete being expressed as phrase-oriented communications. Some aspects of these will be explained in more detail later.

1. *Traffic, especially frequent traffic jams (which is certainly due in part to the post-Katrina experience invoking a temporary presence of New Orleans population, as has already become clear in this section).*
2. *Poor urban planning, particularly excessive urban sprawl, partly linked to traffic because of the congested road system (here the traffic issue from point 1 reappears, this issue is dealt with in more detail in Sect.* 6.1.3*).*
3. *The dominance of restaurant chains (which is certainly also due to the low-income level of the population in Baton Rouge; this is discussed in more detail in Sect.* 5.5*).*
4. *The presence of LSU, especially the 'LSU Tigers' football team, combined with the widespread use of the color combination 'purple and gold' in and around Baton Rouge (this may also be because of the fact that in cities with under a million residents the local to regional sporting identification anchors tend to appear in the singular; a topic that is taken up repeatedly in the book, but specifically in Sect.* 6.3*).*
5. *Katrina criticism, where respondents were disappointed by the distrust displayed towards them by the people of Baton Rouge in the post-Katrina era*

(although the author points out that the help from the Baton Rouge people was considerable, as this section also shows).

6. *Smelly and polluted, in Baton Rouge an aspect of the strong presence of the oil and petrochemical industry (especially in the northern parts of the city; this topic runs throughout the book, but a focus is given in Sect.* 6.2*).*
7. *The high rate of HIV infections, Baton Rouge is to be found in second place among the new infections in the nationwide comparison, New Orleans, according to the author's own indication, is barely behind it in fourth place (the reasons for such high infection rates are manifold and are often connected with a low endowment of symbolic capital; a topic that appears again and again in the book, but especially in Sect.* 5.5*).*
8. *A dead downtown, which is characterized by a predominance of office workers in particular and less so by residents (a trend that can also be found in downtowns of other cities, but which takes on specific characteristics in Baton Rouge; this topic also runs through the book, but with a focus in Sect.* 6.1*).*
9. *Baton Rouge is a city without culture, lacking 'distinctive character', as is to be found in New Orleans or Lafayette (although, according to the author, parts of the city, such as Spanish Town, could certainly set new cultural accents; this topic is also dealt with in different places, with a focus in Sect.* 6.1*).*
10. *Treatment of LGBT* residents, Baton Rouge is not known for its tolerant treatment, especially not by the police (although, according to the author, the city legislature has adopted regulations against discrimination).

When the anger of the Baton Rouge residents (both residents and 'refugees') began to turn against the authorities, Louisiana was hit by a second hurricane: Rita. Like towns in eastern Texas, Lake Charles was evacuated, and also from there were many of the people who had fled Katrina and its aftermath. In Baton Rouge, Rita brought a heavy rainfall that lasted from noon on 23 September until noon on 24 September, causing more damage than Katrina's storms (Rodrigue and Phillips 2011).

5.5 Baton Rouge Today: (Spatial) Statistical Approximations

The population of Baton Rouge has stagnated since the economic boom of the 1970′s on into the early 1980′s with population varying between just under 220,000 and just under 230,000 inhabitants (Fig. 5.2). A major reason for this is the ongoing suburbanization, which is evident in the ratio of the population of Baton Rouge and the East Baton Rouge Parish minus Baton Rouge (Fig. 5.15): While in 1950 (after extensive incorporations) approximately 80% of the population of East Baton Rouge Parish still lived in the city of Baton Rouge (after extensive incorporations), this was only approximately 60% in

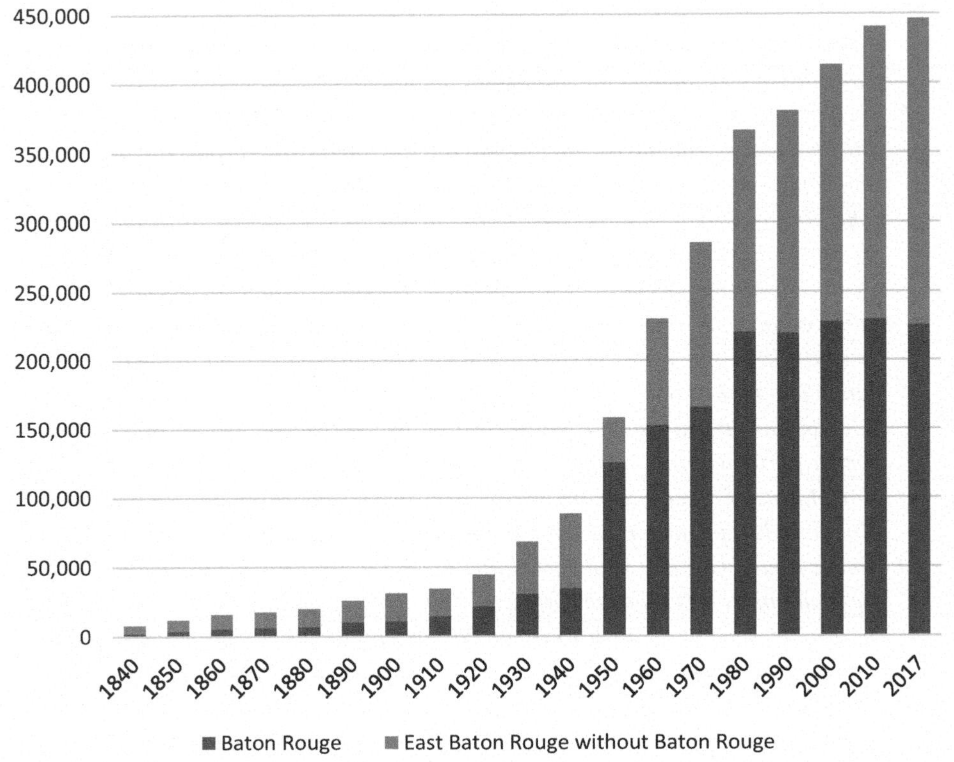

Fig. 5.15 The ratio of the population of Baton Rouge to the population of East Baton Rouge Paris without Baton Rouge (own presentation based on: World Population Review 2019a, b)

1980 and only 50.2% in 2017. The surrounding communities also increased more from the influx of Katrina victims (which is rather small in the long run) than the city of Baton Rouge itself. Historically, the data graphically presented in Fig. 5.15 clearly show the more rural–urban function of Baton Rouge up to the industrialization since 1909: up to that time, the population of Baton Rouge in the Parish was at most about 40 percent. More abstractly formulated, the data illustrate the relatively late onset of the trend towards industrialization and urbanization, while also indicating the trend towards reurbanization, which will be discussed in more detail later (Sect. 6.1) and which has not yet fully taken hold.

Compared to the population density distribution of Baton Rouge in 1958 (Fig. 5.12), the population density distribution in 2015 can largely be described as condensed (Fig. 5.16). The population concentration between the Exxon Refinery and LSU, with a bulge east of downtown, has largely disappeared. There are some more densely populated areas, for example, in the western part of the triangle involving Florida Street/Boulevard, Interstate 110, and Airline Highway, in the area of LSU, and also in suburban neighborhood sections, such as found around Siegen Lane and along Florida Boulevard, east of the intersection with Airline Highway. The low density of the resident population in the

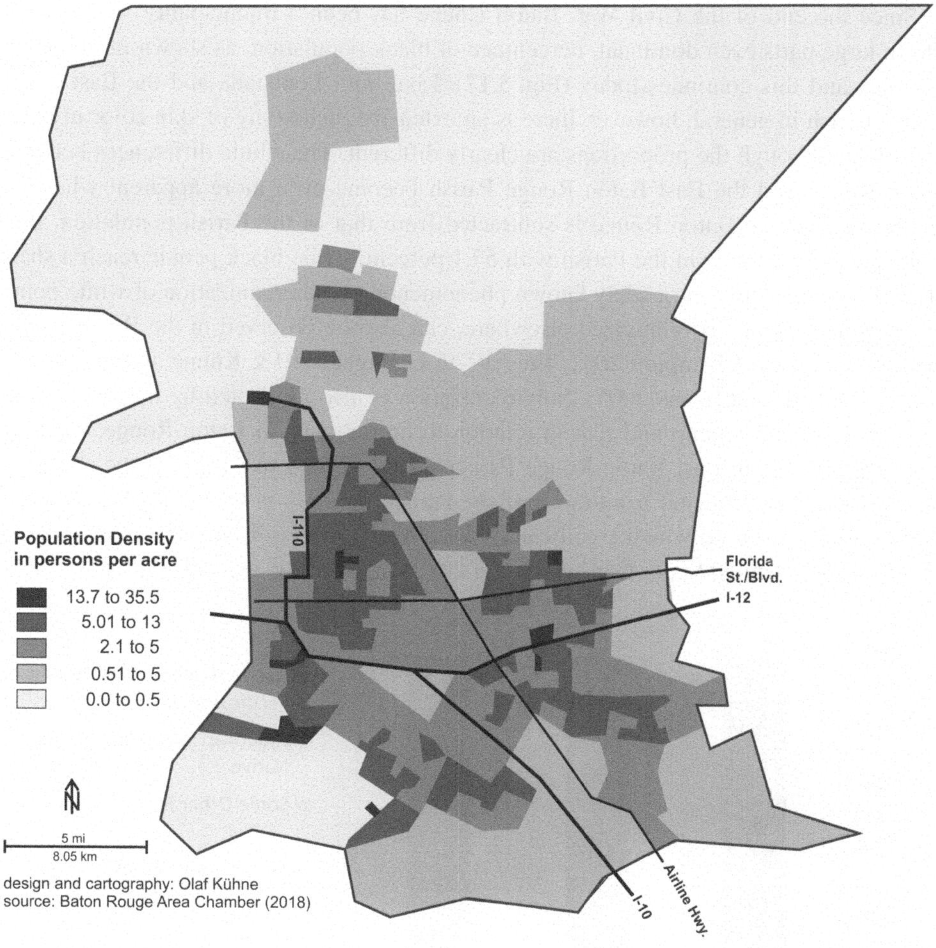

Fig. 5.16 Population density in the East Baton Rouge Parish, 2015 (after: Baton Rouge Area Chamber 2019)

vicinity of the Central Business Districts is due to the small number of multi-story (also historical) residential buildings in this area. The settlement of Baton Rouge was largely based on single-story (suburban) buildings, so that the population density here remained quite low overall. A higher population density can be found where the density of buildings is higher (east of the I-110), in particular due to the low level of economic capital, or where multi-story buildings (often apartments) are being constructed, although this is a more recent trend that is manifesting itself in areas eccentric to the city center, apart from a few buildings in and around the city center. To put it succinctly: even the trend towards reurbanization (in terms of architecture and lifestyle) is taking place in the spatial pastiche of Baton Rouge, primarily in a decentered suburban setting.

Since the end of the Civil War, Baton Rouge has been a municipality with a high, and in large parts even dominant, percentage of black population, as shown in the previous one, and this continues today (Fig. 5.17). Like both Louisiana and the East Baton Rouge Parish in general, however, there is an extensive dichotomy of skin color in black and white, although the proportions are clearly different. The ethnic differences between Baton Rouge and the East Baton Rouge Parish become even more apparent when the respective share of Baton Rouge is subtracted from that of the Parish population, then white people dominate in the Parish with 57.1 percent, while black people reach a share of 36.3 percent. Thus, the widely known phenomenon of suburbanization of white population groups, which is relativized elsewhere, can also be observed in the Baton Rouge metropolitan area (Champion 2001; Frey 1979; Gallagher 2013; Kühne and Schönwald 2015c; Renzulli and Evans 2005). Suburbanization is also illustrated by the proportions between home ownership and rent in relation to households: in Baton Rouge, the ratio is 49.1 to 50.9%, in East Baton Rouge Parish a total of 58.8 to 41.2%, if the shares of Baton Rouge are deducted from those of the Parish, the ratio is 69.1% for homeownership to 30.9% for rent, which significantly exceeds the ratio in Louisiana as a whole of 65.5 to 34.5% (United States Census Bureau 2019a, b, c; partly own calculations).

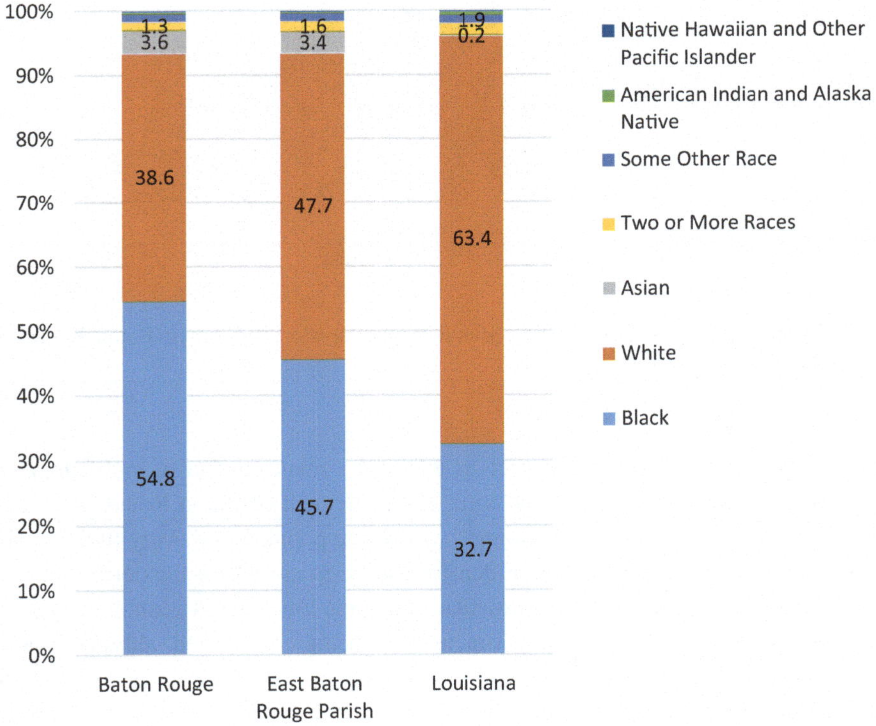

Fig. 5.17 The ethnic population composition of Baton Rouge compared to East Baton Rouge Parish and Louisiana in 2017 (own presentation according to: United States Census Bureau 2019a, b, c)

The suburbanization of Baton Rouge is taking place in the metropolitan area of Baton Rouge according to a special path of suburban development. This is due to the specific physical bases: the Baton Rouge settlement area is divided by numerous water bodies, bayous, canals, small creeks, and marshy areas. Development and expansions were and are only possible on certain – exaggeratedly formulated – 'spatial-islands'. Between these islands there are corresponding physical barriers, which - as we will show later (Chap. 8) – also result in a strong internal relationship within these 'spatial-islands'. The suburbanization of shopping was also encouraged by these structures: low-lying, marshy areas were initially kept free from development, but later became available as large contiguous areas for the construction of shopping malls (such as the Corporate Mall at College Drive, now a restaurant complex).[1]

In the Baton Rouge metropolitan area, the concentration of ethnic minorities (especially blacks, who, as shown, make up the majority population in Baton Rouge, but are referred to as a minority by national statistical conventions; Fig. 5.18) lies north of Florida Boulevard, especially in the downtown and Exxon Refinery areas. In comparison to the 1959 map of the ethnic division (Fig. 5.11), on the one hand, the significant increase in areas inhabited by colored people, a consequence of their increasing share of the population, is evident. On the other hand, the comparison shows the increase in colored population in the triangle between Interstate 110, Airline Highway, and Florida Street/Boulevard. In other words, an area that in large parts was previously populated by white industrial workers. In the Parish, beyond the city limits of Baton Rouge, ethnic minority neighborhoods are particularly common in these areas, although with increasing distance from the Baton Rouge urban area, their share is lower, especially in the northeast and south/southeast.

The population of Baton Rouge, in particular, and the East Baton Rouge Parish, in general, has a high level of formal education compared to Louisiana: 87.7 percent have a high school diploma, 36.6 percent possess a bachelor's degree, in the Parish as a whole it is even higher at 90.0 and 38.0 percent, while Louisiana as a whole has 84.0 and 27.6 percent (World Population Review 2019a, 2019b, 2019c). This is a general trend towards higher educational qualifications in metropolitan areas, due to the concentration of jobs requiring higher educational qualifications, especially in Baton Rouge, and particularly in higher educational institutions and administrative functions. Since suburbanization is mainly supported by higher earning families, and the higher availability of economic capital is in turn strongly dependent on certain cultural capital (specifically formal education), the educational level of the Parish is generally higher than in the city.

The educational differences mentioned above between Baton Rouge, East Baton Rouge Parish, and Louisiana as a whole, are also evident in the high school graduation

[1]We thank Craig Colten for this information.

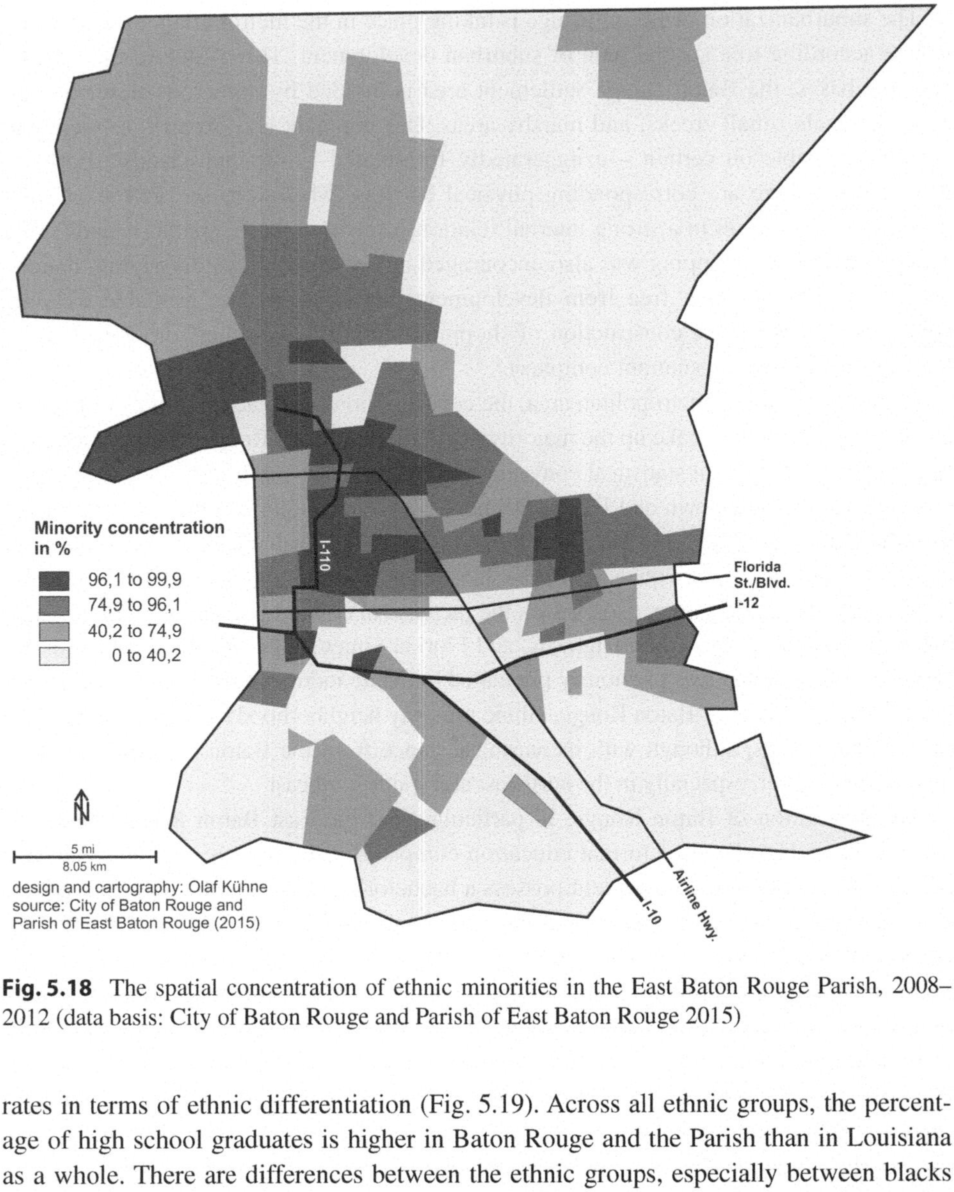

Fig. 5.18 The spatial concentration of ethnic minorities in the East Baton Rouge Parish, 2008–2012 (data basis: City of Baton Rouge and Parish of East Baton Rouge 2015)

rates in terms of ethnic differentiation (Fig. 5.19). Across all ethnic groups, the percentage of high school graduates is higher in Baton Rouge and the Parish than in Louisiana as a whole. There are differences between the ethnic groups, especially between blacks and whites (the Baton Rouge and the Parish dominate). These ethnic differences in terms of education are evident in relation to the proportion of people with a bachelor's degree (Fig. 5.20): specifically in the comparison between black and white populations, there is a considerable divergence, particularly in Baton Rouge and in the Parish as a whole. In Louisiana as a whole, people of Asian origin have the highest bachelor's degree rate. Regarding the relationship between Baton Rouge and the Parish, the higher level of education of the black population in the Parish compared to the city is striking, while the ratio of the white population is reversed. This may be due to a stronger suburbanization

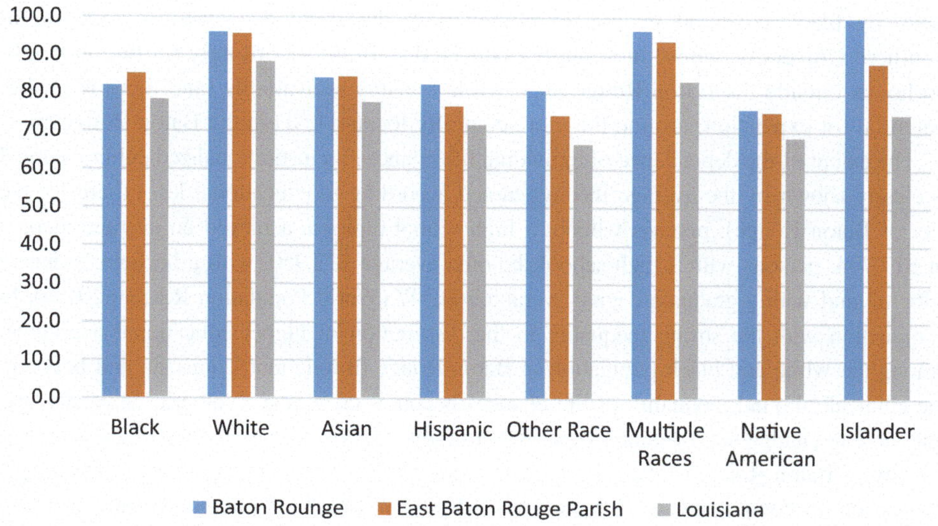

Fig. 5.19 The comparison of the rate of inhabitants with high school diplomas of different ethnicities, in Baton Rouge, the East Baton Rouge Parish, and Louisiana (own presentation according to: World Population Review 2019a, b, c)

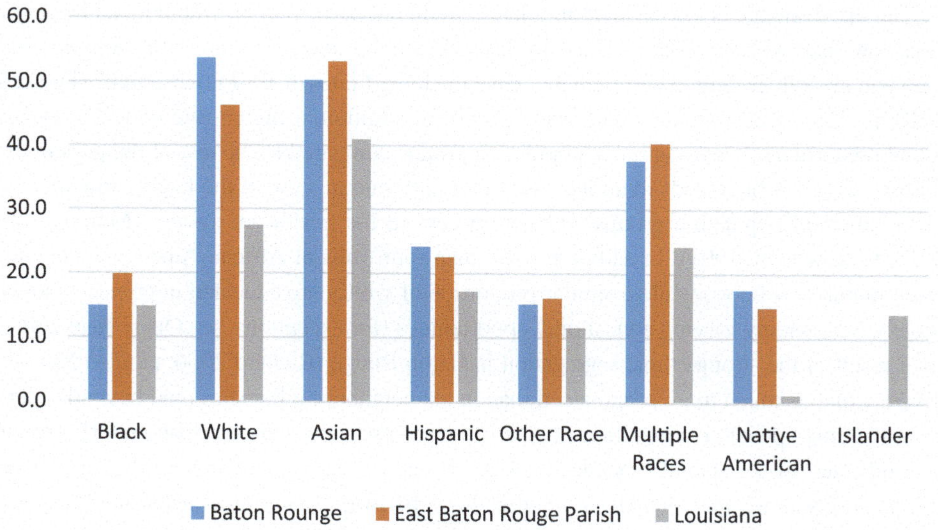

Fig. 5.20 The comparison of the rate of inhabitants with Bachelor's degrees of different ethnicities, in Baton Rouge, the East Baton Rouge Parish, and Louisiana (own presentation according to: World Population Review 2019a, b, c)

drive by black educated people, combined with the more decentralized location of Southern University, whose more highly educated staff are accordingly settling in neighborhoods outside the Baton Rouge area, while housing is available (and there is further potential for expansion) around the more centrally located LSU in the Baton Rouge area.

The extent of the dependence of individual life chances on institutionalized cultural capital is clearly shown by the average income earned according to educational level (here for the city of Baton Rouge): persons without a high school diploma achieved an average income of $17,774, persons with a high school diploma average $22,950, with a bachelor's degree $46,763 and with a graduate degree average $60,447 (World Population Review 2019a). In conjunction with the strong inequality in the distribution of higher educational attainment among the white and black population in Baton Rouge (throughout Louisiana and beyond), the ethnically-bound inequality of life chances becomes clear, and Southern University was only able to change this to some extent. Accordingly, the poverty rate among black people in the city of Baton Rouge is also 31.6 percent, compared to 17.9 percent among white people (Louisiana for comparison: 32.9 percent to 12.1 percent (World Population Review 2019a)). This socio-spatial context is also clear from the comparison of Fig. 5.18 (The spatial concentration of ethnic minorities in the East Baton Rouge Parish 2008–2012) and with Fig. 5.21 (The spatial concentration of poverty in the East Baton Rouge Parish 2008–2012). Particularly in the Baton Rouge urban area, a clear congruence of black and poverty-stricken neighborhoods is evident. Overall, the two maps also show a higher degree of detail in the neighborhood-specific distribution of ethnicity and the risk of poverty than the other Parish areas.

The disadvantage is also evident in relation to digital participation, where it can be stated "that low income households headed by a black, single, female parent with multiple children and no college degree will probably have the lowest rates of Wi-Fi ownership" (Driskell 2010, p. 26; see also Driskell and Wang 2009). In addition, a higher rate of HIV cases is found predominantly among black population groups compared to the rest of the population (Gibson 2002), which leads to an increased mortality rate in view of the largely low income of the affected population groups with less access to the healthcare system (Marcus et al. 2016; Ostrow et al. 1991). In addition, more than four-fifths of African-Americans are educated in public schools, while a similar proportion of whites are educated in private schools, usually with widely varying educational opportunities (for exceptions, see Olson Beal 2008). As a result of the strong ethnic segregation in Baton Rouge (Richard 2006; cf. also Xypolia 2003), which is much more pronounced than in the neighboring Parishes (and beyond), there is an extreme spatial concentration of low life chances in the sense of Dahrendorf. This is also reflected in the spatial breakdown of the Human Development Index (HDI) in Baton Rouge. To establish the context: in a national comparison, Louisiana is listed in 49th place out of 50. In Louisiana, Baton Rouge contains the area with the highest HDI (South) and the area with the lowest HDI (North and Center of the City; Burd-Sharps et al. 2009).

The pastiche spatial structure of Baton Rouge is based less on postmodern spatial developments, which are recognizable but have by no means taken on the dimensions of growth regions, and was already laid out by the specific logic of local urban development with its three spatially separate development paths of state administration, petrochemical industry, and higher education institutions, combined with a rather unambitious urban development policy.

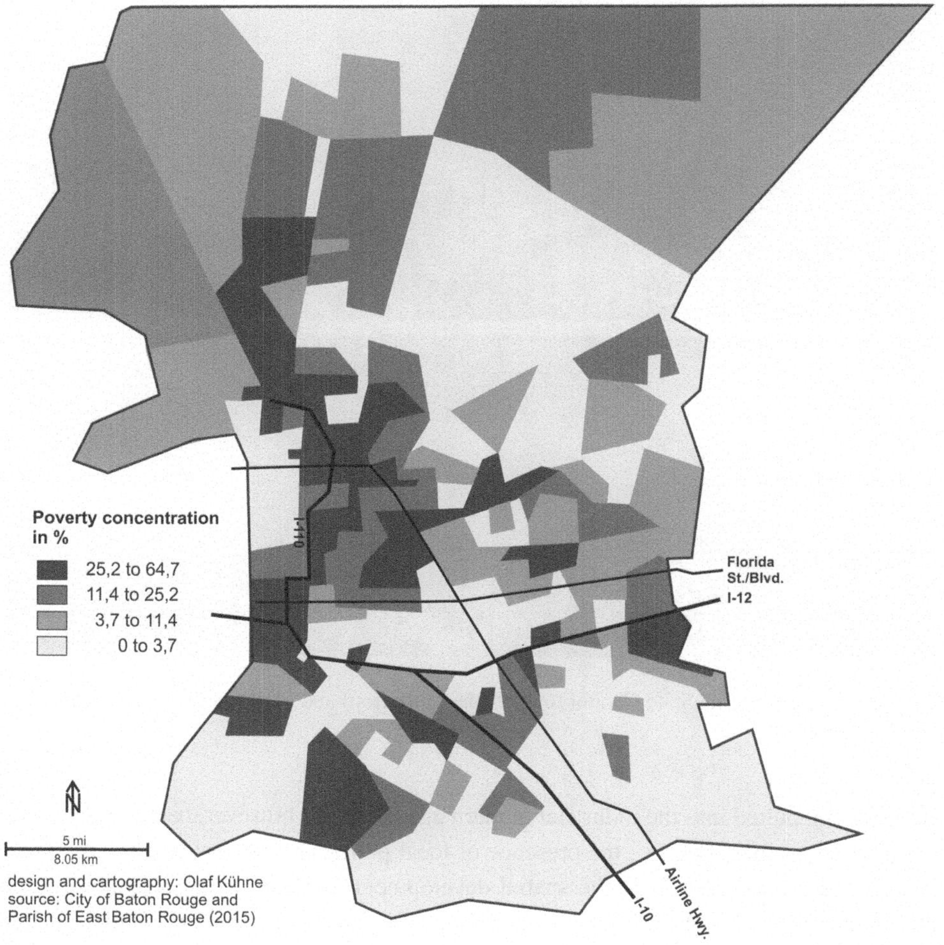

Fig. 5.21 Spatial concentration of poverty in the East Baton Rouge Parish 2008–2012 (Data basis: City of Baton Rouge and Parish of East Baton Rouge 2015)

5.6 Historical Aspects of the Urban-Rural Hybrid Landscape Biography of Baton Rouge – An Interim Conclusion on the 'Tripole City'

The neighborhood development of Baton Rouge began modestly and was marked by numerous political and cultural ruptures until Louisiana's reintegration into the Union after the Civil War. Only since the middle of the nineteenth century did the southern state capital develop a regional trading function as well as a political and administrative significance that shaped the State of Louisiana through the assumption (initially with setbacks) of the capital function of New Orleans. The second pole of Baton Rouge development appeared in 1909 in the form of the Standard Oil Refinery. Baton Rouge was

Fig. 5.22 View from the State Capitol towards the south (top) and north (bottom) of the city; own photographs 2019

suddenly catapulted into the industrial modern age. Without, however, being prepared for this in terms of local politics: the presence of local politics in the spatial order remained restrained at best (Fig. 5.22). The spatial development of Baton Rouge has always followed and continues to follow economic interests (with a few current exceptions, such as the safeguarding of historical buildings), regardless of whether human lives are endangered as a result (as the fire in the refinery in 1989 demonstrates). The third pole of spatial development took shape in the mid-1920s south of the then relatively self-contained built-up area around the LSU campus, which is still located there today. Despite the two growth spurts in the 1940s (World War II) and 1970s (OPEC crisis) within the city of Baton Rouge, due to surges in demand for oil products, the growth of the metropolitan region increasingly went beyond the administrative boundaries of Baton Rouge. In conjunction with the loss of economic importance of the oil refining industry, which also affected Baton Rouge, the ethnic composition and availability of symbolic capital changed roughly north of Florida Street/Boulevard: the percentage of the population of people of color increased, as did the percentage of the population suffering from poverty, without an increase in life chances through access to higher education.

6 Compartments of Development in the Spatial Pastiche of Baton Rouge

The spatial structure of Baton Rouge is very differentiated. This is not only the result of general developments and logics, but is also based on specific physical-spatial arrangements in Baton Rouge as well as specific logics in dealing with spaces. In this chapter we will follow these different influences and developments.

6.1 The Spatial Structure of the Downtown Area and Its Spatial Links

It is difficult to draw a clear line between downtown Baton Rouge and Baton Rouge because of the considerable pastiche patterns of differentiation. As a rough approximation, the area between the Mississippi to the west, Interstate 10 to the south, Interstate 110 to the east, and south of Capitol Lake to the north can be summarized (see also Duany Plater-Zyberk and Company 1999). In this classification, however, parts of neighborhoods are found under the designation 'downtown', which were genetically independent neighborhoods or have a clearly suburban structure, but are in part undergoing intensive functional transformation. This is particularly true of the 'Beauregard' neighborhood, where - due to the proximity to the courts - the structures originally built for residential purposes now increasingly house law firms. To the east of the Capitol complex in the north of the downtown area, the 'Spanish Town' is also a gentrified quarter of suburban buildings. South of the Capitol, in turn, an extensive government district has

Electronic supplementary material The online version of this chapter (https://doi.org/10.1007/978-3-658-30716-5_6) contains supplementary material, which is available to authorized users.

O. Kühne and C. Jenal, *The Multivillage-Metropolis Baton Rouge*, RaumFragen: Stadt – Region – Landschaft, https://doi.org/10.1007/978-3-658-30716-5_6

been developed, which is adjoined to the south by the classic Central Business District. Overall, the downtown of Baton Rouge has a comparatively low density and disperse spatial structure. In addition to the accumulation of law firms in the Beauregard quarter, the concentration of authorities south of the Capitol (state) as well as on the western North Boulevard (especially municipal authorities), there is a certain accumulation of office buildings (some with a larger vertical extension) in the area between North Boulevard and North Street (south and north directions) as well as Lafayette Street to the west and Seventh Street to the east.

6.1.1 Downtown Development in the Second Half of the Twentieth Century and the Beginning of the 21st

Since the last decade of the nineteenth century, the marked changes in the downtown and adjacent areas have been clearly expressed through the change in the ground plan structure and also in the construction of Interstates 10 and 110 (Fig. 6.1), which in the Zoning Plans of 2001 and 2019 mark linear undeveloped areas in the south and east. In the period under consideration of almost thirteen decades, an increasing coarseness of the building structure becomes clear: even north of North Boulevard and west of Fourth Street, 1891, as the most densely populated part of the section under consideration, was dominated by buildings with a comparatively small floor plan; by 1951, these small-scale basic structures had already been replaced by larger ones. However, the most significant revision of the building floor plans took place in the second half of the twentieth century. After a densification of the building structure beyond the area north of North Boulevard and west of Fourth Street had also been carried out in the previous sixty years, a comprehensive restructuring of large portions of the area under consideration took place, which

Fig. 6.1 Zoning plans of the downtown area and partially adjacent areas for the years 1891, 1951, 2001, and 2019 (for the years 1891, 1952 and 2001 according to Desmond et al. (2005); for the year 2019: own result based on google maps orthophotos)

only a few rather peripherally located communities resisted (especially Beauregard in the south and Spanish Town in the northeast). The trend towards increasingly coarse-grained floor plans continued after the turn of the millennium, although the efforts to protect historic buildings considered valuable were referred to as 'islands of fine-grain' (in German 'Kleinkörnigkeit', an expression for small structures; along and west of Third Street, as well as Beauregard and Spanish Town; see also Sect. 5.4). A particular change in the floor plan structure can also be observed between First and Lafayette Street: as a result of the loss of importance and even the abandonment of the port facilities close to the city center, the construction of buildings with a large floor plan is accompanied by the demolition of buildings without replacement (together with the construction of large parking lots). The conversion of the floor plan structures in favor of buildings with a larger floor space not only follows the economic structural change, but also how the public authorities have fundamentally restructured the downtown area north of North Street and south of North Boulevard (in each case in the area approximately west of Fifth Street), which will be examined more closely in the following.

In the mid-twentieth century, the downtown area was characterized by multi-story reinforced concrete and brick buildings (Fig. 6.2) and began to develop from a diverse usage structure, with a focus on shopping and office work, to a usage dominated by offices (Brill 1963; Gleason and Brockway 1991). Third Street remained the central shopping street (Fig. 6.3). In general, the building height structure in the central area of the downtown in 1959 - especially on the other side of Third Street - shows a large dominance of flat buildings, as well as undeveloped land that was often used as parking lots. An exception to the otherwise rather restrained vertical development is the luxurious 10-story 'Heidelberg Hotel', completed in 1927, located on the south-western corner of the intersection of Lafayette and Florida Street. The hotel became a symbol of power in the city, for example Huey P. Long and his brother, Earl Long, used "the hotel bar as an unofficial office" (Rodrigue and Phillips 2011, p. 37) and Huey Long "became infamous for meeting his guests while wearing his pajamas" (Rodrigue and Phillips 2008, p. 62). The structure of the building heights and the small floor space of the buildings - which can partly be deduced from this - illustrates the minimally planned development of the depicted part of the downtown area. However, it also illustrates the low profitability of a vertical building development, since firstly, sufficient free space is available for a (cheaper) horizontal building development, secondly, a low land price level does not force a vertical building orientation, and thirdly, no headquarters or branches of nationally or internationally important companies can be found here, which would like to symbolize a high presence of power, influence, and significance via a high vertical building extension (Hasse 2012). The land use of the central area of Baton Rouge downtown in 1959 (Fig. 6.3) shows, in addition to the assortment of official uses in the north and south of the study area - already mentioned in other contexts - and the dominance of shops on Third Street, as well as their accumulation on Main Street. Warehouses are located primarily near the Mississippi River and the train station. Especially in the southeast, commercial and public use is losing importance, with open spaces and residential

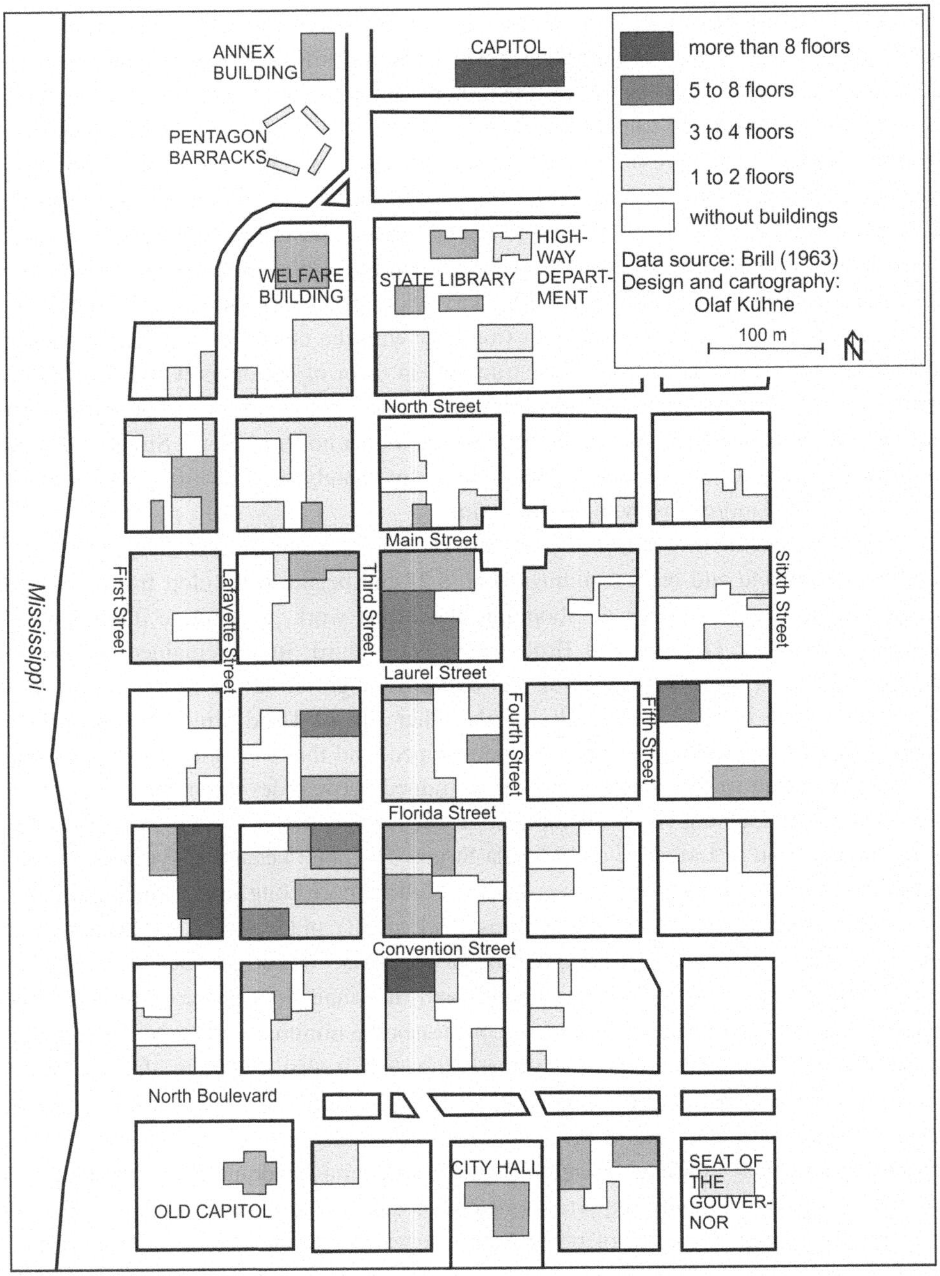

Fig. 6.2 Building heights in the central part of downtown in 1959 (own representation according to Brill 1963)

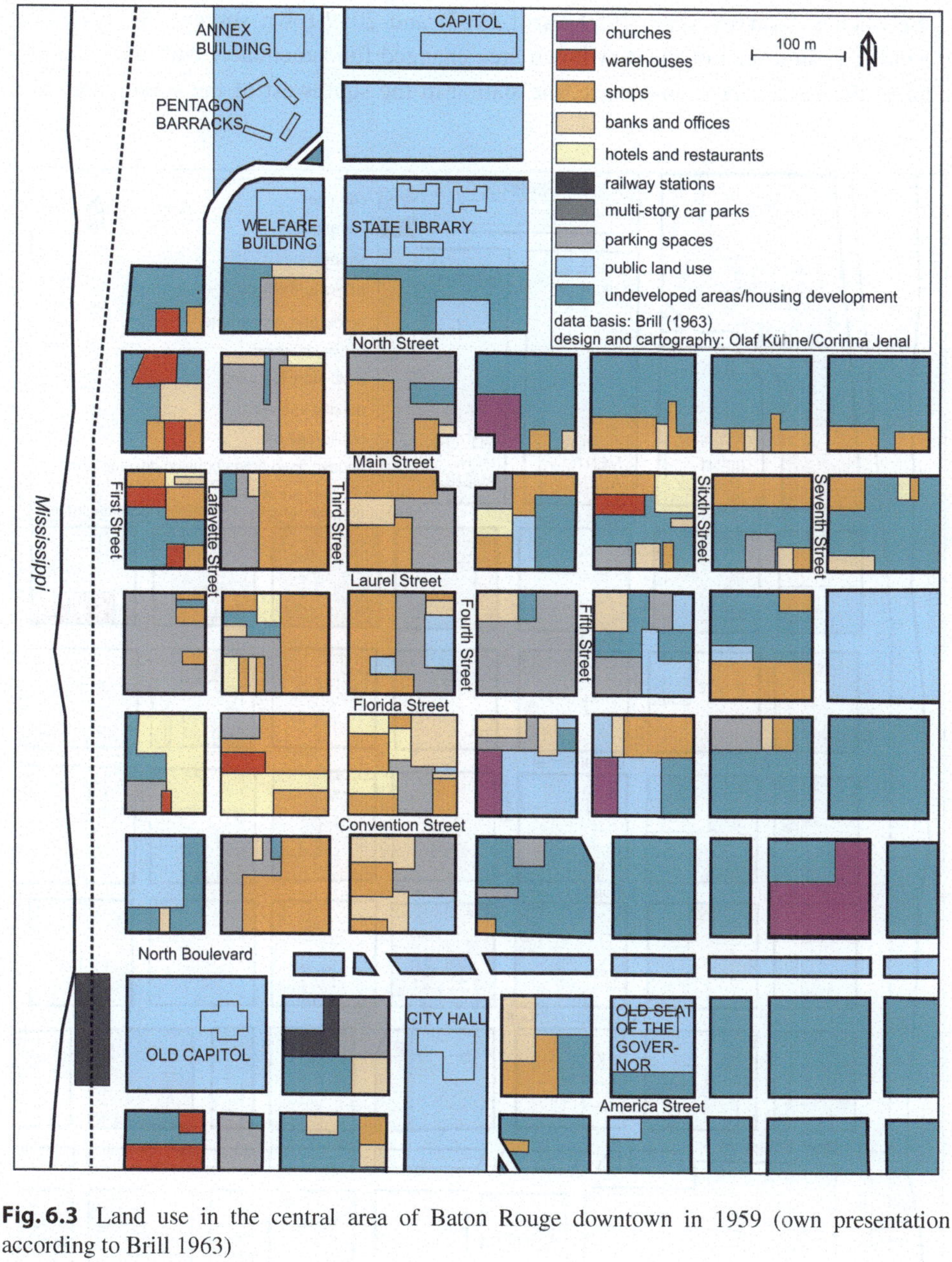

Fig. 6.3 Land use in the central area of Baton Rouge downtown in 1959 (own presentation according to Brill 1963)

buildings dominating. In addition to these conurbations, however, there is a large degree of mixed use. A strong functional differentiation of the downtown area has not taken place.

Between the survey years 1959 (Brill 1963) and 2019 (own survey; Fig. 6.4), land use in the central area of the downtown area changed fundamentally: with the abandonment of the passenger train service, the station in the southwest of the survey area also

Fig. 6.4 The land use in the central area of Baton Rouge downtown in 2019, vacant buildings were attributed to the last use (own survey and presentation)

disappeared. The track running along the Mississippi is only used for freight traffic (especially shunting traffic). The disappearance in the use of warehouses in the area surrounding the former railway station is connected with its closure as well as its abandonment along with the former port facilities along the Mississippi River bank near downtown. In addition, the downtown area described above has undergone a fundamental change of use: the use of the buildings by shops, which was specifically localized on Third Street in 1959, has largely disappeared, giving way in particular to use by banks and offices. A significant expansion of public use can be found in both the north and south of the study area. Completely new - in comparison to 1959 - are multi-story car parks, which - as in the case between North and Main Street - can take up entire building blocks.

The functional change of the downtown area is also reflected in its vertical structure (Fig. 6.5). Not only has the construction of taller commercial buildings (such as One America Place or the Chase Tower) in the CBD contributed to this increased vertical orientation of the downtown area, but also the expansion and redesign of the building structure through government buildings north and south of the CBD. The already mentioned multi-story car parks also contribute to the increased vertical extension of the buildings in the examination area. The heavily transformed parts of the portion in the downtown area under investigation also show the tendency towards an increased footprint area of the (newly constructed) buildings, as shown elsewhere (in the context of Fig. 6.1). With the exception of the administrative districts to the north of the investigated area, a noticeable preeminence of undeveloped areas can be observed. Most of these are used in the form of parking lots (even on weekdays when they are sparsely used). This low intensity of use of space in the central parts of the downtown area indicates a low (relative as well as absolute) economic importance: while in the mid-twentieth century a certain concentration of (tertiary) jobs were still to be found in the downtown area (Brill 1963), the downtown area is now just one location area among many (Antipova 2010), whereby the trend towards an Edgeless City type development is clearly discernible (cf. also Brian 2003; Wang et al. 2011).

The weakening of the downtown area in terms of retail trade since the Brill survey (1963), as well as its low significance in comparison with downtowns of cities possessing comparable economic and political importance and population numbers, may also be due to its very decentralized location on the western edge of the urban area: historically located on the Mississippi River, and thus conveniently situated for both river-based passenger and freight transportation, the transformation of transportation to - in Baton Rouge, less important - rail and automobile has created a distinct location preference, especially for the hospitality industry (which is strongly focused on business customers) and retail. The decentralization of the downtown area increased with every suburban settlement activity. In line with this decentralization of the downtown area, inner-city retail trade lost importance early on (Brill 1963) and increasingly lost its competitive edge relative to the shopping malls that developed in rapid succession – and also increasingly the residential and working locations (for example around the Airline Highway and Siegen Lane) of affluent customers. The development of shopping

Fig. 6.5 Building heights in the central part of downtown in 2019 (own survey and presentation)

structures in Baton Rouge has been and continues to be subject to transformations and spatial shifts (see Fig. 6.6). This development is characterized by an ever stronger focus on the accessibility of individual shops by automobile. While the traditional shopping

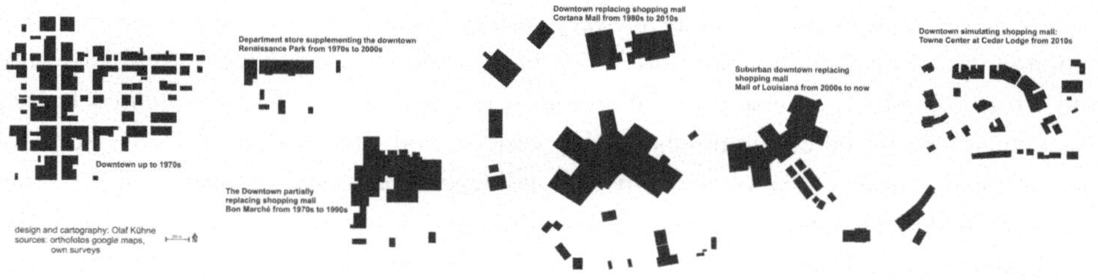

Fig. 6.6 The floor plan development of dominant physical shopping structures in Baton Rouge over time (own survey and presentation)

downtown is still pedestrian-oriented (in combination with the flowing car traffic) by shifting the car parking traffic to surrounding parking lots, the parking lots are moving closer to the Renaissance Park (now converted department store) on Florida Street east of Interstate 110. Located further east on Florida Boulevard, at the intersection with Lobdell Avenue, Bon Marché Mall's parking lots are partially enclosed by the shopping buildings. At the same time, the range of goods and services on offer has grown to a degree that has largely replaced the classic downtown area. After its closure in the late 1990s, the mall was converted into a technology center. Its function being replaced by the Cortana Mall (opened in 1976), which had far surpassed Bon Marché in terms of sales area. This is located in the northeastern corner of the intersection of Florida Boulevard and Airline Highway. The parking spaces are located between the central mall and additional shops. In the first months of 2019, this mall was in a state of advanced decline, with only around 10% of the shops still in operation, mostly holding closeout sales. While the development of out-of-town shopping structures up to the Cortana Mall was linked to a constant growth in floor space, this trend has been reversed with the Mall of Louisiana (opened in 1997) located in southern Baton Rouge: the two-story design allows for a more compact construction and reduces travel times between the car and the mall as well as between the individual shops. The mall is centrally located between parking lots, offering footpath connections to the adjacent neighborhoods (compared to the previous shopping structures without footpaths). An open-air pedestrian area stretches out in a south-easterly direction, which takes its design cues from a downtown area of a medium-sized urban community. This motif comes to the fore even more clearly in the Towne Center at Cedar Lodge, located between the Mall of Louisiana and the former Bon Marché Mall. In order to increase the accessibility of the shops from the parking lots, however, they are arranged in single rows, with access along arcades and not in a hall that is centrally located and unaffected by weather conditions. Complementary shops scattered across the central parking area further reduce travel times. In comparison to the Mall of Louisiana, the brand structure of the shops, the presentation of the goods as well as the price level is geared to an audience with a higher availability of symbolic capital. However, here too - also in the wake of increasing competition from Internet trade and possibly the construction of shopping malls that are more closely geared to future consumer needs - a more-or-less rapid filtering down can be assumed.

Moreover, the downtown area is not very well integrated into the transportation infrastructure, rail-bound public transport has been shut down, Interstates 10 and 110 only affect it at the edges, diverting traffic around it rather than directing destination traffic, the connections to the communities on the western bank are located far to the north or beyond the downtown area to the south of it and depend on motorized means of transportation (no pedestrians, no passenger transport by rail) for the transporting of people. The transportation axes leading from the eastern suburbs towards the downtown area, especially the Florida Corridor and Government Street, taper off significantly before reaching the downtown area. The abandonment of the Downtown Airport (finally closed in 1976) in the area of the Airline Highway past Florida Boulevard and the shifting of air traffic from the jet-engine airfield (which generally depends on longer runways due to its greater weight and higher take-off and landing speeds) to the 'Municipal Airport' located in the north of the city, has further extended the traffic connections of the Downtown Region. This airport was expanded in the mid-1990s and now has a renovated terminal and new parking facilities (Brian 2003; Rodrigue and Phillips 2011). The decentralization of transportation infrastructures, of residential locations, of working locations as well as of supply locations, is - in connection with an increases in employment - associated with an increase in vehicle traffic. The resulting traffic, in turn, is mainly handled by private cars - also a result of the low density (see Antipova 2010), regardless of whether the infrastructures were designed for such a high load. In order to reduce travel times, interlinked transportation is used (different locations are served on one route, shops are visited on the way to work; Antipova 2010). Switching to other means of transport is made more difficult by low availability (local public transportation), poor infrastructural suitability (incomplete footpaths, hardly any bicycle paths), but also by climatic specifics (sultriness in summer, changeable weather, frequent precipitation; see also Sect. 6.4).

While the gastronomic offerings in the downtown area are strongly oriented towards the needs of the CBD and government employees, for example with lunch, small snacks, and coffee-to-go as well as delivery services, and sometimes closes before evening hours, Government Street (and south of it) is increasingly developing as a gastronomic location (Fig. 6.7). Today there is a diversified offering of different restaurants (different price levels and kitchen traditions) and a larger number of bars, but as a result of the low density structure caused by the incomplete local public transportation system, there is a clear focus on motorized individual transport. This development continues a long tradition of linear development of commerce (not only in Baton Rouge). Government Street, like Florida Street, had already formed such districts in the 1950s (Brill 1963). These districts can still be found functionally - as can those along Highland Road (Brill 1963) - while this development along Main Street extends to the area within the downtown district and to the east of Interstate 110 only offers isolated commercial areas. The functional change along Plank Road and the Scenic Highway north and northeast of downtown, which Brill (1963) still called 'shopping streets', is even more apparent. Today, empty commercial buildings dominate here, and there are often manifestations of a 'poverty economy', such as liquor stores, one-dollar shops, and pawn shops. Here, the decline of the North as a result of the loss of industrial jobs in Baton Rouge is clearly reflected in physical space.

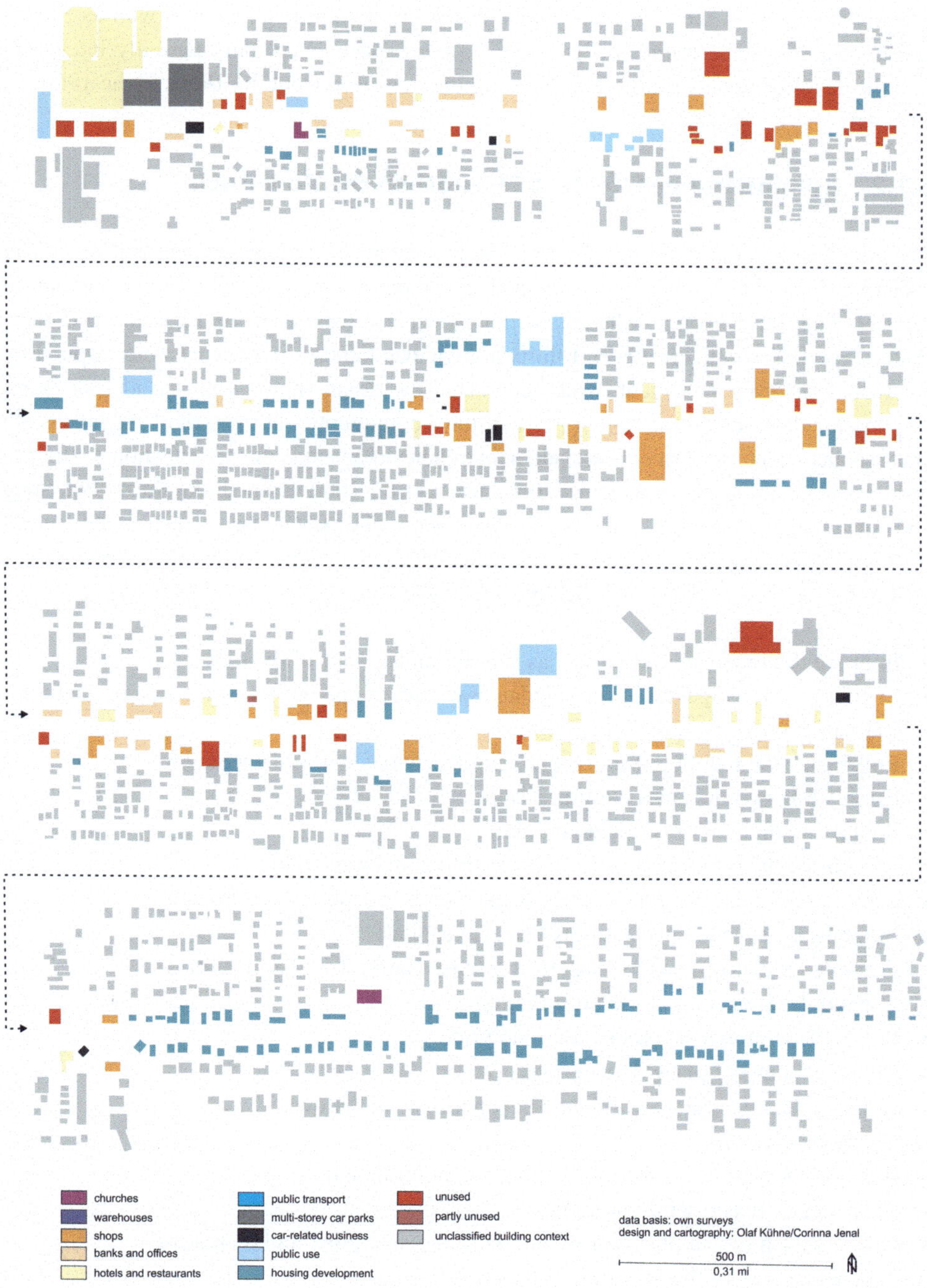

Fig. 6.7 The building use on Government Street, starting from downtown, in the west (Own presentation)

The development of the 'Edgeless Cities' along the Airline Highways as well as Siegen Lane is not yet to be found in Brill (1963), these spaces were only partially integrated into the development of Baton Rouge during the transition from the 1950s to the 1960s.

6.1.2 Administrative Efforts to Strengthen the Downtown

The efforts to develop Baton Rouge, that was less determined by economics as it was along social lines, had led to a highly fragmented spatial development which found expression in the 1998 Plan Baton Rouge, wherein the downtown area was given special attention (Duany Plater-Zyberk & Company 1999). Efforts to revitalize the downtown area were inspired by the New Urbanism movement, which countered the urban sprawl with the idea of dense neighborhoods based on the design of small-town communities at the turn of the nineteenth and twentieth centuries. An essential element of new urbanism is the participation of (future) residents in planning (Speights-Binet 2004; general Ellis 2002; Grant 2006; Katz 1994; Song and Knaap 2003). Civic participation, in which about 500 people participated in the context of 'Plan Baton Rouge', identified the following needs and wishes for the downtown area as well as for other parts of the city (Duany Plater-Zyberk and Company 1999, p. 11):

- Improve the image of Downtown as the common center of the city.
- Encourage municipal and state regulators to become proactive rather than procedural.
- Capitalize on work already in progress.
- Emphasize the identity of individual neighborhoods within Downtown.
- Establish Downtown as a cultural center.
- Retain and maintain historical landmarks.
- Streamline the permitting process for building and renovation.
- Create an ongoing forum for citizens to discuss and control their destiny.
- Emphasize Downtown's status as the state capital.
- Promote Downtown as the place to showcase Baton Rouge to visitors.
- Expand the areas of historic preservation.
- Present Downtown as a focus for civic pride.
- Create a 'sense of place' Downtown.
- Confirm the city and state alliance in support of Downtown.
- Celebrate diversity.
- Promote Downtown as the city's spiritual, cultural, and economic center.
- Make use of existing infrastructure.
- Create a 24-h city: a place for living, working, shopping, and recreating.
- Take advantage of the Mississippi River.
- Create a pedestrian experience in contrast to the suburban surroundings.
- Set an example for planning beyond Downtown.
- Promote the economic impact of urban revitalization.
- Maintain a focus on the needs of Downtown residents.

- Focus on history and heritage as a marketing theme.
- Showcase the regional architectural style.

The plan identified six areas of interest: Catfish Town (around the Rouge River Center), Old State Capitol, State Capitol, Central Business District, Beauregard Town, and Spanish Town Neighborhoods. In addition, it specified development corridors between these areas of action, in particular the Downtown Parks Corridor, which was to connect Catfish Town with the State Capitol District, the Riverfront Parkway Corridor between Catfish Town along the Mississippi River, and the Seventh Street Corridor, which was to connect the two historic quarters of Spanish Town and Beauregard. In addition, the plan addressed general issues such as increasing commercial development, improving the traffic situation, design proposals, etc. In total, the plan included more than 100 projects and proposals for action (Duany Plater-Zyberk and Company 1999; Speights-Binet 2004).

The revitalizing developments of Baton Rouge downtown, which received a new urbanistic focus in the course of 'Plan Baton Rouge', can be interpreted as a synthesis of conservative, middle-class values of a well-ordered and understandable world, for instance in the sense of Jackson's 'Landscape Two' (Burckhardt 2004a; Jackson 1984). Thus, the call for the conservation of historical elements or their reinvention can be understood as "active return to the design elements of the past, and in doing so, implicit is a larger promise for the lifestyle of the past, the community of the past, and a 'simpler time'" (Speights-Binet 2004, p. 183). However, the focus on the historical, the local and the attempt to emphasize the specific is also done - in the sense of Bourdieu (Bourdieu 1991; 1998) - in order to create distinctive characteristics (for more details: Höfer and Vicenzotti 2013; Jenal 2019a; Kühne 2007; Schönwald 2015). The focus of the development will be on a synthesis of economic and social 'upgrading'. Increased accessibility for pedestrians and cyclists is also more likely to meet the needs of urban milieus with an upscale endowment of symbolic capital beyond those of marginalized residents in the downtown and adjacent areas, including the interest in benefiting from a general increase in property values through 'reurbanization successes' (Speights-Binet 2004).

The path of a historicizing revitalization of downtown that began with 'Plan Baton Rouge' was continued, expanded, and refined in the following two decades. In the first years of the twenty-first century, the focus was on 'smart growth' as an overall strategy - a thematic specification, for example, on bicycle and pedestrian traffic, an integrated traffic concept, the renaturation of water bodies or a stronger integration of East Baton Rouge Parish (for example: City of Baton Rouge 2019). The plans of the past two decades have in common a strong citizen orientation.

6.1.3 Downtown Developments Today

The plans for the development of downtown Baton Rouge in recent years have a strong orientation towards the needs of a population with an upscale endowment of symbolic

Fig. 6.8 Modest reurbanization tendencies along North Boulevard, in the St. Louis Street confluence area: restaurants and bars in restored building fabric; own photograph in 2019

capital, on the one hand, and a strong economic link, on the other. Planning should serve the economic development of Baton Rouge, or at least not hinder it. With the exception of increased citizen participation, they are thus in the tradition of planning that is strongly oriented towards economic needs, as was also the case in the context of the development of the oil and petrochemical industry in Baton Rouge - with quite problematic consequences for the inhabitants of the city and the Parishes (see Sect. 6.2 below). The focus of the current economic fixation in the downtown area is on the cultural and creative industries, which in turn should act as a nucleus for increasing the attractiveness of tourism and for residents with an urban lifestyle.

The current downtown development includes – albeit comparatively modest – reurbanization tendencies (Fig. 6.8). Apartment buildings that are being erected, especially between the CBD and Interstate 110, and the neighboring Spanish Town, Beauregard, and Garden District, are also subject to URFSURBanization of varying intensity (Fig. 6.9). While in the area peripheral to the more inner-city development there are also more intensive over-developments, for example in relation to the construction of apartment blocks instead of a smaller building structure, the over-development of large parts of these historical suburban structures is limited to modernization, gentrification, and in some cases the tranformation of purposes (especially housing by offices).

Fig. 6.9 A document of the (comparatively modest) construction of URFSURBs southeast of downtown: newly erected urban apartment buildings in suburban neighborhoods; own photograph taken in 2019

Overall, the downtown development of Baton Rouge can be characterized as unambitious and - compared to communities of similar size in the United States and beyond - relatively modest. There is a degree of impetus where state actors can have a direct impact (i.e. in the case of state-owned real estate). This applies not only to the government district around the Capitol in the north of the downtown area, but also to the south, particularly where municipal authorities and state museums (such as LSU) are located, and also Catfish town. The development of Third Street into an inner-city shopping and nightlife street - following its own historical model - remains just as restrained as the riverfront development or the connection of Spanish Town and Beauregard along Seventh Street. These two neighborhoods are subject to a process of URFSURBanization because of their extensive fit with the spatial ideas of a 'good life' by urbanophiles possessing an upscale configuration of symbolic capital. In comparison to most European societies, the limited influence of the spatial impact of social processes and especially relative to economic processes in the United States, in general, and in the political tradition of Louisiana and Baton Rouge, in particular, becomes clear here: spatial planning has a function that promotes economic processes, but hardly controls them and certainly does not exclude them in certain locations.

The developments in and around the downtown area can be interpreted as an expression of a differentiating postmodern society. In addition to the continuing need for a suburban life simulating a rural community, there are the needs of urbanophile population groups manifested in physical space in the form of apartment buildings, but also bars, cafés, and restaurants. These developments differentiate the compartments of the postmodern spatial pastiche - also towards the urban-cultural poles of urban–rural and cultural-natural hybridities.

6.2 From the Janus Face of Accelerated Industrialization: Oil and Petrochemical Industry in and Around Baton Rouge

The opening of Standard Oil's refinery in Baton Rouge marked the beginning of the accelerated industrialization of Baton Rouge and thus the third path of a 'Tripole City', as discussed in Sect. 5.3. However, this third path was not only associated with economic growth – which induced population growth, intensified integration into the regional, national and, especially regarding the port, international transport system - but also considerable socio-ecological burdens. These will be focused on in the following.

The construction of the Baton Rouge Refinery acted as a nucleus not only for the industrialization of Baton Rouge, but also for a corridor of oil and petrochemical production between Baton Rouge and New Orleans and beyond. Fueled by investments during and after World War II, 177 refineries and chemical plants were built in Louisiana in 1947, rising from 211 in 1962 and 284 in 1981 to 320 in 2002 (Colten 2006b). In 2017, oil refining had become concentrated at 18 sites in Louisiana that processed more than 3.0 million barrels of crude oil per day (LMOGA - Louisiana Mid-Continent Oil and Gas Association 2019b). In terms of crude oil processing, the Exxon refineries in Baton Rouge rank fourth in national comparison, the Marathon refinery in Garyville third (LMOGA - Louisiana Mid-Continent Oil and Gas Association 2019a). Overall, the spatial distribution of the refinery sites, and particularly the production in Louisiana, shows a clear focus in the south of the state, especially between Baton Rouge/Port Allen and Belle Chasse (Fig. 6.10). With considerable consequences for the appearance of the affected areas of Louisiana: "A landscape once dominated by sugarcane fields had been thoroughly transformed" (Colten 2006b, p. 96). This transformation took place through massive support from politicians and their administrations: the State of Louisiana promoted the growth of the oil and petrochemical industry through massive tax breaks, the disposal of waste by industry was facilitated by the authorities, the maintenance of the dikes and keeping the Mississippi Fairway open for ocean-going ships can be interpreted as subsidizing industry, and, in addition, the urban planning of cities with major oil and petrochemical plants followed the primary goal of maintaining reserve land for companies (Allen 2003; Burby 2000; Colten 2000, 2014a). This policy has been (and continues to be) widely supported by large sections of the population of Louisiana, as it enables people with a low level of education to earn a comparatively high income, which is why

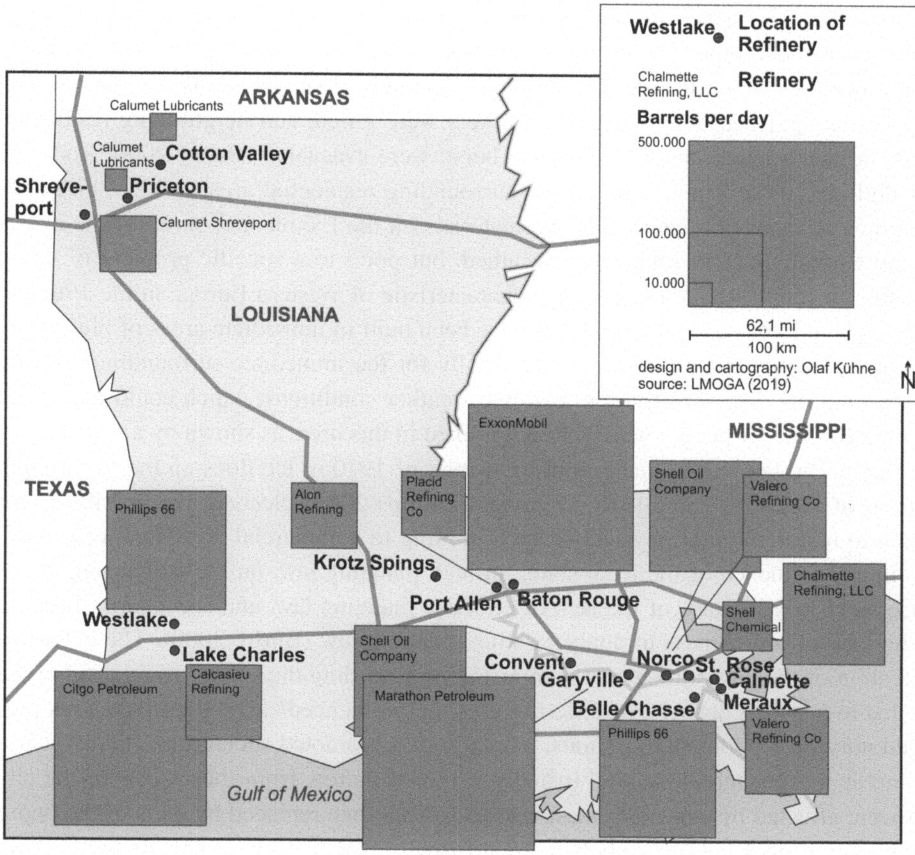

Fig. 6.10 The refinery sites and daily production of the refineries in Louisiana in January 2017 (own presentation)

growing environmental problems have been widely accepted (Colten 2006b; Hochschild 2016). This acceptance existed despite numerous industrially induced environmental damages, such as the fish mortality in the Mississippi River in the winter of 1963–64, the designation of 40 active or inactive hazardous waste deposits along the petrochemical corridor, the emergence of a significantly increased cancer rate along this corridor, and explosions at several refineries, one of them the Exxon refinery in Baton Rouge, which will be discussed in more detail in the following (Allen 2000; Burby 2000; Colten 2000).

The fire at the Exxon refinery and its spatial impact can be understood as a combination of lack of operational diligence, extreme weather conditions, and spatial planning that is not very sensitive to environmental issues. Thus, unusually low temperatures (the second lowest ever recorded in Baton Rouge) between December 21 and 25, 1989 indirectly had profound consequences for the spatial development of Baton Rouge: after the air temperature of eight degrees Fahrenheit (-13.3 degrees Celsius) on December 23 reached its lowest level in the frost period, the cold caused an explosion in a corroded pipe. The explosion could be heard and felt as far away as six miles from downtown Baton Rouge. The fire with its clouds of smoke visible from afar could only be extinguished after 14 h, two workers were killed, and neighboring residential areas, in which numerous windows also burst, were evacuated (Burby 2000; Rodrigue and Phillips 2011). The effects on the surrounding residential areas, which, given the quantities of flammable and explosive materials on the Exxon site, survived the events around Christmas 1989 relatively unscathed, but point to a specific problem of urban planning in Baton Rouge (as was also characteristic of Western Europe in the 19th and early 20th centuries): residential areas have been built in immediate areas of high emission and potential risk of accidents, especially for the immediate surroundings (Burby 2000). First, the low sensitivity to extreme weather conditions, which could have been anticipated even by people who were not trained in this area, as shown by a well-known photograph by Fonville Winans from the winter of 1940 of ice floes on the Mississippi in front of the State Capitol (Rodrigue and Phillips 2008). Second, the possibility for Exxon to restructure the physical space according to commercial considerations; third, the limited influence of the local state, through planning law, on the structuring of the space, primarily because of the lack of effective planning law, and secondarily because of the lack of willingness to apply existing planning law (Burby 2000). The extensive abstention of the city politics and administration regarding the spatial structuring of the city led to a spatial pastiche of differing uses that influenced each other. Industries produced noise and pollutant emissions, which in turn promoted social (and ethnic) segregation, as more symbolic capital (usually white) withdrew from those parts of the city that were affected by emissions these residents were then replaced by parts of the population with a low level of symbolic capital (usually black; Burby 2000; Richard 2006). In response to the 1989 explosion and the growing sensitivity of the Baton Rouge citizens (and, more generally, large sections of the population of Louisiana and the United States) to the environmental and health hazards associated with the oil and petrochemical industry, Exxon began buying up and demolishing the buildings surrounding the plant (Burby 2000). Today the area is characterized by the extensive disappearance of the buildings, while the streets have been preserved (Fig. 6.11). This has resulted in a physical-spatial structure that - due to the state of care - evades the widespread classification patterns. A hybrid park-street settlement space of limited accessibility has been created, which is a singularity even in the very complex and multiform spatial pastiche of Baton Rouge.

But accidents are not the only health hazard posed by the oil and petrochemical industry to Baton Rouge residents (Anderson 2011). The methanol, styrene, dioxin, benzene, toluene, xylene, and many others emitted by the chemical industry are carcinogenic and, despite mostly decreasing emissions since the 1980s (e.g. for benzene, dioxin,

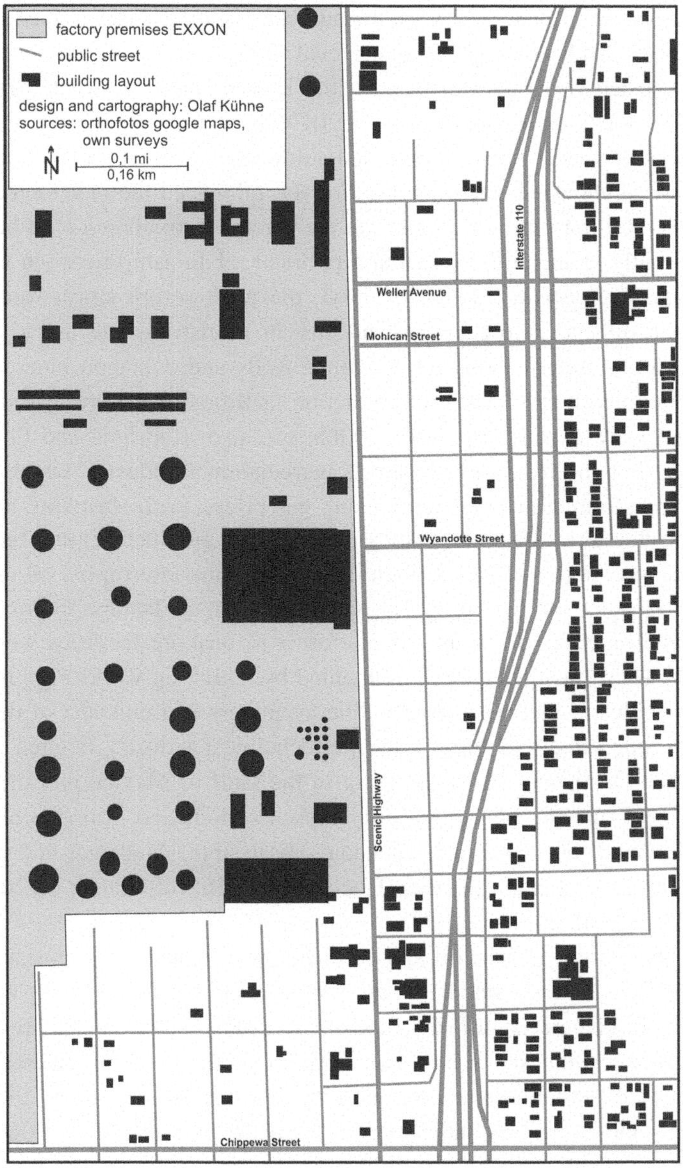

Fig. 6.11 The ground plan layout west of the Exxon plant site in February 2019. As a result of the acquisition and demolition of buildings, the ground plan west and east of Interstate 110 contrasts sharply. Also noteworthy is the naming of the 'Scenic Highway', immediately east of the factory site, which - especially the part of the surface terrain shown in the map section - differs greatly in terms of stereotypical ideas of beauty or picturesqueness. The name of the street was given before the industrial plants were built and has not been revised

formaldehyde, weaker xylene), increase the number of cancer cases in East Baton Rouge Parish (Anderson 2011; see also: Huebner et al. 2004), in a state that has the seventh highest age-adjusted rate of all cancers in the United States between 2011 and 2015 (Centers for Disease Control and Prevention 2019).

Hurricanes Katrina and Rita not only had a devastating effect on the coastal cities of Louisiana (and neighboring states) and indirectly influenced the development of Baton Rouge, and as well the physical impact on the oil and petrochemical industries, but it also fundamentally changed the social and political relationship between them (Colten 2006b). Hurricane Betsy in September 1965, the most severe storm since the establishment of the oil and petrochemical industry in Louisiana, had already interrupted production at around eight thousand active oil wells and damaged numerous onshore pipelines. Subsequently, the offshore production facilities which are less susceptible to storm damage, were expanded (Colten 2006b; see also: Rodrigue and Phillips 2008). Hurricanes Katrina and Rita hit Louisiana's petrochemical industry, which was already in a state of transformation because of rising gas prices, with shrinking methanol and ammonium production (affecting plastics and fertilizer production) and increasing liquid natural gas production (LNG; Colten 2006b). Katrina interrupted oil refining at 15 facilities in Louisiana and Mississippi, Rita caused serious damage to thirteen oil platforms, and nine pipelines from the oil platforms to onshore facilities were temporarily shut down. These incidents are accompanied by shrinking workforces in the oil and petrochemical industry and a growing public awareness in Louisiana of the social and environmental side-effects of the oil and petrochemical industry (Colten 2006b). The explosion of the Deepwater Horizon oil rig in the Gulf of Mexico and the subsequent 87-day outflow of oil into the sea not only caused serious and lasting ecosystem damage in the Gulf and the coasts of Louisiana, Mississippi, Alabama, and also Florida, but also had clear social consequences (Hochschild 2016; Silliman et al. 2012; Sumalia et al. 2012; White et al. 2012).

Thus, the vulnerability of the oil and petrochemical industry in Louisiana is increasing in a double sense: on the one hand, the ecological threat posed by the consequences of global climate change is growing, and on the other hand, social pressure is also increasing; environmental and health damage is increasingly no longer seen as a natural side effect of the wealth-generating industry, but is seen not just by the public, but increasingly by politicians and authorities, as a call to action (Colten 2006b; Karl et al. 2009; van der Wiel et al. 2017; see also Fig. 6.12).

Fig. 6.12 A sticker widely used in public spaces in Baton Rouge, picking up on the aesthetics of historical film posters, addressing the high air pollution in Baton Rouge, Artist: Brad Jensen; own photograph 2019

6.3 The Centers of Academic Life: Louisiana State University and Southern University

The description of the development of Baton Rouge has already made clear the importance of the universities (Louisiana State University and Southern University; see Fig. 6.13) for the biography of the community (Chapter 5). In the following, the main developments and interrelationships of the universities in and around Baton Rouge (and beyond) will be presented. In the order in which they were founded, the focus will first be on LSU and then on Southern University.

The Louisiana State University was founded in 1860 under the name 'Louisiana State Seminary of Learning and Military Academy', whose students were called cadets due to their connection to the military (Cowan 2013). Due to the chaos of the war, however, teaching began haltingly and with interruptions, although the Civil War had an effect on LSU that is still felt today: a battalion, the Louisiana Fighting Tigers, gave it its mascot, the Tiger (Brill 1963; Cowan 2013; Jenkins 2013). On October 15, 1869, the buildings of the seminary's original location in Pineville burned down, the cause of the fire

Fig. 6.13 Impressions of Southern University (top); impressions of LSU (bottom), own photographs 2019

remaining unexplained (Ruffin 2006). Only a few days after the fire, on November 1, 1869, the seminary was moved to Baton Rouge, where it was initially housed in the buildings of the Institute for the Deaf, Dumb, and Blind, where it remained until the campus was built in the former military buildings. Already in 1877 it had been united with the 'Agricultural and Medical College', which was extended in 1887 by an agricultural research institute. However, the early years of LSU were often marked less by a new beginning than by uncertain financing and correspondingly poor planning (Cowan 2013; Fleming 1936).

The bipolarity of the two sites in the north and south of the city was ended by moving the northern parts of the university to the southern site, this was completed in 1932 (Brill 1963; Cowan 2013; Gleason and Brockway 1991; Rodrigue and Phillips 2011; Ruffin 2006). The development of LSU in the 1920s was greatly accelerated by the election of John M. Parker to the office of governor in 1920 (as a Democrat). Parker, who had already tried in vain as a member of the (short-lived) Progressive Party to be elected governor in 1916, advocated a program to promote road construction, especially in rural areas, on the one hand, and the expansion of the educational infrastructure, especially LSU, on the other hand, through tax increases (for example, on unprocessed natural resources; Miller 2000; Rodrigue and Phillips 2011). The design of the new LSU campus followed the design of the famous architect Frederick Law Olmsted, Jr., who had already designed 250 public parks and 60 university campuses, including Stanford and Cornell. On Thanksgiving Day 1924, the new (though not yet fully completed) Tiger Stadium was also opened in front of more than 17,000 spectators with a defeat against their eternal rival Tulane. Despite this, the LSU Tigers became a very successful college football team, winning national championships in 1958, 2003, and 2007 (Rodrigue and Phillips 2011; Ruffin 2006). The rapid expansion of LSU resulted in financial bottlenecks in 1924. But Parker's successor, LSU graduate and brother-in-law of the long-time LSU President, Thomas D. Boyd, also supported the LSU expansion program (LSU 2019a; Ruffin 2006).

The increase in importance of LSU, which was physically manifested in the new campus, also took place in research and teaching. In November 1928, LSU was granted full Class A accreditation by the Association of American Universities, which honored the credits earned at LSU throughout the United States, and in 1935 the first doctoral degree was awarded (Ruffin 2006). Governor Huey P. Long also supported the development of LSU, not just the football team and the marching band, but also academic education, not only interfering in the affairs of the first two, but also trying to work with them towards academic development, although President James M. Smith, certainly one of his confidants, slowed him down in these efforts (Rodrigue and Phillips 2011; Ruffin 2006). He pursued the goal of making LSU one of the leading universities in the United States and provided additional funds to increase the salaries of professors in order to attract renowned scientists (Rodrigue and Phillips 2011). One expression of LSU's new self-image was the founding of the LSU Press in 1935, the campus changed physically through the construction of buildings and a park structure, and also acquired national funds to drain a swamp area east of the campus plus create University Lake near City Park Lake, which was built in the 1920s (Rodrigue and Phillips 2011). At the same time, with the construction of the new Capitol complex in the north of downtown, not only the institutional presence of the LSU had disappeared, but also LSU building structures: they were largely demolished in favor of the new government and administrative headquarters for the State of Louisiana (Ruffin 2006). With the rise of LSU as an academic research and educational institution, the influence of the military, which was still dominant until the 1920s, declined until the dress code for students was finally abolished in 1969 (Cowan 2013).

After a decline in enrollment during World War II, the prewar growth of LSU continued after the war. The G.I. Bill entitled veterans to a college education, which "led to a huge influx of former soldiers on campus" (Rodrigue and Phillips 2011, p. 50). This entitlement to college education provided the basis for social advancement for many veterans who would otherwise never have been able to afford such an education, such as sons of underprivileged immigrants and African-Americans (Mettler 2005). The G.I. Bill provided for one year of education or training for every veteran who had served at least 90 days, with one additional month of education for every additional month of service, up to a maximum of 48 months (Mettler 2005). Accordingly, the number of students rose from the low of 3,105 in 1943 to 8,705 in 1946, which posed a challenge for the spatial and personnel capacities of LSU (Ruffin 2006). However, the education of African-Americans at LSU took place only shortly before the Supreme Court ruled in 1954 that separate educational institutions were inherently unequal, making Roy S. Wilson the first

African-American to graduate from LSU, and in 1961 Pearl Andrews became the first African-American to receive a Master's degree in Education from LSU (Rodrigue and Phillips 2011).

LSU grew in the following decades, new buildings were erected, others were expanded, including several times the stadium of the LSU Tigers, the curriculum was diversified. Since the 1960s, the expansion to an 'LSU system' has taken place by successively establishing campuses at other locations, such as Shreveport and Alexandria (Ruffin 2006). The crisis in Louisiana in the 1980s also affected LSU, where enrolment exceeded 30,000 in 1982 for the first time, fell below 25,000 in 1988 (in conjunction with changes in enrollment standards) and at the same time government allocations fell as a result of the oil price crisis. Savings were made across the LSU campuses, which also affected essential courses of study, and fees were increased at the same time (Ruffin 2006). At the end of the 1980s, however, there were signs that the situation was easing: since 1988, LSU has ranked among the top one percent of U.S. colleges and universities, although government funding remained scarce despite a slow rise in oil prices (Rodrigue and Phillips 2011; Ruffin 2006). Currently, LSU of all universities within the United States is ranked 115th (QS World University Rankings 2019). Even though these rankings suggest a quantitative comparability that needs to be questioned in detail, they do allow a rough classification. At the turn of the millennium, the financial situation of LSU also eased, "higher oil prices combined with a supportive governor, M.J. 'Mike' Foster [in office from 1996 to 2004; authors' note], and his legislature helped LSU to catch up with overdue pay raises, needed course offerings, and delayed maintenance" (Ruffin 2006, p. 133). The extensive separation of the LSU from the downtown area was surmounted with the opening of the LSU Museum of Arts in the south in 2005. The LSU system was directly affected by Katrina, especially on the campuses beyond Baton Rouge. The indirect effects also affected the campus in Baton Rouge both as a refuge for storm victims and because numerous students from other university locations continued their studies at the Baton Rouge location because their home campuses were so badly damaged by the storms that a regular course of study could not be guaranteed there (Ruffin 2006). The number of students at LSU has fluctuated around 30,000 for the decade, and by fall 2018, 30,987 students were enrolled at LSU (LSU Office of Finance and Administration 2019; see Fig. 6.14).

The spatial influence of LSU extends far beyond the campus, which is particularly striking to its north. The space between the LSU campus and downtown has undergone an intensive transformation in recent decades: this area along the Nicholson Drive and Highland Road was much more connected to campus life until the mid-1970s, when alcohol laws required that places where spirits (hard liquor) were served be at least a mile away from campus. So here was a cluster of bars a mile north of campus on Highland Road - until the huge Tigerland apartment complex in the late 1970s offered another drinking opportunity. This is where low-income neighborhoods were, and still are, located to the north just off Highland. Along Nicholson Drive there were shops serving students in the north of the campus, and then a group of sprawling apartment

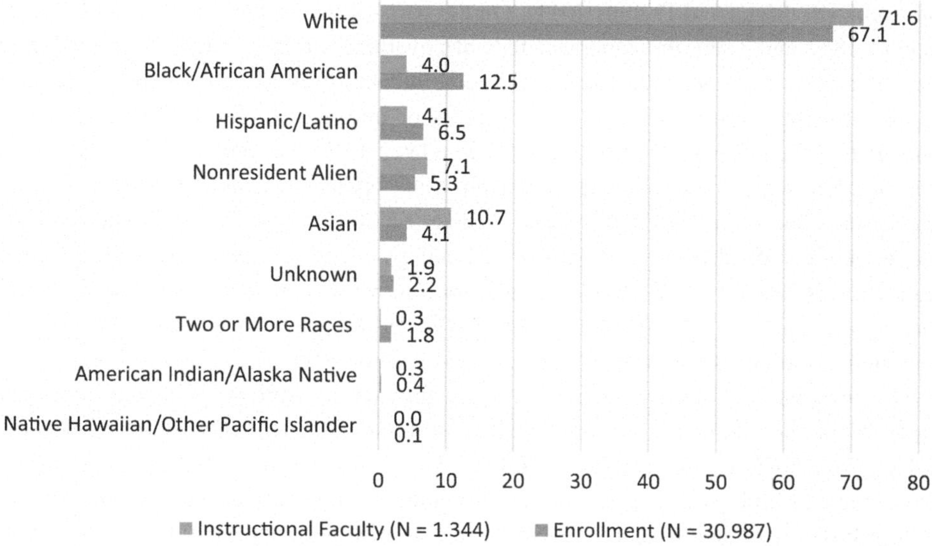

Fig. 6.14 The ethnicity of enrollment and instructional faculty at LSU in autumn 2018 (own presentation and calculation according to LSU 2019b)

complexes near Magnolia Mound, which has now been demolished (since 2000).[1] Currently, a physical and social restructuring is taking place north of the campus: former suburban development is being replaced by university-related buildings (especially apartment buildings) - a development that can be described as decentralized URFSURBanization.

The LSU is still a white-dominated university, both in terms of the ethnic composition of the student body and - even more pronounced - of the teaching staff (Fig. 6.14). Blacks, who make up the majority of the population in Baton Rouge (Fig. 5.17), are a minority among students and even more so among faculty, even falling behind faculty from outside the United States, of Asian or Hispanic descent which in Baton Rouge are a very small minority. The low presence of black people at LSU is (in part) due to the strong alignment of LSU with the Southern University, whose development is briefly outlined below.

Southern University and the Agricultural and Mechanical College System are the result of the efforts of some delegates at the Louisiana State Constitutional Convention 1879 to increase the educational opportunities for colored people in higher education and is one of the largest mainly black universities in the country (Gleason and Brockway 1991). Southern University opened in New Orleans with twelve students, five faculty, one building, and a budget of 10 thousand dollars. In 1890 it was converted into a land-grant

[1]Many thanks to Craig Colten for this information.

university (Rodrigue and Phillips 2011). On the basis of the Morrill Land-Grant Colleges Acts of 1862 and 1890, the American federal government transferred land ownership to the American states under the condition that the income from the sale or use of the land be used for the construction and running costs of universities (Ruffin 2006). Southern University's inclusion in the 1890 system is related to the second Morrill Land-Grant Colleges Act, which required states to establish and operate special colleges for African-Americans (Anderson 1976; Christy and Williamson 2017; Cross II 1999), whereas LSU had already benefited from the first Morrill Land grant and also participated in the 1890 grant (Ruffin 2006). In 1912, the legal basis for a move was created: Southern University was granted the right to sell the property and in 1914 opened in Scotlandville, a small community north of Baton Rouge (Rodrigue and Phillips 2011).

The mission of Southern University is to provide its students with the necessary expertise in the respective disciplines of a professional qualification and to develop skills to contribute to the solution of cultural, social, and political challenges of society (Rodrigue and Phillips 2011). In this sense, Southern University developed into a major driving force of the civil rights movement in Baton Rouge (and beyond) in the 1960s, with Southern University students repeatedly using sit-ins to fight racial segregation and discrimination. The clashes with the police and the courts reached a sad climax on November 16, 1972, when - after several clashes - two black students were shot dead by a police officer during a protest on the campus of Southern University, probably because the officer had misinterpreted an order (Rodrigue and Phillips 2011).

Although the focus of education is on the bachelor's level, Southern University also offers the opportunity for education up to the doctorate level and, especially in technical and scientific subjects, has become one of the leading national institutions in the higher education of the African-American population with a budget of more than 200 million dollars (Rodrigue and Phillips 2011). Today, Southern University has more than 15 thousand students at its five campuses in Louisiana, coming from 46 states and 40 countries outside the United States (Southern University and Agricultural and Mechanical College 2019). Since 1999, buildings bordering the Mississippi River on the Southern University Campus in Baton Rouge have been listed on the National Register of Historic Places. The buildings, including Riverside Hall, are former plantation buildings built in the second half of the nineteenth century and have since served the university in various capacities (such as conference center, dining hall, president's office, dormitories; Rodrigue and Phillips 2011).

Still, there persists a strong ethnic division in higher education between white (LSU) and black (Southern University) in Baton Rouge. Despite the presence of Southern University, as shown in Fig. 5.20, Baton Rouge has a fraction of its black population with a bachelor's degree relative to the white population.

6.4 The Distribution of Physical-Spatial Indicators of the Unequal Distribution of Symbolic Capital and Life Chances in Baton Rouge

Baton Rouge, as shown elsewhere (Sect. 5.5, particularly Fig. 5.18 and Fig. 5.21), has large spatial concentrations of both the ethnic composition of the population and the distribution of poverty. In the following, we will examine to what extent these distributions are also reflected in physical space. To this end, both physical-geographical aspects, such as the distribution of trees, urban climate aspects, and the distribution of noise will be studied. These are not systematic complete surveys, but rather indicator-based approaches. Subsequently, the spatial distribution of physical-spatial indicators of poverty is examined, as well as those of restaurants having different price levels. Finally, the distribution of public and private high schools in Baton Rouge is examined.

6.4.1 Cultural Hybridization

Communities modify, shape, and displace physical objects that are strongly inclined towards the pole or core of the 'elemental' within their conception of a nature-culture hybridity. In the following, we will focus on nature-culture hybridizations that in their differentiation have a retroactive effect on social references, especially the availability of symbolic capital.

The first example of nature-culture hybridization is the spatially differentiated implementation of Baton Rouge with trees. Aerial photographs in particular show the strong degree of penetration of trees throughout Baton Rouge, with residential areas in particular showing a high stock of trees, which can be interpreted as a greater tendency towards the pole of 'naturalness' of such a component in the spatial pastiche of Baton Rouge.

Trees are associated with different physical effects as well as symbolic connotations. While, in a metaphysical sense, they are regarded as symbols of life, which can sometimes also be the subject of religious-mythical interpretations, in a purely physical sense, they provide shade and thus lower the air temperature near the ground, bind carbon dioxide, and contribute to an improvement in air quality through the filtering effect of their leaves. Trees are not only perceived positively by the urban population, there are also inverse correlations between the presence of urban trees and the crime rate (read more at: Heisler 1974; Henninger 2011; Jenal 2019b; Lohr et al. 2004; Nowak et al. 2000; Sullivan and Kuo 1996). Following such connections, we have examined for comparative purposes the tree population along Government Street (Fig. 6.15) and Plank Road (Fig. 6.16) which were already largely built-up in the middle of the twentieth century and several residential streets (the other two streets examined more intensively in our study of Baton Rouge either have only isolated residential buildings, in the case of Florida Street/Boulevard, or have been developed more recently and are also strongly influenced by commercial use, as in the case of Siegen Lane). This examination is

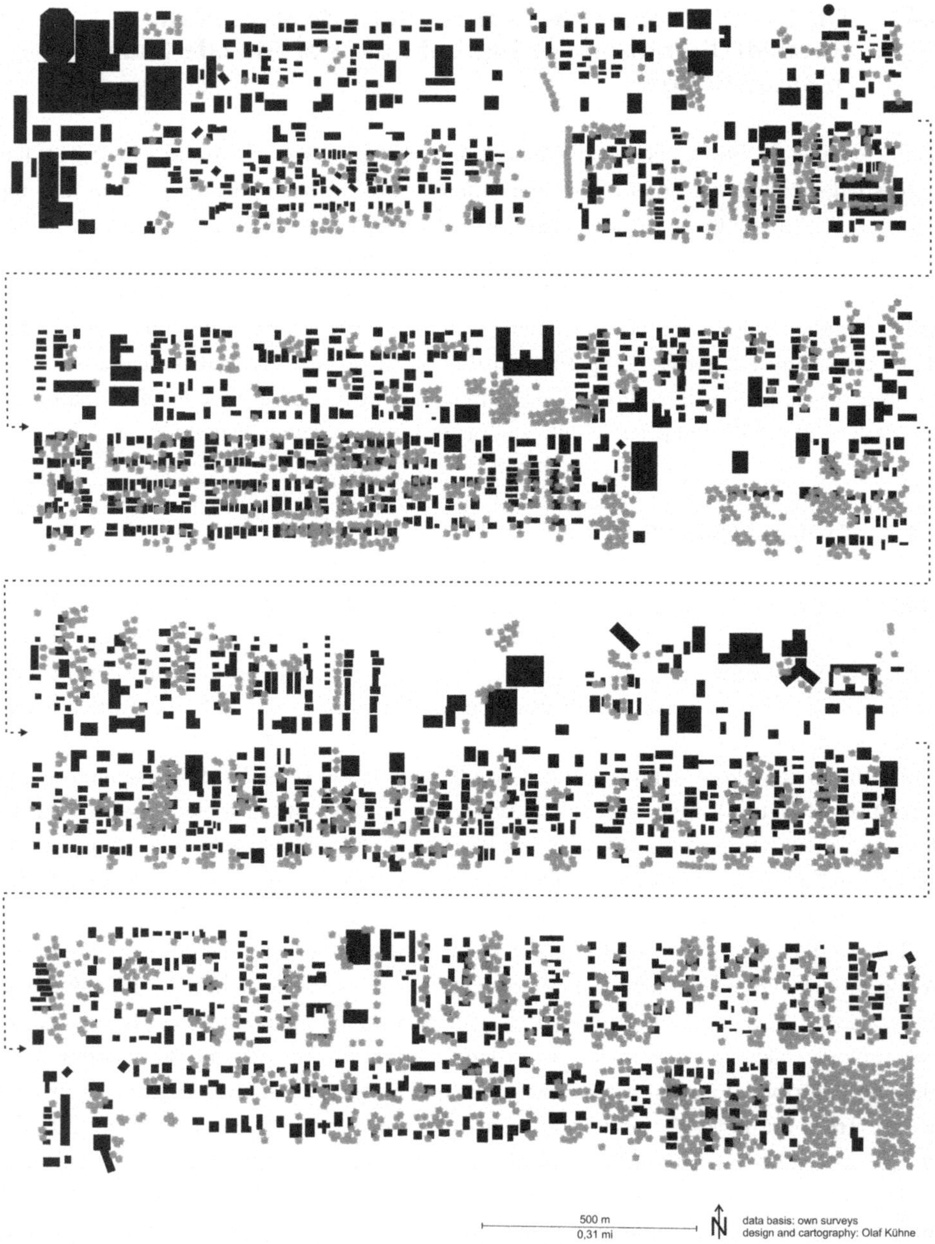

Fig. 6.15 Figure-Ground Diagram with tree locations along Government Street (own survey)

supplemented by the presentation of the downtown and adjacent areas, whereby the different tree populations in the context of different building structures and also their functions become better elucidated (Fig. 6.15).

Along Government Street, a higher total tree population south of the reference street is evident. A small collection of trees can be found in the downtown area, also along the Mississippi River south of Government Street. A smaller tree population, especially to the southern neighboring residential areas along Government Street, can be found to the east after crossing Interstate 110 as well as in the area of the Foster Drive intersection. While this area is characterized by commercial uses, the area east of the Interstate 110 is characterized by a mixture of used and unused residential buildings as well as commercial buildings in use and those abandoned. In this area, as highlighted in Sect. 5.5, the proportion of ethnic minorities with a low level of symbolic chapter is particularly high. In the tree-rich areas south of Government Street, east of Park Boulevard, on the other hand, a primarily white population with a high level of symbolic capital dominates.

Plank Road has a low tree population compared with the residential areas east of Park Boulevard, especially those to the south of Government Street (Fig. 6.16). Along Plank

Fig. 6.16 Landuse plan with tree locations along Plank Road (own surveys)

Fig. 6.17 Figure-Ground Diagram with tree locations in the city center and adjacent areas (own surveys)

Road there are residential and commercial buildings, each with a different intensity of use. The residents of Plank Road are particularly characterized by people of color, and the proportion of people living in poverty is generally above average.

The figure-ground diagram with tree locations in the downtown and adjacent areas (Fig. 6.17) shows a very clearly differentiated distribution of trees: the Central Business District has a very low tree population, as does the area north of Government Street, along Interstate 110 (already mentioned in the context of Government Street); a more extensive tree population is found around the public buildings in the north and south of the CBD. Beauregard and Spanish Town are strongly marked by trees. Here, too, a denser tree population can be seen in parts of the community inhabited by people with a higher endowment of symbolic capital (which are also usually white).

The presence and absence of trees have impacts on the local climate. During the day, the short-wave radiation on the earth's surface (or parts of buildings) is reduced, which leads to a reduced radiation turnover and thus a comparatively lower air temperature. Transpiration, especially through the leaves, reduces the air temperature, but since only tangible heat is converted into latent heat, the total heat content of the air does not change, more on this later in the context of the equivalent temperature. At night, trees reduce the long-wave radiation, thus counteracting the cooling of the air close to the ground. However, this effect can be more than compensated for by reduced heat absorption, especially by buildings and paved surfaces during the day (see in particular Akbari et al. 2001; Georgi and Zafiriadis 2006; Kühne 1999). In this respect, inner-city tree stands are capable of significantly modifying the expression of local climatic phenomena, especially air temperature. This, too, can be interpreted against the background of considerations of nature-culture hybridity: the modification of the local climate through the collective activities of those in the community (in particular the construction of buildings and infrastructure facilities, especially roads and parking lots) leads to a stronger tendency towards the side of culture regarding the scale of nature-culture hybridity.

Of far-reaching bioclimatic importance is the equivalent temperature t_{eq}. This is a fictitious quantity that cannot be measured directly but can only be calculated. It is regarded as an essential parameter for the thermal load on humans and is calculated as follows (Bründl et al. 1986):

$$t_{eq} = t_a + \frac{0.622 * r' * e}{p * c_p}$$

where by:

t_a = air temperature in °C
r' = heat of evaporation of water in Jkg^{-1}
e = vapor pressure in hPa
p = air pressure in hPa
c_p = specific heat of air at constant pressure in $Jkg^{-1}\ K^{-1}$

The equivalent temperature represents the total heat content of the air. This is composed of tangible and latent heat. This is the temperature that a volume of air would assume if all the water vapor contained in it condensed and the heat of condensation released in the process was fed into the air. The combination of high air temperature and high air humidity hinders the body's ability to dissipate heat, resulting in heat accumulation and a feeling of sultriness. For a wind speed of 1 m/s applies (Flach 1957):

51.0 °C (123.8°F)	t_{eq}		sultry
47.0 °C (116.6°F)	$t_{eq<}$	51.0 °C (123.8°F)	slightly muggy
34.0 °C (93.2°F)	$t_{eq}<$	47.0 °C (116.6°F)	neutral (comfortable)
27.0 °C (80.6°F)	$t_{eq}<$	34.0 °C (93.2°F)	slightly cool
	$t_{eq}<$	27.0 (80.6°F)	cool

The underlying air temperature and relative humidity were measured at the bioclimatic effective altitude of 1.5 m above ground level with a calibrated thermometer, the air pressure was taken from the National Weather Service data (converted to sea level) and reduced to the altitude level (above sea level) of Baton Rouge. The diurnal temperature and humidity curve of the atmosphere for Baton Rouge was taken from the National Weather Service data and the measured values were adjusted according to these curves (more on this: Kühne 1999). The measurements took place under cloudless skies and an allochthonous wind with a speed of less than one meter per second.

The distribution of the equivalent temperature in the downtown and adjacent areas of Baton Rouge on September 12, 2019 (Fig. 6.18) exceeds the above threshold for sultriness everywhere. The maximum area of the equivalent temperature is located east of Interstate 110 in the area of mixed residential and commercial buildings consisting of both a different usage intensity and a small tree population. Both the area of the Central Business Districts and the areas with a larger tree population show – as a result of the shading - significantly lower equivalent temperatures. The location and thermal level of the maximum equivalent temperature area in a neighborhood whose inhabitants are disproportionately affected by poverty is particularly problematic, as the households located here often have no air conditioning.

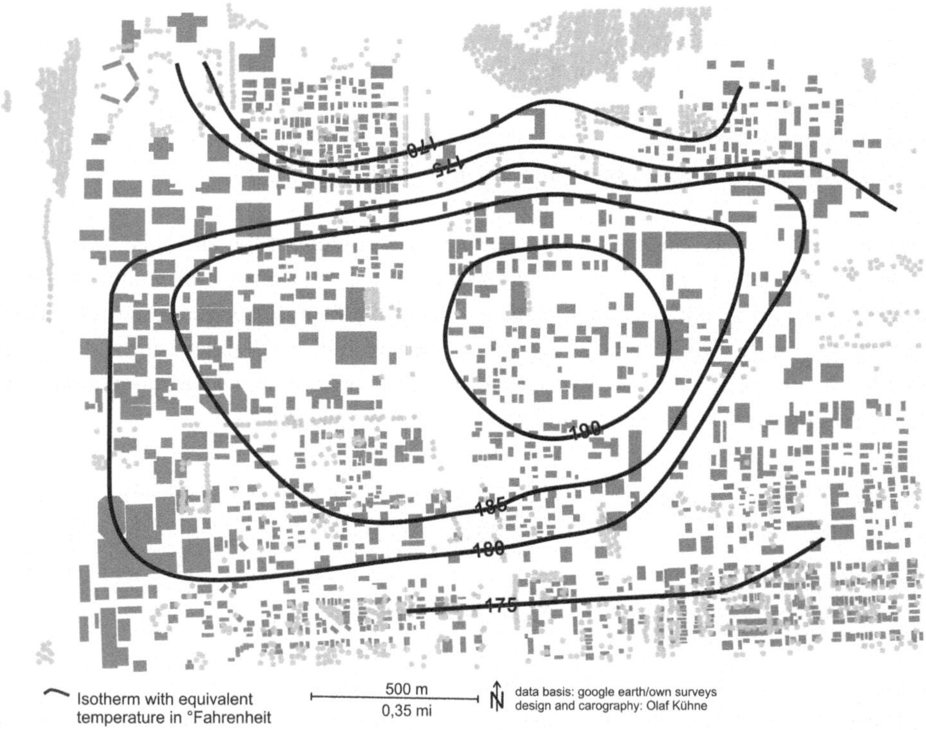

Fig. 6.18 The distribution of equivalent temperature in downtown and adjacent areas of Baton Rouge on September 12, 2019, measured between 13:50 and 15:45 (own measurements)

Not only during the day, but also on the night of September 11–12, 2019, there is a maximum area of equivalent temperature covering the above-mentioned area east of Interstate 110, in this case extending over the Interstate because of the heat storage effect of the building structures and sealed surfaces into downtown Baton Rouge (Fig. 6.19). This heat storage effect is particularly apparent around the former Bon Marché Mall and its extensive parking spaces, the area around the intersection of Lobdell Avenue and Florida Street, and the intersection of Florida Street/Boulevard and Airline Highway, where extensive sealed surfaces (including the Cortana Mall) are also found. The lowest

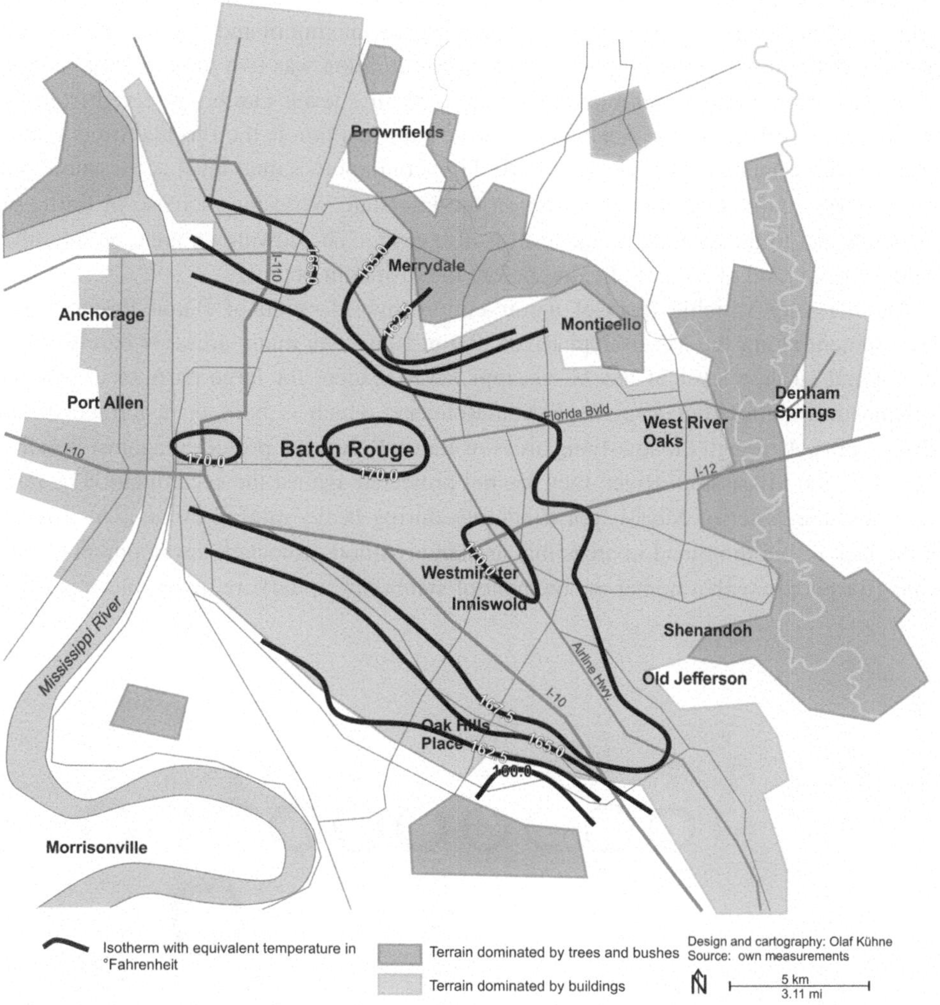

Fig. 6.19 The distribution of the equivalent temperature in Baton Rouge on September 12, 2019, measured between 0:50 and 2:50 (own measurements)

equivalent temperatures are found - although still well above the threshold of humidity - in low-lying areas lacking in taller vegetation. Cold air that is produced collects here and the humidity contained in the air condenses out when the dew point is reached and is correspondingly no longer bioclimatically effective above the equivalent temperature.

Although the distributions of equivalent temperature are only snapshots, classical patterns of equivalent temperature distribution can be found in Baton Rouge on these days. In this respect, it can be assumed that these distributions also occur in similar large-scale weather situations.

Another snapshot that can be generalized is the result of the measurements of sound pressure at noon on September 13, 2019 (Fig. 6.20). The measurements were made with a calibrated sound level meter. Due to the dominance of road traffic noise, two traffic light switching cycles were always measured and the maximum and minimum noise levels were determined (the minimum measurement duration was two minutes). In general, an increase in minimum and maximum sound pressure levels can be observed from the Mississippi riverbank to the Acadian Thruway. An exception is the Florida Street underpass beneath Interstate 110, where a fairly high continuous sound level is measured due to the fairly continuous flow of traffic on the Interstate. With the increase in traffic on Florida Street to the east, the noise level is also rising. The downtown area, on the other hand, forms a kind of 'island of calm' before the rush hour begins.

Another very variable form of nature-culture hybridization of Baton Rouge is the development along the Mississippi River: Baton Rouge, in many areas, is barely above the normal surface level of the Mississippi River (except for large parts of downtown and the universities, but also the Exxon Refinery, which is built on a terrace of the Mississippi). Although the low-lying areas of Baton Rouge are protected against flooding caused by the Mississippi River, they are not protected against the accumulation of rainwater and the water of Mississippi tributaries during heavy rainfall events). As a result of the lack of building land in areas that are largely flood-protected, development expansions (for people with a higher endowment of symbolic capital) are taking place in areas

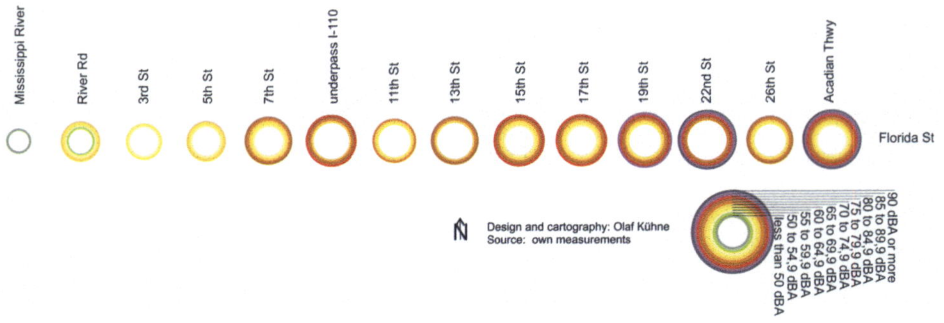

Fig. 6.20 The noise level along Florida Street from the Mississippi to the Acadian Thruway, measured on September 13, 2019, between 13:15 and 14:30 (data basis: own measurements)

at risk of flooding, whereby an increase in the risk of flooding is predicted as a result of anthropogenically intensified climate change (another nature-culture hybrid; Black 2008; Mathur and da Cunha 2001; van der Wiel et al. 2017).

6.4.2 Geographies of Consumption Depending on the Availability of Symbolic Capital

A differentiated endowment of symbolic capital manifests itself in physical spaces in the form of specific structures relating to use and supply. Because of the widespread use of automobiles (even by people with little symbolic capital), physical-spatial distances are losing importance, but specific spatial patterns of use and supply structures still exist. With the goal of identifying specific spatial patterns, we studied the distribution of pawn shops and one-dollar stores in Baton Rouge, as well as interior design and antique shops. In addition, the distribution of restaurants with different price levels was also examined.

Pawnshops in particular can be understood as an expression of a low level of symbolic (especially economic) capital, since these are used primarily by people who have no or very limited access to alternative credit (whether private or through banks; for more information see: Bos et al. 2012; Carter 2015). Cheap mixed-goods stores, which usually have 'dollars' in their name (Family Dollar, Dollar Tree, Dollar General) can also be seen as an indicator of consumers with special price awareness, whether forced or voluntary (for more details: Carpenter 2008; Carpenter and Moore 2009). Conversely, antique dealers are aimed at persons with a higher endowment of symbolic capital who are financially able to pay higher prices for objects with the same or less practical value than newer objects, based on their own taste or need for distinction. The same is true for specialized interior design shops. Here, the focus is not on pure function, but on the 'refined' design (see also Bourdieu 1985b; Combs 2003; Dimaggio and Useem 1978). The spatial distribution of pawn shops shows an accumulation along the entry and exit roads Airline Highway and Florida Street/Boulevard, especially in the area north and west of their intersection (Fig. 6.21). Cheap mixed-goods shops are more common north of Florida Street/Boulevard, with samples showing that the range of goods on offer in those parts of the city with a high proportion of people living in poverty was limited compared with other parts of the city in terms of the supply of items not needed for immediate physical survival. Interior design shops are more common south of Florida Street/Boulevard, as well as in the suburban areas east and north of the airline highway, a distribution that is also found for antique dealers, with a clear concentration in Denham Springs, where the relevant products are offered in a small-town, historic atmosphere that is set apart from the mere function of a shopping mall, thus enabling a gain in distinction for those with relevant specific tastes.

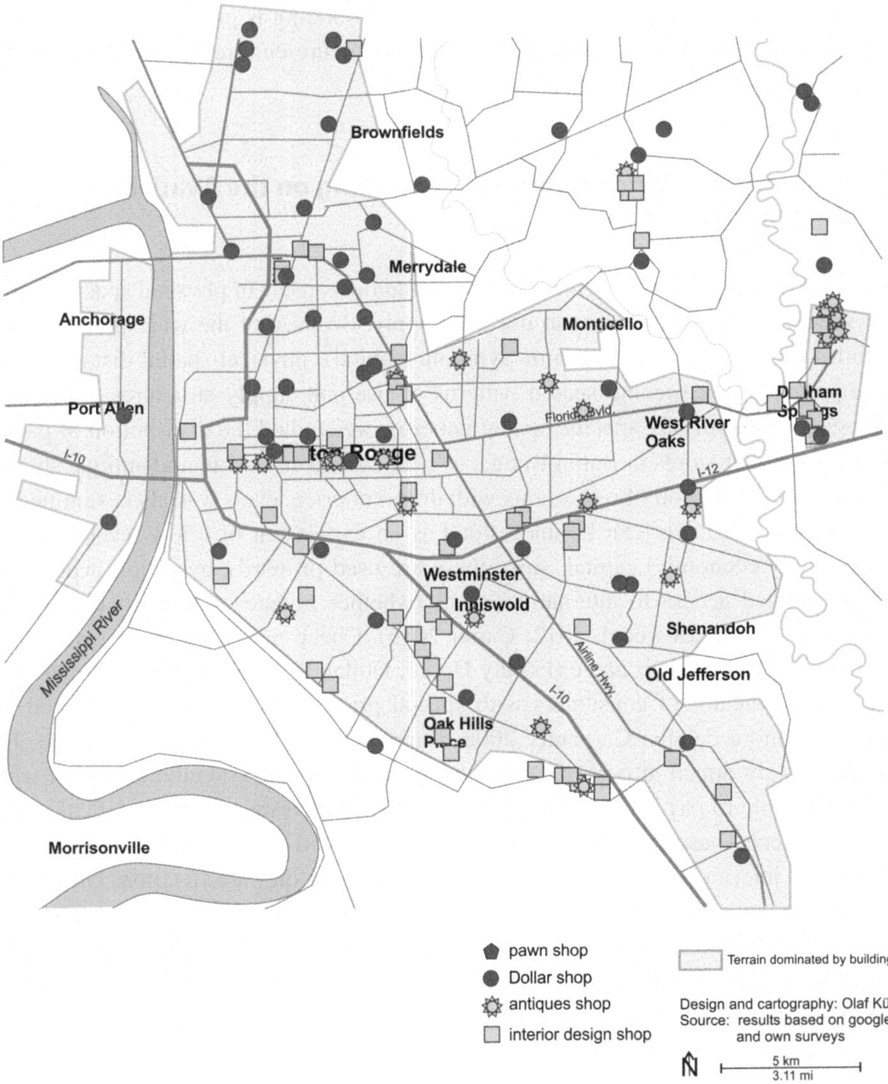

Fig. 6.21 The spatial distribution of pawn shops, cheap mixed-use shops (Dollar shop), antique dealers, and interior design shops (own survey in September 2019)

The distribution of restaurants as well as their price categorization follows, supplemented by a random tracking in the area - via Google Maps, the most frequently used search engine. The restaurants found in Baton Rouge largely serve the need to purchase food at a low price (Fig. 6.22). On the one hand, there are clusters linearly along the already mentioned entry and exit roads, but on the other hand, there are also rather flat

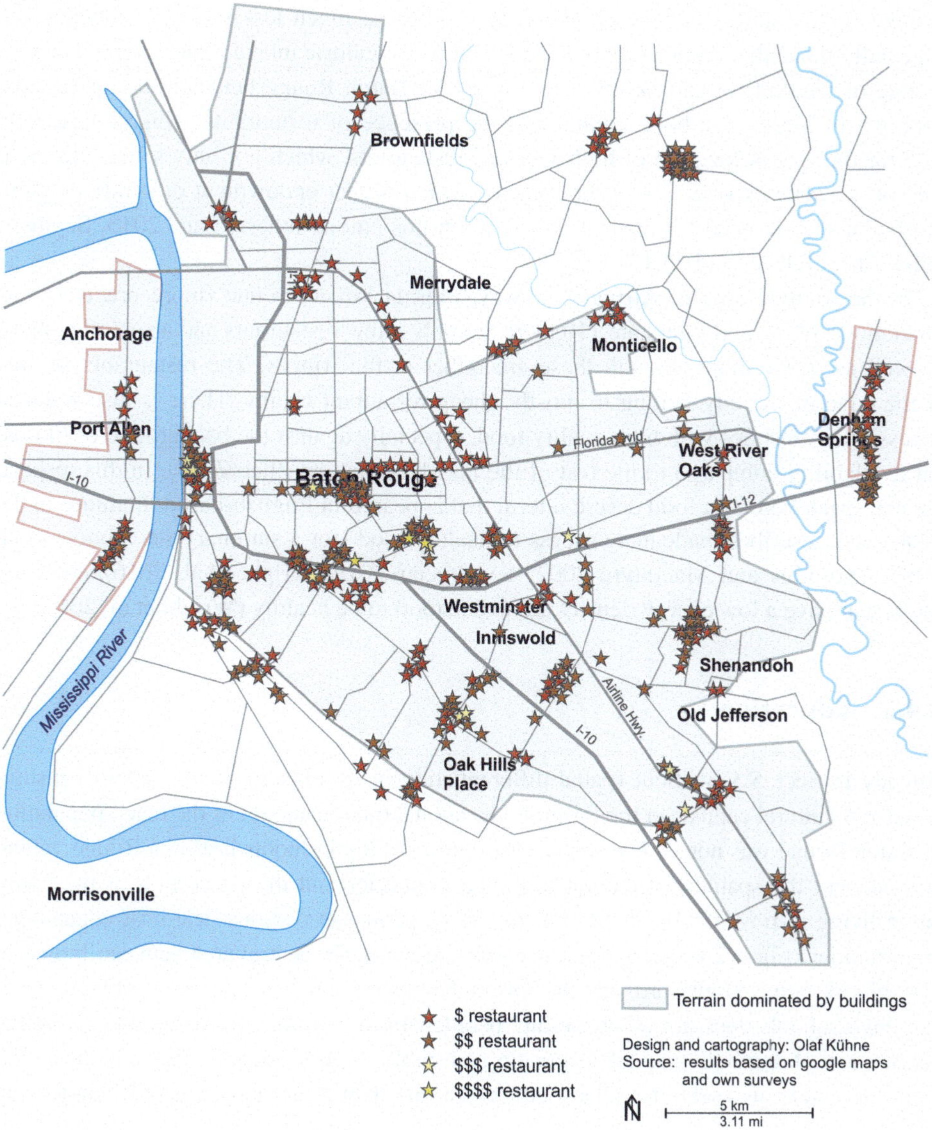

Fig. 6.22 The distribution of restaurants in different price categories (own surveys)

conglomerations, especially in the downtown area, in the neighboring city of Central, in the area of the overpass of Interstate 10 via Perkins Road, the middle area of Government Street, and somewhat more linearly in Denham Springs, as generally in the southern suburban residential areas of Baton Rouge. Higher-priced restaurants are usually located in these conurbations. In general, the area north of Florida Street/Boulevard (except for the above-mentioned areas) appears to be much less served by restaurants, especially those that do not satisfy the basic need for calorie intake alone. Overall, a high degree of parallelism can be observed (also) for Baton Rouge between the spatial distribution of higher symbolic capital and the presence of restaurants. This is especially true for the spatial location of higher-priced restaurants, which not only satisfy the need for special taste experiences of their guests with a higher endowment of symbolic capital, but also their need for distinction (more on this among others: Lane 2013; Seymour 2004; Stringfellow et al. 2013).

In the triangle area of Airline Highway, Florida Street/Avenue (more precisely one block north of it), and Interstate 110 there are only a few restaurants and also dollar shops as well as pawn shops, though these are rather at the fringes. The restaurants located at the edge of the city belong to mostly cheap restaurant chains. There is also reduced accessibility to shops with high quality food, especially against the background of limited car availability compared to the rest of the Parish (Kuai and Zhao 2017). In this respect, we can speak here of a food desert, a term that appeared in English-based literature in the 1990s and describes inadequate access to healthy food (for a summary see Alviola et al. 2013; Cummins and Macintyre 2002; Sedelmeier 2019; Wright et al. 2016), whereby foods that have a low energy density are understood to be healthy (Wright et al. 2016).

6.4.3 Education

Already in Sect. 5.5 (without spatial differentiation) a significantly lower level of institutionalized cultural capital (in the form of educational qualifications) of the black population of Baton Rouge was noted. The spatial distribution of high schools in Baton Rouge, taking into account the spatial distribution of ethnic minorities and the proportion of the population living in poverty (Fig. 5.18 and Fig. 5.21), shows an obvious dearth of educational opportunities (Fig. 6.23) due to the low concentration in the central area between Interstate 110, Florida Street/Boulevard, and the Airline Highway. The distribution of public and private high schools does not allow for any parallelism in principle between ethnicity or the distribution of symbolic capital to private and public high schools, except for the dominance of private high schools in the southeastern suburban expansion areas of Baton Rouge.

6.4.4 The Distribution of Physical and Spatial Indicators of the Unequal Distribution of Symbolic Capital and Life Chances - A Conclusion

The physical manifestations of life chances possibilities in Baton Rouge are spatially dispersed in very different ways. Areas of cumulative preference, especially to the southeast of Florida Street/Boulevard, but also north of it, east of the Airline Highway, show greater accessibility to higher education, a higher degree of greening and less climatic stress, but also increased accessibility to healthy food. The areas of culminating disadvantage, on

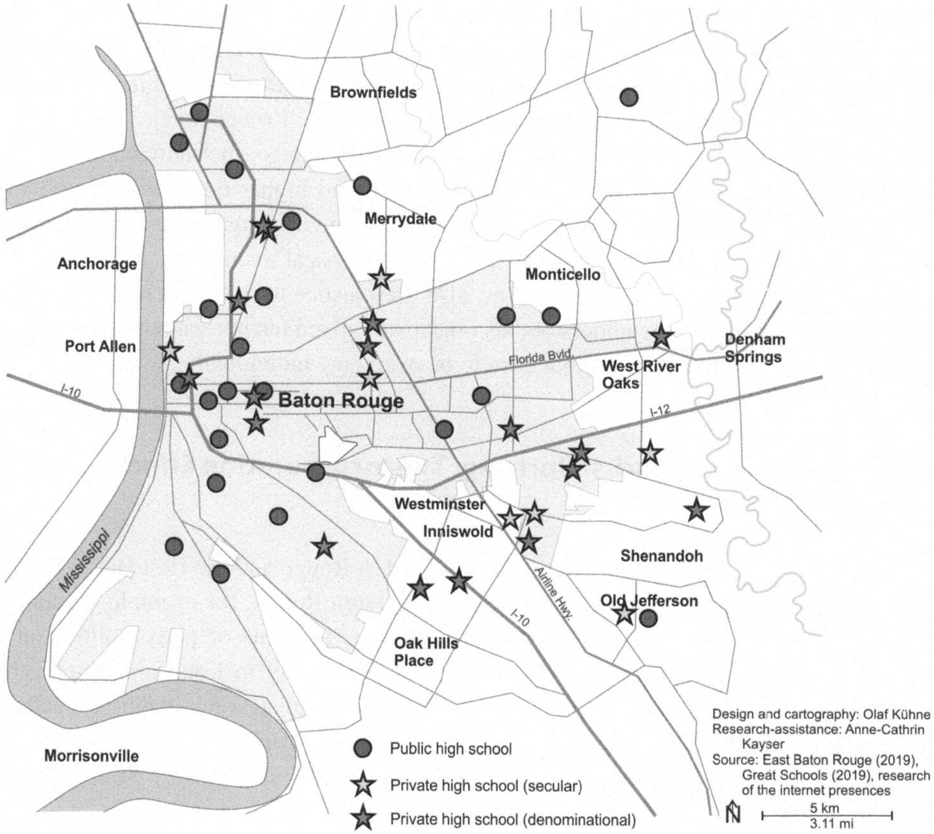

Fig. 6.23 The distribution of public and private high schools in Baton Rouge. We would also like to thank Michael Stein, who in a presentation of our first results at LSU recommended that we map the school locations, separated into public and private

the other hand, are characterized by lower tree populations, higher equivalent temperatures, and, along the arterial roads, by higher noise pollution as well as a lower supply of (healthy) food and higher education. The low endowment of the inhabitants with symbolic capital in this space increases their vulnerability (on this concept see: Adger 2006; Cutter et al. 2003; Lankao and Qin 2011): low economic capital prevents, for example, individual reduction of the consequences of the sultriness 'island' through availability of air conditioning systems. Accessibility to higher education institutions is not only limited by their lack of physical existence in the area between Interstate 110, Airline Highway, Florida Street/Boulevard, it is also less likely to be compensated for by participation opportunity via transportation through reliable cars (lack of economic capital) or by moving around on foot as a result of limited accessibility of footpaths (see Sect. 6.6 for details). As a result of such limitations, this area can be described as a 'triangle of diminished life chances'. There is also - due to the lack of social, economic, and cultural capital - considerable vulnerability to the consequences of climate change.

The studies presented in this section deal - in the positivist tradition of thought - with the unequal distribution of physical objects as indicators of the unequal distribution of life chances and symbolic capital. These inequalities induce further-reaching questions, which - framed theoretically in a constructivist manner (see also Sedelmeier 2019) - deal with questions of the awareness of advantage and disadvantage, the individual and collective coping strategies in relation to life in a 'food and higher education desert', the heat island, etc., but also the political and administrative handling of these inequalities. Such a framing, in combination with the mapping of physical manifestos of the distribution of symbolic capital and of life chances, also does justice to the claim made here that these 'deserts' should not be understood as objectively given territories, but as indicators of disadvantageous life situations that have to be overcome individually.

6.5 The White Suburbs' Uprising: The Incorporation Efforts Around St. George

This ethnic and economic division in the city of Baton Rouge and the East Baton Rouge Parish is not only reflected in the election results: Baton Rouge, for example, is ranked third on the list of the country's most polarized cities in terms of party politics on a national scale after New Orleans (Shreveport is the third city in Louisiana among the most polarized cities, Shreveport is ranked fifth, with the greatest polarization reaching Jackson, Mississippi; Dottle 2019), it is also noticeable in terms of centrifugal territorial tendencies, the current example of which is the incorporation efforts of St. George in southeast Baton Rouge parish.

As is often the case with such endeavors, subsequent efforts at incorporation are based on the desire of white, wealthy parents to establish their own school district. The reason given for this was that East Baton Rouge public schools are generally considered to be among the poorest in the state and their situation is unlikely to improve in the foreseeable future (Harris 2019; The City of St. George 2019). In 2012, when the parents failed to achieve the necessary quorum to establish the Southeast Community School District (Harris 2019), preparations began to create a community in the previously unincorporated area (White 2019). But even for the establishment of a separate municipality, the organizers failed in 2015 because of the necessary quorum for a vote on incorporation (25 percent of the voters registered in the area) fell short by 71 signatures (Jones 2019a). This failure was due in no small part to the activities of the 'Better Together' initiative, which conducted a withdrawal campaign that led 1,000 eligible voters to withdraw their support signature for the holding of a referendum on the founding of St. George (Harris 2019). In view of the dominance of private high schools, however, a comparison of the maps of the planned boundaries of St. George (Fig. 6.24) and the sponsorship of high schools in Baton Rouge (Fig. 6.23) leads to the question of whether the initiators are primarily concerned with a better public school system in the area of a St. George municipality or whether the primary goal is to largely withdraw from financing the public school system in Baton Rouge in order to be able to reduce taxes in St. George, for example.

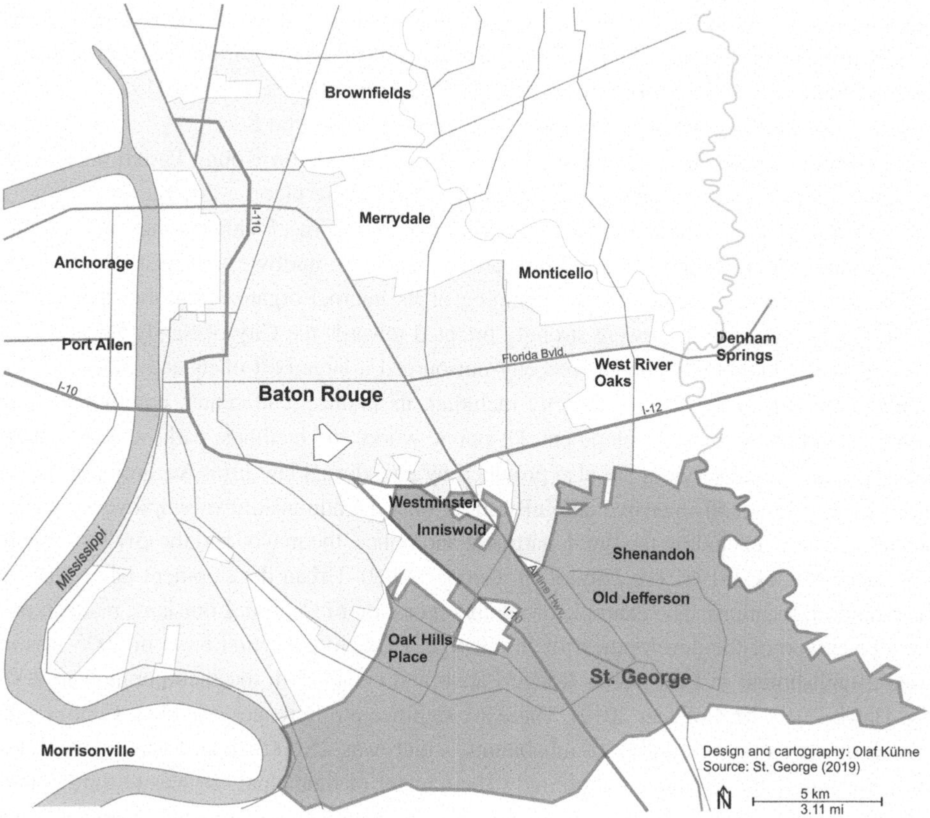

Fig. 6.24 The demarcation of the planned city of St. George in the 2019 design

The proponents of the incorporation of St. George justified the failure of the project by citing too large an area and population of the area to be incorporated (Jones 2019a; The City of St. George 2019). Proponents of incorporation reduced the territorial scope and initiated a new petition for the establishment of the municipality (Fig. 6.24). According to this, the city of St. George would have a population of more than 86,000 people. Despite a reduction in the 2019 proposal by 15 square kilometers of land area and 20,000 inhabitants compared to the 2015 proposal, St. George would become the fifth largest city in Louisiana with its incorporation, ahead of Lake Charles (Harris 2019; White 2019). More than 70 percent of the population of the proposed area is white and less than 15 percent is black (Harris 2019), while in the Parish just under 46 percent of the population is black (see Fig. 5.17: The ethnic population composition of Baton Rouge compared to the East Baton Rouge Parish and Louisiana in 2017 (own representation according to: United States Census Bureau 2019a, b, c). The backdrop of the planned city of St. George includes not only residential quarters with a predominantly white, affluent population, but also prosperous commercial areas. St. George's tax revenues could develop accordingly: while the budget of the proposed city could include expenditures of $33 million in its first year of independence, this income would be offset by $58 million in revenue. Fed mainly on sales tax revenues within the limits, which were estimated at $53 million (Jones 2019a; The City of St. George 2019). Two other figures illustrate the capacity of the area that could one day be St. George: this segment of the community is where about two-thirds of the East Baton Rouge Parish tax base is generated, but only about one-third is spent (The City of St. George 2019).

As exemplary benchmarks for the incorporation of St. George, Sandy Springs, in the vicinity of Atlanta, Georgia, and Central, Louisiana, a city in the northwestern East Baton Rouge Parish, are referenced. Regarding the planning of its internal organization, the operators of the St. George incorporation were strongly oriented towards the City of Sandy Springs with its more than 100,000 inhabitants. The city outsourced a large part of its activities, with the contractually agreed services of the city including its finance, call center, communications, economic and community development, IT, public works and facilities, parks and recreation, and city court. Legal services are also provided by an external law firm. Accordingly, Sandy Springs has a low staff-intensive administration: only 14 administrative employees manage the city's payroll, excluding the fire department and police, the mayor and the city's six-member council (Jones 2019a; The City of St. George 2019). Urban development takes place at quasi-random locations. The example of the incorporation of Central, Louisiana, reads in part like a prototype of the development of St. George. Here, too, the first issue in 2005 was its failed establishment in East Baton Rouge Parish and ended with the incorporation in 2007 (e.g. Harris 2019; McCollister 2019). There are significant differences regarding Central, particularly in terms of the number of inhabitants, which was 25,000 when it was founded, but also regarding Sandy Springs. Accordingly, the transfer of municipal services to third parties means the absence of public servants in the event of a disaster, such as the floods of 2016, which also affected large parts of the planned area of St. George (Jones 2019b). In addition, it is to be assumed that the budget plan of the planned city of St. George will be additionally burdened by pro rata liabilities of the Parish from the past, including pension and retirement liabilities (The Hayride - Southern Politics & Culture 2019). Of additional note, there are liabilities from the renovation of the Woman's Hospital on the Airline Highway and new offices of the local criminal prosecution department (Karlin 2019b). On October 12, 2019, the citizens of the potential city of St. George voted 54 to 46 percent in favor of incorporation as an independent municipality. A majority of the western constituencies bordering the city of Baton Rouge voted against incorporation, while the eastern constituencies voted in favor (Jones 2019b). In these parts of the voting precincts, however, there are also discernible preferences to be annexed by the city of Baton Rouge rather than becoming part of the St. George Community, which means that incorporation is now geared towards questions of exact demarcation (Jones 2019b; Schmaltz and Casserleigh 2019). With the book entering the production process at the beginning of December 2019, this was the state of development.

As mentioned earlier, both the proponents and opponents of the Baton Rouge incorporation are engaged in intensive public relations work. This also includes the operation of internet hosted 'homepages' on which their own position is presented plus their own activities and news are shared. The evaluation of the text content of the respective homepages illustrates the different focuses of the two competing initiatives (Fig. 6.25 and Fig. 6.26).

Fig. 6.25 The textual content of the homepage of the initiative 'The City of St. George' in a word cloud, for technical reasons 'Baton Rouge' was abbreviated as 'BTR' (The City of St. George 2019; created with www.wordwolken.com)

Fig. 6.26 The textual content of the homepage of the 'Together Baton Rouge' initiative in a word cloud, for technical reasons 'Baton Rouge' was abbreviated to 'BTR' (Together Baton Rouge 2019; created with www.wordwolken.com)

The central reference point of both initiatives is 'Baton Rouge', while 'St. George' is added to the incorporation initiative (the 'St.' was not considered in the presentation for technical reasons), which, interestingly enough, only has an implicit meaning on the homepage of 'Together Baton Rouge', because here the emphasis is placed on community involvement; accordingly, the thematic diversity is also higher here, which is also reflected in the word cloud. In the word cloud 'The City of St. George', the indicator for the future ('will') is clearly marked by abstract terms such as 'incorporation', 'taxes', 'budget' or 'fund'. Terms that have a much lower frequency on the homepage of 'Together Baton Rouge'. Here - due to the pronounced sponsorship of the initiative by churches - words with church connotations dominate. Much more than on the homepage 'The City of St. George', words are used here that characterize social relationships - far beyond the name of the initiative 'Together', for example in relation to 'share' or 'believe'. In Max Weber's terminology, the homepage of the proponents is dominated by arguments that suggest rational social action, while the opponents of incorporation seem to focus more on value-rational or affective dimensions of social action (Weber 1976 [1922]).

As the examples of Sandy Springs and Central already suggest, efforts to incorporate St. George are part of a long tradition of economic and ethnic polarization of communities, which is now manifested in increased administrative separation. In addition to this general trend in large parts of the United States, which is based largely on the massive expansion of the highway infrastructure and, thus, in combination with cheap home construction based on Fordist principles and mass motorization, laid the foundation for the 'White Flight', a major reason can also be found in Louisiana's legal history (detailed Chapter 3): in the late 1960s, the state's public schools were legally forced to comply with the requirements of ethnic integration. The unintended side-effect of this equal opportunities policy was not only the establishment of a considerable number of private schools (White 2019), but also the efforts of wealthier school districts, such as those in the Central and 'St. George' areas, to revise the system to create greater economic and ethnic homogeneity. The result of these efforts to improve the life chances of young people with a higher endowment of symbolic capital, as well as the hope of lowering municipal taxes, has as a side effect the reduction of life chances in those areas outside the incorporated municipalities, where the only way to maintain the qualitative level of public schools is to increase municipal taxes, which seems difficult to implement in view of the low endowment of symbolic, especially economic, capital of the inhabitants. This in turn is linked to the consequence of a further deterioration of the school system, which in turn is a driving force for further active segregation of people with a higher endowment of symbolic capital. The result is the above-mentioned almost dichotomous division between a white, wealthy Baton Rouge and a poor, black Baton Rouge, whose demarcation line is roughly formed by Florida Boulevard (see also Harris 2019). With considerable consequences: "The persistence of racial segregation in American cities continues to define those cities' politics" (Dottle 2019, o.S.).

As put forth by the Polish-English sociologist Zygmunt Bauman (2008, p. 130) this development can also be described as an expression of "mixophobia". Mixophobia is thus expressed "in a striving for islands of similarity in the middle of a sea of diversity and difference" (Bauman 2008, p. 130). Resulting from the extreme social, ethnic, political, and largely increasing territorial antagonisms, Baton Rouge in a special way only incompletely forms "a broken whole" in functional terms and otherwise forms "a 'unity without spatial context'" (Hasse 2000, p. 32 f.). Social homogenization, manifested in distinct spatial units with unintended side effects, "lowers the tolerance of the inhabitants for difference and multiplies the causes for mixophobic reactions, which leads to urban life appearing more 'risk-prone' and thus more tormenting than safer and thus more relaxed and pleasant" (Bauman 2008, p. 136). This - following the logic of mixophobia - heralds another round of de-differentiation of urban functions and structures (using Los Angeles as an example: Kühne 2015).

6.6 Walkability

The automobile 'orienting' of traffic in Baton Rouge has been addressed at one point or another (e.g. in Sect. 6.1.1), in the following, the consequences of automobile orienting for the automobile traffic itself will be addressed first and the accessibility of Baton Rouge will be examined in more detail using selected example roads.

6.6.1 The Traffic Framework of Walkability

The traffic structures of Baton Rouge show - as already discussed elsewhere, especially in Sect. 6.1.1 - a clear focus on automobile traffic. This in turn is characterized by an exceptionally high degree of delays (Lomax et al. 2019): the amount of annual hours of delay per car commuter in Baton Rouge has risen since 1982 from 12 to 58 in 2017 (Fig. 6.27). Since the second half of the 1980s, Baton Rouge has been on a par with New Orleans, which is significantly more populous. Compared to the smaller densely populated areas in Louisiana, Baton Rouge, on the other hand, shows a significantly higher degree of delays. For comparison: nationally the most delayed metropolis of Los Angeles, Anaheim, and Long Beach measures 119 h. Relative to population, however, Baton Rouge has 90.6 annual hours of delay per car commuter per 1,000 inhabitants, Los Angeles, Anaheim and Long Beach only 9.4 and New Orleans 58.9, while smaller cities such as Shreveport 91.8 or even Lake Charles 216.2 have similar or greater population-based traffic loads. In terms of metropolitan areas with a population of 500,000 to 1,000,000, Baton Rouge, along with New Orleans, ranks second only behind Honolulu, HI, with 64 annual hours of delay per car commuter (Lomax et al. 2019). Such a development of traffic flow in Baton Rouge cannot be attributed to a single cause (see also the comments in Sect. 6.1.1),

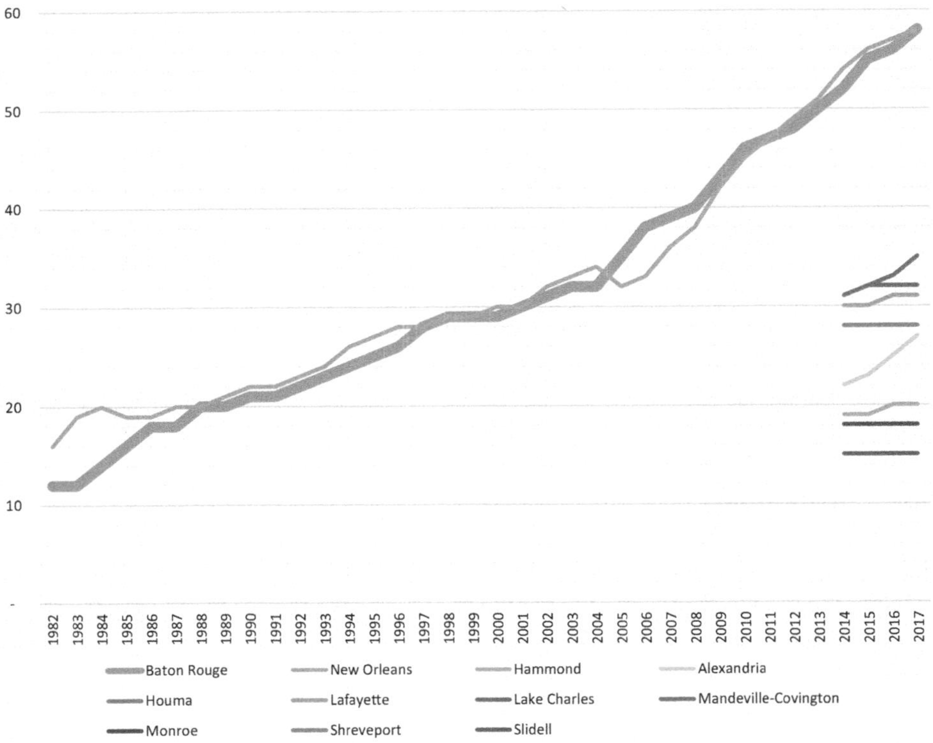

Fig. 6.27 The number of annual delay hours per car commuter from Baton Rouge compared to other densely populated areas in Louisiana (own presentation based on data from: Lomax et al. 2019)

but is the result of complex interrelationships, a community structure that has developed along with the decentralized nature of the downtown area and the associated low level of central-location goods and services, a low level of development of public transport (including the frequent, time-consuming shunting of rail-bound freight traffic over level crossings, a strong tendency towards suburbanization, and a practice (as in large parts of the United States) of socializing spatial mobility wishes over generations without reflecting on alternatives other than satisfying them with the private car (see Los Angeles, for example: Bottles 1987; Bratzel 1995; Hanich 2007; Kühne 2015), in conjunction with the often poor state of preservation of public roads. For example, 36 percent of Baton Rouge's roads were found to be in poor condition and 25 percent in moderate condition, resulting in additional vehicle wear, additional accidents, and delays that would cost the average Baton Rouge motorist $2,300 annually (Paterson 2019; TRIP 2019). Avoiding automobile traffic jams by using the bicycle seems risky due to the large number of road deaths among bicyclists in Louisiana (Sentell 2018). And also, the development of Baton Rouge with the most basal form of locomotion, i.e. on foot, can at best be described as challenging, which will be discussed in more detail in the following.

6.6.2 Walkability - Criteria and Procedure

With the aim of exploring the walkability of Baton Rouge, it was inductively divided into the following categories after exploratory walks:

- Very good walkability: no other means of transport in the traffic area, even road surface, passage of other pedestrians (also with prams) possible without problems.
- Good walkability: other means of transport in the traffic area, but no conflicts (clear separation), no danger of stumbling, passage of other pedestrians possible (also with prams)
- Medium walkability: other means of transport in the traffic area, conflicts possible (no clear separation), danger of tripping if walking carelessly, passage of other pedestrians possible (prams only at alternative points).
- Low Walkability: other means of transport are dominant in the traffic area, conflicts probable (footpath passes close to danger spots), danger of stumbling requires excessive care walking, passage of other pedestrians only possible at certain points (prams only with great effort).
- Minimal walkability: other means of transport very dominant in the traffic area, conflicts are inevitable (footpath leads over danger spots), danger of stumbling extreme care while walking, passage of other pedestrians is difficult (prams not at all)
- No foot traffic: foot traffic only possible at the risk of life.

With the help of this categorization, the following streets were subjected to a double-sided investigation based on Google Street View and (where possible without danger to one's own life) our own inspections:

1. Florida Street/Boulevard from the Mississippi River to the now closed Cortana Mall, as the traditional connecting axis of the downtown area with the eastern suburbs and flanked by central-location facilities.
2. Government Street, as the gentrifying little sister of Florida Street.
3. Plank Road, as one of the central streets in North Baton Rouge, which once connected the northern districts with downtown.
4. Siegen Lane, as the central tangential transport and business axis of the suburbanizing south of Baton Rouge.

The approach pursued here accordingly combines a social constructivist view (the construction of categories) and its positivist operationalization in the form of the classification of physical-spatial arrangements (referred to as footpaths) into these categories.

6.6.3 Florida Street/Boulevard

The walkability of the Florida Street sidewalks is generally above average in the downtown Florida Street area (Fig. 6.28). However, the passage of Interstate 110 is only

Fig. 6.28 Walkability of Florida Street/Boulevard, as of September 2019 (own surveys)

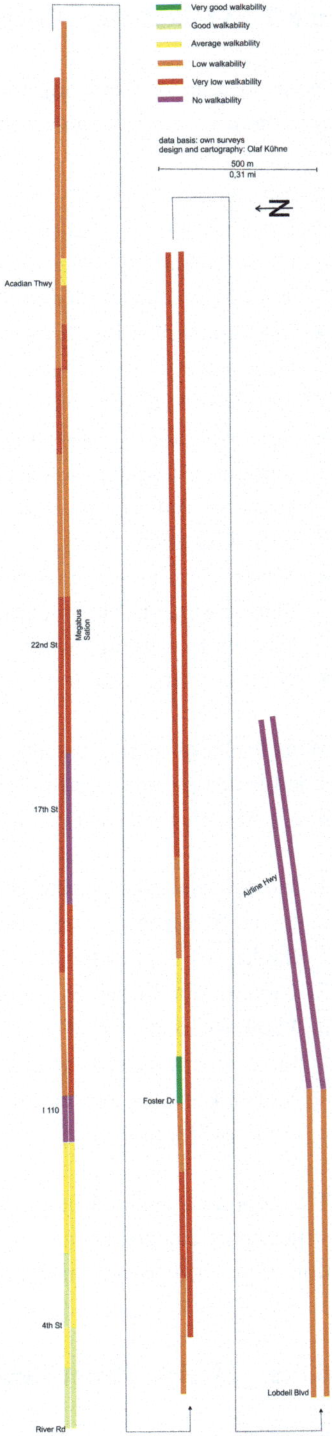

possible at the risk of one's life, as the feeder roads to the Interstate do not have any pedestrian crossings. Further to the east, however, foot travel is mostly below average. This situation becomes clearer both in the area of the long-distance bus station, which is hardly accessible on foot, and also for the area between Interstate 110 and the Acadian Thruway, since here (in comparison to the other parts of Florida Street beyond downtown) pedestrian traffic actually takes place. A section of the northern sidewalk east of Foster Drive appears remarkable, which - clearly separated from the highway - has the highest walkability not only of Florida Street, but of all the streets investigated in Baton Rouge. This section also appears to be particularly remarkable, in that a filling station is passed here, and that the adjacent road sections (especially to the west) show a clear decline in walkability. Accordingly, access to this high-quality section is limited. Numerous bus stops, along Florida Street, also have similar limited walkability. Often there are no built up road connections to these. In general, the walkability of the eastern part of the unexplored section of Florida Street/Boulevard is characterized by a poor to life-threatening condition for pedestrians. The latter is east of the junctions of the on-ramps and off-ramps to the Airline Highway, which have no crossings. The underpass itself and the area to the east of it are also free of traffic jams. A limited possibility to cross Florida Street is problematic - especially east of Interstate 110. This also avoids bypassing problematic sections of road on one side by having to change sides of the road.

6.6.4 Government Street

In comparison to Florida Street/Boulevard, the walkability of Government Street is significantly higher overall (Fig. 6.29). However, two patterns can also be found here: first, the Government Street underpass under Interstate 110 is shown to not be pedestrian-friendly because of the subsequent lack of pedestrian lights or crossings over the access and exit roads to the Interstate. Secondly, here too, the footpaths where they are used most intensively east of the Interstate overpass are in a less-than pedestrian-friendly condition, again an area of town with a predominant black population possessing a limited supply of symbolic capital, see Sect. 5.5. However, the low level of pedestrian traffic in this area is due more to its state of repair than a lack of sidewalk pavement.

6.6.5 Plank Road

Plank Road (Fig. 6.30) is generally less pedestrian-friendly than Government Street (although the pedestrian routes here are subject to greater use overall). Apart from a generally poor condition of the footpaths, three specific features are conspicuous: first, the life-threatening crossing of Interstate 110, also found along Florida Street and Government Street, including the Airline Highway's Plank Road (like Florida Boulevard) also representing a life-threatening barrier for pedestrians. Secondly, especially around

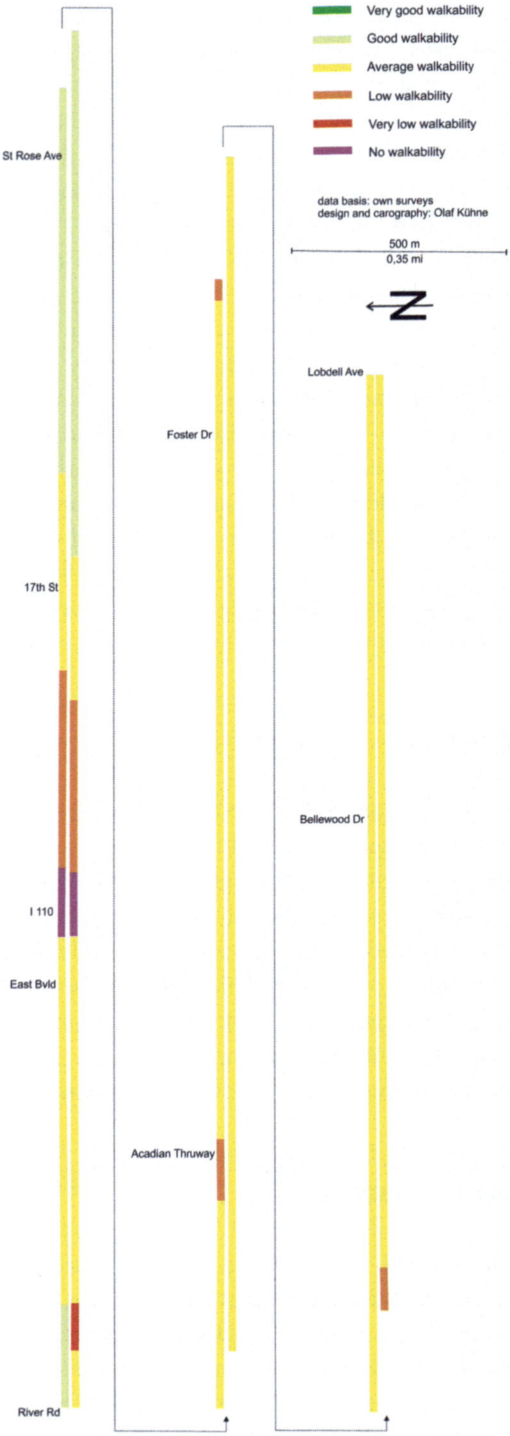

Fig. 6.29 Walkability of Government Street, as of September 2019 (own surveys)

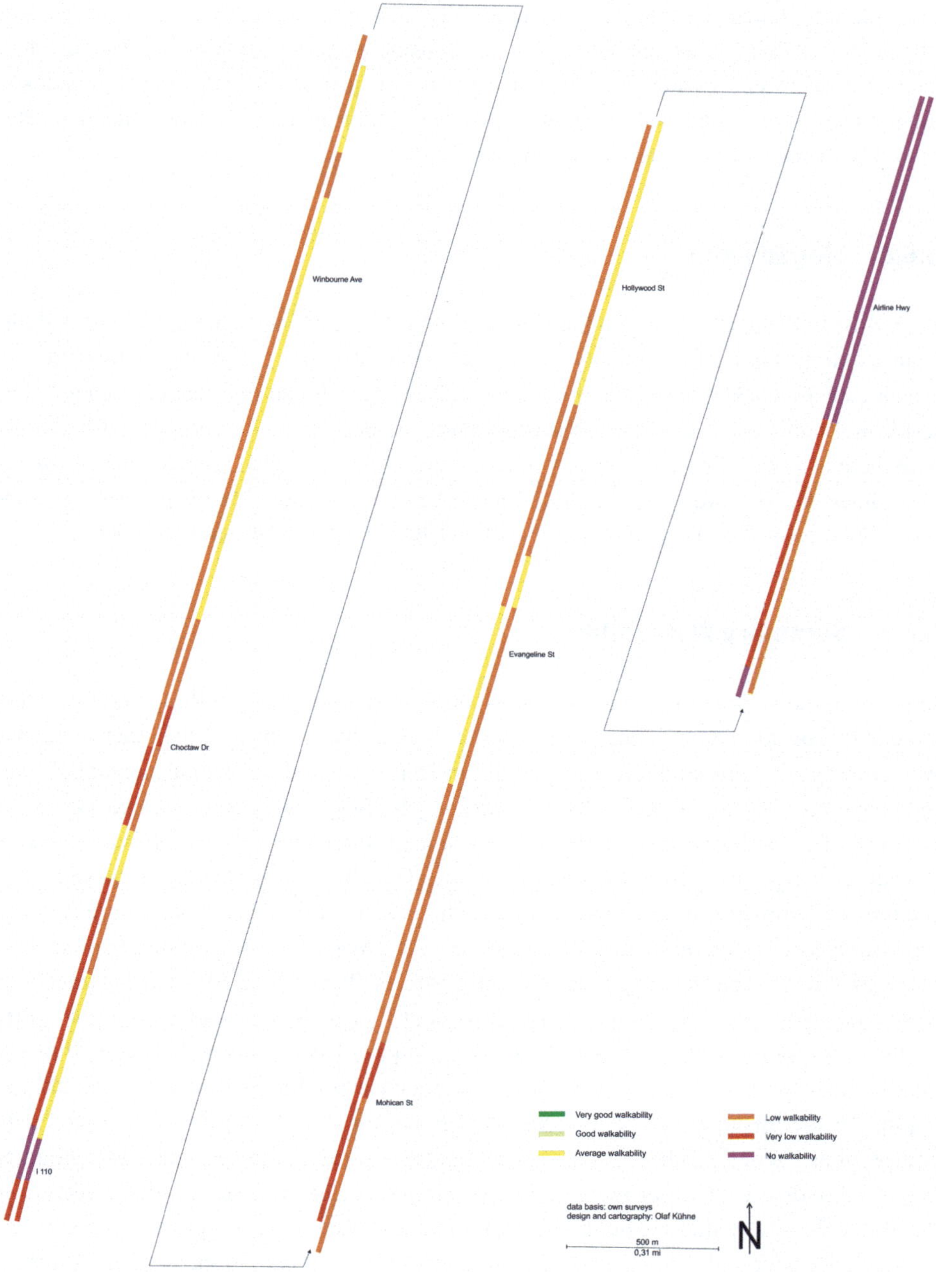

Fig. 6.30 Walkability of Plank Road, September 2019 (own surveys)

the underpass under the Interstate, footpaths are missing completely or only on one side of the street. Thirdly, walkability alternates considerably (but mostly at a low level), a lack of pedestrian crossings on Plank Road makes it just as difficult to change sides of the road so to use a sidewalk of better quality and, unfortunately, the sidewalks that often run on both sides of the road are of similar quality.

6.6.6 Siegen Lane

Compared to Plank Road, Siegen Lane presents a fairly continuous structure, the walkability changes rarely, but is of low quality - except for a longer section before its south-southwestern end (cf. Fig. 6.31). This area of medium walkability is found in a section dominated by residential buildings, the north-northeastern parts are dominated by commercial buildings, here there is a lack of paved footpaths in large parts. Also, at Siegen Lane the undercrossing of an Interstate Highway, the 10, along Siegen Lane is only possible for pedestrians at the risk of their lives, this also applies to the Airline Highway, where Siegen Lane ends.

6.6.7 Summary Walkability

In general, the walkability of the footpaths along the roads of Baton Rouge that we studied can be considered moderate at best (which is also true for most of the other roads that were not specifically studied). For pedestrian traffic beyond individual neighborhoods, highways (Interstates 10 and 110 and Airline Highway) represent massive obstacles. Because of its pathway through the downtown area, Interstate 110 is a primary cause of downtown's separation from the rest of the city. The situation is striking in these related sections of Government and Florida Street where foot traffic takes place in areas where the population is predominantly black and in less possession of symbolic capital: here the walkability is particularly bad. This also affects Plank Road to a large extent. The low walkability here, less so in the eastern parts of Florida Street/Boulevard or the north-northeastern Siegen Lane, is caused by the lack of paved sidewalks and often by the poor condition of the sidewalks, as well. The consequence for Baton Rouge is - which also applies to the unsuitability of the examined roads for bicycle traffic - the forced generation of automobile traffic. The use of alternative means of transportation beyond most neighborhoods is difficult or impossible. This, in turn, reinforces the 'insular' structure of the Baton Rouge spatial pastiche, the individual compartments are primarily connected by machine, which also leads to a discontinuous spatial construction, since the physical space is not consciously experienced in continuity, but is selectively observed (especially in relation to traffic) within a controlled enclosure of private space. In order to make possible a spatial construction that is contingent on this, we have nevertheless set out, insofar as it is possible without massive endangerment of our own lives (this applies to Siegen Lane), to approach the streets that have been examined in terms of walkability in a phenomenological way. The results are presented below.

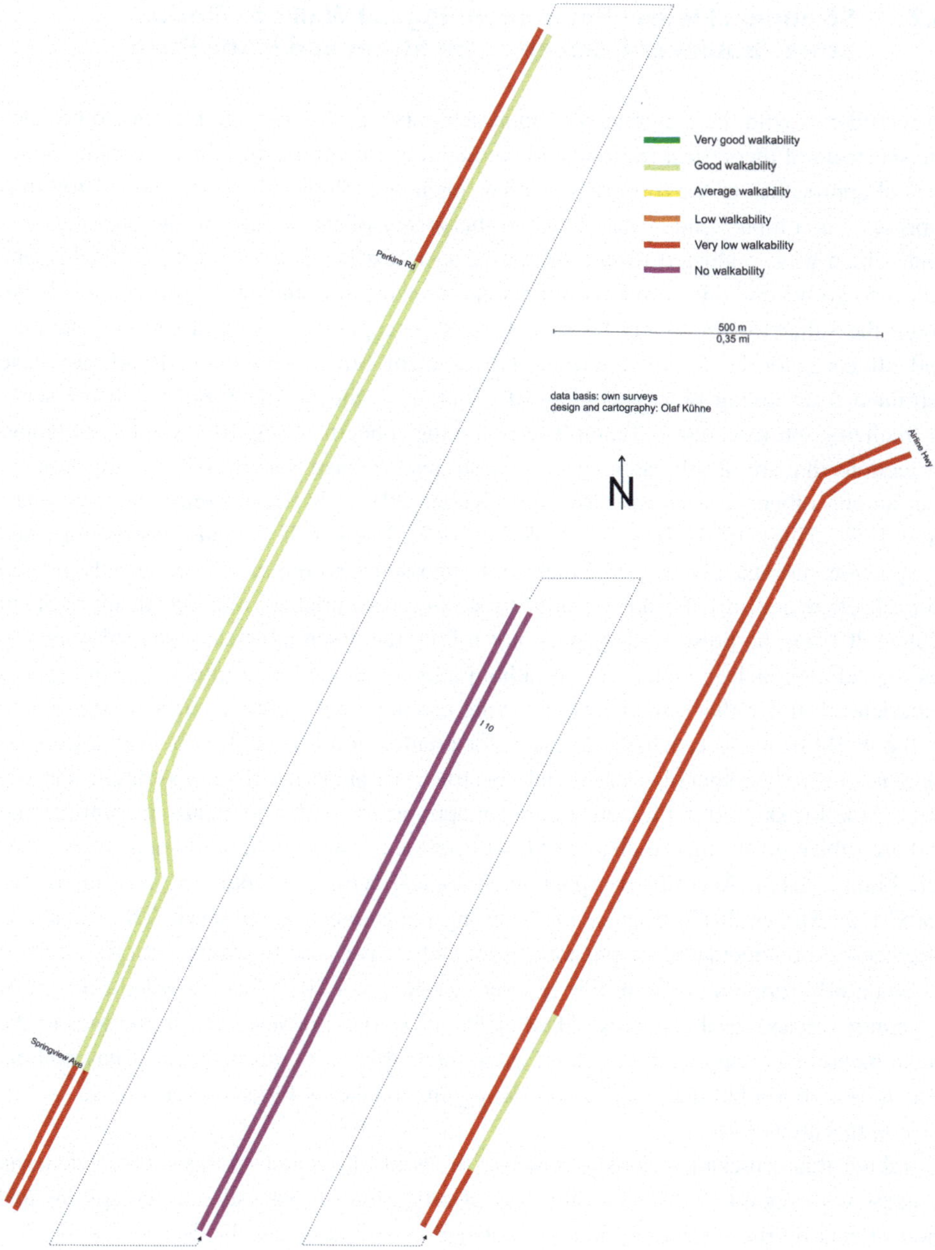

Fig. 6.31 Walkability of Siegen Lane, as of September 2019 (own surveys)

6.7 Strangers Fields - Phenomenological Walks in Florida Street/Boulevard, Government Street and Plank Road

In accordance with the neopragmatic approach pursued, this section attempts to broaden the spectrum of theoretical framing and methodological approaches through phenomenological approaches in the form of so-called 'phenomenological walks'. The section thus represents a complementary parameter to the survey of walkability in the previous section, which was conducted from a primarily positivist perspective. In the following, the aim is to go beyond the collection of data and to achieve a subjective gain of knowledge about the subject of the study by accentuating perception - in contrast to calculation - and affects (Florida Street/Boulevard, Government Street, and Plank Road; we have refrained from giving phenomenological attention to Siegen Lane, mainly in the sense of finalizing our research on Baton Rouge and the concern arising from the large number of sections that are life-threatening for pedestrians (see also Rapier 2019); in this context see, among others, Barrett and Ko 2009; Kühne 2019c; Merleau-Ponty 1945; Nogué i Font 1993; Tilley 1997; Tuan 1989; Wylie 2019). Even if the social constructivist and the phenomenologically shaped (scientific) approaches to the world are closely related to each other and in parts have a mild to strongly essentialist color (cf. in more detail Kühne 2019c), the present study does not follow the essentialistically shaped phenomenological approaches, which aim to fathom a hypothetical being inherent in the spaces experienced. It is rather a matter of exploring and approaching the (sensory) experiences of the world or a space, which are taken for granted and accepted without question, on the basis of a phenomenological-social constructivist approach (Henriksson and Friesen 2012; Sokolowski 2000). Phenomenological approaches are by no means uncontroversial and are subject to strong criticism, not least because of vague methodological approaches (cf. Finlay 2012). Accordingly, phenomenological writing is akin to 'Writing in the Dark' (van Manen 2015) because of its strong intersubjective influence. The vagueness of what can be understood by phenomenology also gives rise to strong criticism, even if a certain arbitrariness arises in this respect (see Seamon 2019 for a detailed account of the controversies). In this context, Hard (1995, p. 133) expresses: "The limitations of the phenomenological approach are clear: as an intersubjective empirical test it has nothing else at its disposal but the reader's consenting understanding based on related life experience in this area of life".

Taking this criticism seriously, however, it should be noted - a research-related gap remains with regard to the recording and investigation of synaesthetic perceptions and their effects on the interpretation and explanation of spaces and landscapes, which has not (yet) been resolved. Landscape-related research results indicate that so-called sensory perceptions and impressions, such as brightness, smell, and the acoustic backdrop of a room or landscape, play a central role in the interpretation, explanation, and categorization of a room as pleasant/unpleasant, beautiful/ugly etc. (see, among others, Jenal 2019b; Kazig 2013, 2016; Kühne 2018c; Linke 2018; Schönwald 2017; on the methodological challenge of their recording and analysis, see also Edler et al. 2018).

Accordingly, the present phenomenological walks represent an attempt to approach this research gap with the help of phenomenological approaches and to document and analyze the practical experience of space with all the senses (hearing, smelling, feeling, seeing). For the synaesthetic impressions that result from the experience of walking and experiencing of the spatial environment – often unconsciously – a setting that triggers further associations, emotions, and cognitive references and can lead to corresponding interpretations and explanations - with the associated world of everyday life implications in everyday practical actions such as avoidance, targeted searching, etc.

The phenomenological walks ultimately represent a subjectively shaped, consensually validated approach of two people, each with a specific 'professional' deformation. Taking seriously the above criticism of phenomenological approaches, they form a complement in the context of the other analytical approaches to Baton Rouge. This complement appears essential in so far as space is not only subject to construction within the consciousness and intersubjectively negotiated, but also to experience (see, among others, Hasse 2017; Kazig 2013, 2019; Kazig and Weichhart 2009). Accordingly, places and spaces are "not only experienced as *ego* and *alter ego*, but also [...] as bodily practice" (Dörfler and Rothfuß 2018, p. 100; emphasis in original). Therefore, according to Dörfler and Rothfuß (2018, p. 99; emphasis in original) "*bodily experiencing subjects* are needed who communicate these effects of space as structures of meaning, because otherwise they cannot be generalized and understood".

Thus, in the present context, 'spatial experience' refers first of all to all those phenomena - i.e. events outside the subject - to which the individual is exposed with all his senses (hearing, smelling, feeling, seeing) in the chosen spatial environment. A central point here is that in the present case it is a question of the perspectives of non-natives, who (can) update less emotional references to the selected spatial environments, for example, because of personal social contacts, but tend to dominate aesthetic-cognitive approaches. In the following, the focus will be on the narration of a day's walk along the most important and largest access road from the downtown area towards the Suburbs and the parallel roads running south of it, Government Street, as well as Plank Road, the connecting axis of the neighborhoods in the north that diagonally crosses the north of the city. The two main access roads running directly parallel from east to downtown are currently in different stages of development: while Florida Street/Boulevard, which appears to the viewer to be partially abandoned in terms of development, Government Street is in the midst of a restructuring process.

The Phenomenological Walks are a 4.5 h (Florida St/Blvd, Government St) and 2.5 h (Plank Road) walk in the spring (February; Florida St/Blvd, Government St) and fall (September, Plank Road) of 2019, with significant variations in the external conditions: While Florida Blvd had temperatures around 61°F and moderate humidity alternating between cloudy and partly cloudy, temperatures on the day of the Phenomenological walk to Government Street rose to about 81°F in combination with almost 90% humidity, also alternating between cloudy and partly cloudy whereas Plank Road was approaching 97°F and about 95% humidity with cloudless sky and sunshine.

6.7.1 Florida Boulevard/Florida Street: On the Hybridity of Past and Present and its Reflection in Uselessness-Low Use Hybrids

Even as we approach the starting point of the Phenomenological Walk to Florida St./Blvd. at the intersection of Florida Blvd. and Lobdell Blvd. in the east of the Baton Rouge metropolitan area to the west towards downtown and the Mississippi River, the horizon only a few hundred meters away with its unbroken lines of cars is not very encouraging, even in view of the chosen mode of transport of walking: in the beginning, the sidewalks - which are made up of poured pieces of aging, uneven, small-square, exposed aggregate concrete slabs and indicated paths across mown lawns along Lobdell Blvd. - point the way to a crossroads where twenty-four lanes will meet, without traffic lights or crossings for pedestrians as can be observed on arrival.

When walking along the six-lane roadway (three lanes in each direction), we experience a feeling of being lost, which results in particular from the out-of-place nature of the chosen mode of locomotion through the area, because the paradigm of the area is not oriented towards the non-motorized traveler. Instead of bicycle paths or footpaths, generously asphalted entrances for motor vehicles alternate with short, wide lawns, over which the walker or bicyclist must find his or her way, whereby the lawns, due to their lack of worn footpaths, refer to the rarity of passing pedestrians and bicyclists. Also, the density of the traffic rushing past at up to 50 mph, and more, hardly allows a thought contemplating changing sides of the road - crossing the road is rather a danger to life and limb, so leave it be and try your luck along this side of the road - in this case the northern side, where you are currently walking, thus turning the road, which does not provide any crossings for anyone other than motorized individual traffic even at traffic lights, into an asphalt wall. When passing a so-called 'Ghostbike', it becomes clear once again how dangerous the route designed solely for motorized individual traffic is for pedestrians and bicyclists alike (cf. Fig. 6.32).

But the acoustic backdrop of traffic is also not very beneficial for people on foot, as it generates a noise level that abruptly prevents any kind of conversation and only allows it to take place with great effort on the part of the vocal and hearing organs. Rather, a carpet of sounds covers the area, which is dominated by the humming of powerful, multi-cylinder reciprocating piston engines, occasionally drowned out by tinny music from inexpensive loudspeakers of passing cars. A thin layer of flattened garbage has accumulated along the roadways and the smell of combustion engines lacking catalytic converters mixes with the vapors of used cooking fat. There are hardly any pedestrians on the road except for a few African-American people with an apparently small endowment of symbolic capital, who appear here and there somewhat lost along the roadside, which offers a cheap range of goods and services and has a considerable number of pawn shops and shopping malls left open. The omnipresence of car-related trades such as gas stations, garages, tire, and used car dealerships is contrasted by a lack of cultural manifestos - for example, no creative graffiti can be found on the surfaces of the often fallow buildings and their walls, apart from a few sprayed-on words and sentences, as is usually the case in many other urban metropolitan areas of similar structure. An attempted valorization of the physical arrangement is often achieved via polyvalence through national symbolism - for example

Fig. 6.32 Ghostbike (a bicycle left as a Memorial designating the location where a bicycle rider was hit and killed by an automobile) on Florida Blvd

in representations of the individual car trade with US national symbols such as the US flag and the placement of pick-up trucks in the front line of the car lots.

Overall, the passage of the neighborhood along Florida Blvd. towards downtown is characterized by a replacement of monovalent and nonvalent spatial segments (see Fig. 6.33), which are composed of useless-underutilized hybrids or, in terms of time, of past-present hybrids, in which - except in the case of complete decay - no certain signs of use/non-use can be identified.

Towards the downtown area, the number of lanes in the street decreases, and Florida Blvd. becomes Florida Street, with two lanes in both directions, and closer to the downtown area there are also more people along the street (we won't talk about sidewalks in some sections), the vast majority of whom are black people with a low level of symbolic capital - in parts with obvious limitations in terms of physical or mental health.

Along the Magnolia Cemetery - when stopping at the cemetery with drive-in facilities - the relative calm of the area sets in, which, however, appears utopian for a few hundred meters: when crossing under the Interstate to enter the downtown area, the noise of tires driving over concrete slabs hits the large areas of asphalt hard, which increases the relative volume many times over. The permanent shading of the grey asphalted areas under the overpass construction creates a gloomy, not very inviting structure (see Fig. 6.34).

Fig. 6.33 Past and present hybrids and the replacement of monovalent and nonvalent spatial compartments; own recordings 2019

Fig. 6.34 Transition from mid-city to downtown; own photographs 2019

At the entrance to the downtown area, the noise of engines becomes less present, the increasing number of representative buildings with flags and the increasing presence of white people is conspicuous. There are indications of a restrained reurbanization, such as the increased provision of rental housing through the construction of new buildings or the renovation of existing ones. In addition, the local car repair shops make a much more professional and structured impression than the garage workshops passed further to the east on Florida Blvd resulting from the generous supply of tools and machines. Furthermore, increasingly pre-structured parking spaces can now be found, which were previously only found in isolated cases before the useless-underutilized hybrids. Despite the reduced number of people, wider sidewalks and pedestrian crossings can now be found as well as traffic lights that switch at the push of a button. Isolated sources of noise from non-motorized origins are now also much more audible, such as flags flying in the wind or children romping in playgrounds. Likewise, the perceptible fragrance of food is now more stimulating than dubious - of spices rather than dominated by rancid oil smell.

Having arrived on the Mississippi River, the attempt to upgrade the downtown area can certainly be observed when, for example, an increasing design beyond the individual structure is aimed for and, for example, different materials or path concepts are brought together in a unifying spatial concept. Municipal officials also ensure that the public space is kept clean, which is now also equipped with wastebaskets, while further east open car windows still seemed to perform the wastebasket function. Also, the benches and bus stops of the downtown are not placarded with advertising messages, in contrast to parts of Florida Blvd. further east. However, the ascent to the dam after a walk of almost two hours is the only attraction representative of the differences looking back. Lingering by the river, you can now hear the sound of isolated cars, with the occasional V8 engine now much more present. In addition, booming ship's diesel engines and the sounds of traffic over the bridge characterize the soundscape, the basso continuo of which forms the permanently humming air conditioning system of the nearby hotel complex. The beach promenade, on the other hand, refers in its design form to a leisure-oriented stay in public space: accordingly, this is the first place to find people doing sports or people who have come to the river for the view and a short break, but who move on again after a short stay. Remarkable is the scarcity and even absence of leisure opportunities on and around the river - such as pleasure boats, although the designed space clearly shows the effort for 'leisure on the river'. However, the promenade remains half-hearted in the final analysis, as it ends abruptly in the downtown area and turns into a well-trodden path, at the end of which the people strolling along regularly - in parts also with prams or bicycles in tow - make their way by stepping carefully over the tracks immediately behind it (cf. Fig. 6.35). Thus, the city seems to be turning its back on the Mississippi River, which seems to be primarily a transport route for freight traffic and a traffic obstacle for the city, possibly reinforced by the association of negative aspects such as heavy physical labor, flooding, and disease.

The view of the many parking areas, which are often only partly in existence and can be found in the 'best downtown location' (see for example Fig. 6.35), raises the assumption that the constant availability and provision of omnipresent parking spaces is more

Fig. 6.35 Mississippi Downtown; own photographs 2019

important in urban development than consequential urban development, in which the decisive development phase of the city in a time when the car had risen to become the dominant mode of transport, is still virulent today, even if the aesthetics of space is currently and increasingly oriented towards pedestrians - in some cases even in the form of taking account of mobility-impaired persons - or slow-moving cars in selected areas.

On the way back east along the northern bank of the street, the presence of functional illuminated advertising in the downtown area, which was previously often dilapidated and technically defective along Florida Blvd. In addition, the downtown area is home to image-oriented café chains such as Starbucks Café. Even if the design commitment of the downtown area is clearly evident, the further away from the attempt to revitalize the area, the more the old washed-out concrete slabs reappear. In retrospect, the revitalized inner-city areas appear like the illusion of an urban life beyond the automobile. That it is an illusion becomes clear when residents of the downtown area can only provide themselves with the things of daily life on the basis of a limited supply and are thus 'forced' to live a suburban life with a car in the downtown area for the sake of seeking a wider range of goods.

Back in the second ring suburb along the Florida Blvd., you can hear the loud, bass rap music again, which often comes from customized 1980s cars - often with a set of tires fitted with polished chrome rims. In addition, the smell of rancid fat or the sweetish smell of baked-on fat appears again, and a clear trail of alcohol can be detected from the few people passing by - just like the cannabis smell from passing cars. On the street, in turn, there are more and more people whose symbolic capital obviously has a clear potential for upward mobility. Besides cartridge cases on the street, the limited presence of public transport is striking, which - measured against the high-frequency route - often is hardly occupied or not at all. Once again, the dominance of representations that serve to sell cars or keep them running becomes clear, with the lawns in between often being worked by unskilled workers from private contractors spraying a generous application of pesticides. The sight of loose power lines, broken windows and doors is accompanied by a mix of the sound of engines and the rustling of the flags of flag dealers.

The social function of the gas stations is also clearly demonstrated as a meeting place for the non-refueling local population and those not involved in something other than a fuel purchase. In the process, the gas stations often become lounges where employees and customers gather in conversation, often in the hope of getting some change or leftover food from the previous day.

After leaving the second ring suburb, which is in decline, things become a bit more sophisticated in the following third ring suburb. There are now a few upscale retailers between storage space providers and discounters as well as affiliated gated communities.

In reflection, it can be summarized that the downtown area forces itself upon pedestrians and cyclists alike, but that behind them, due to the 'thorny design' (Flusty 1999), pedestrians and cyclists represent an existence external to the system and that this space is not created for them, nor they are wanted or intended to be there. Inverse landscapes of the Florida Blvd., i.e. also physical arrangements that cannot be found and experienced, but which can be found in comparable physical-spatial arrangements (cf. Kühne 2013a, b), form individually designed concept bars and bistros, owner-managed restaurants or a superior/premium range of goods and services. In addition, a will to design beyond the individual structure is missing, as is the design of public space through art forms of any kind. Florida Blvd, in particular, seems like a space without an idea, which thus also congeals into a meeting place for people who seem to have long since lost the imagination to shape their own lives - for a variety of reasons.

6.7.2 Government Street: On Revitalization and Land Use Cycles

The entrance to the Phenomenological Walk along Government Street is at the eastern end of the street by the intersection of Government Street and Lobdell Blvd heading westbound towards downtown and the Mississippi River. At the entrance to Government Street, there is a view of a four-lane street, which is much less busy at this intersection than Florida Blvd, which was crossed a few hundred meters earlier. Along these streets, rows of residential buildings in standard construction design from the first phase of suburbanization are lined up, each of which stands on generously sized plots of land with extensive lawns. However, attention is also drawn to the busy construction site on the south side of Government Street, where a gated community is being built using construction methods identical with the houses being built - the walls have already been raised - and on whose location, it is assumed, considerably more buildings will be erected than previously. Further to the west over partly newly built pedestrian paths, one passes generous, well-kept, detached lawns that are not enclosed with walls. The residential buildings built in the first phase of suburbanization show an attempt at upgrading and individualization, for example by adding variations to the façade or column constructions, thus also aiming for a design that goes beyond pure form or function (cf. e.g. Fig. 6.36). Thus, in the neighborhood that is passed through, the effort to be creative and a certain willingness to design beyond the individual object becomes clear again and again, for example by staging streets and houses with lighting. At the same time, small houses are in renovated condition and keep a security notice regarding the neighborhood watch program and private security companies in their front gardens. Even if there are only a few people on the way as pedestrians or bicyclists, we do not meet black people under the influence of drugs on the newly laid out pedestrian paths, instead, white people who seem to be inspired by religion.

Fig. 6.36 Government Street residential development; own photographs 2019

Although the Louisiana sultriness is very noticeable even in February and the sun seems to burn mercilessly despite being behind the dense clouds, the conditions for long walks are not really very inviting. Nonetheless, we continue our walk undiminished. After passing through the well-kept residential buildings, it is mainly owner-managed restaurants or regional fast-food outlets that take the place of corporate gastronomy, in whose atmosphere the smell of spices, seafood, and steaks dominates and even the Jack in the Box does not seem to smell of rancid fat. Overall, a high frequency of people using the restaurant can be observed. Similarly, there is a greater volume of business along Government Street, which - often in buildings with a renovated, historic structure - offers a specialized demand for goods and services - such as food, furniture, and interior design, for people with a higher endowment of symbolic capital, rather than a less well-stocked car-related business. In addition, antique dealers can be found instead of pawn shops, as on Florida Street/Boulevard. The efforts, of those with an economic interest to conceptualize this part of Government Street, are made clear by a sign pointing to the entrance of the 'Mid City' (Fig. 6.37). However, with the increasing volume of traffic after passing through the residential area, a similar soundscape as on Florida emerges under the humming of the numerous cars, even if pedestrians feel less out of place and less lost than along Florida Blvd. The public space is also characterized by renovation

Fig. 6.37 'Mid City' sign: The addition 'Growing Heart of Baton Rouge' illustrates the competition between the individual compartments striving for centrality in the polycentric Baton Rouge; own recording 2019

Fig. 6.38 Murals in the public space; own photographs 2019

and repair work, so that failed businesses are torn down and not simply left standing and left to decay. Accordingly, the decline is not experienced systematically here, but merely forms an aspect instead of a theme of the pastiche and the land use cycle. There are also higher-quality schools along Government Street, which repeatedly displays art or murals on buildings or walls (see Fig. 6.38).

As you approach the downtown area, loud rap music and penetrating bass sounds from the loudspeakers of passing cars returns. Also, more and more customized 1980s cars with oversized chrome rims are cruising the Government Street and a similar 'charm' is developing as in Florida Blvd., whereby the transition to downtown seems to be more fluent. The smell of smoked products ranging from tobacco to marijuana also spreads here.

Crossing the downtown area after two and a half hours we arrive at the river, we realize that on the whole way there was no clothing store to find, which would have offered the possibility to buy shorts that would have been more appropriate for the prevailing weather.

6.7.3 Plank Road - of High Potential and the Difficulties of Alternative Interpretations

In order to be able to reach Plank Road for a phenomenological walk as an outsider, the Plank Road corridor must be approached by car. Currently (as of September 2019), there are no overnight accommodations available for travelers and overnight guests in North Baton Rouge, so the likelihood of foreigners landing in North Baton Rouge is extremely low. Finding a parking space on site is extremely difficult, as a large part of the area is privately owned and there are apparently no public parking spaces available. After several extra laps, one finally visits the spacious parking lot of the nearby Memorial Stadium, which is only populated by a few employees of a facility maintenance company, and walks the rest of the way to the actual destination on Plank Road. The path to Plank Road along the stadium and car park facilities is easy to master, despite the lack of footpaths, over the well-kept greenery to the left and right of the stadium feeder road.

However, at the beginning of Plank Road, at the intersection of Foss Street and Plank Road, a noticeably invisible boundary seems to have been drawn: as if drawn with a

Fig. 6.39 Plank Road and Foss Street entrance; own photographs 2019

ruler, flattened garbage can now be seen everywhere, whose labels and packaging, faded and washed out beyond recognition in parts by sun and rain, suggest long periods of lying around. Shopping carts, filled to the brim with rubbish and seemingly parked for eternity in hedges, stand as a greeting at the entrance gate to Plank Road - a deserted burger place, which in retrospect seems to be symbolic of what is yet to come - a middle-class infrastructure that was once recognizably developed in terms of the goods, services, and housing on offer, which over the years has increasingly broken away from its urban core in terms of development and whose own characterization and that of its promising offerings have become muddled (see Fig. 6.39).

In the further course of time, there is an increasing number of significantly damaged or even crashed cars of different degrees of damage with fenders, bumpers and the like fixed with straps, expanders or duct tape - often in colors that deviate from the original color and painted with a conventional paint roller instead of a professional paint job in the original color or without taking any repair measures. The local vehicles are in a much worse condition than those on Florida Street/Blvd. Here and there, fallen bumpers or leftover car seats and old tires (parts) are also simply left at the roadside without any signs of their removal. The visible large oil stains at traffic lights and parking lots also indicate the dominance of older engine-powered cars with high mileage levels in the surrounding area. All in all, the roads and pavements are lined with rubbish and refuse that has disintegrated unnoticed and almost beyond recognition, which seems to be amalgamated with the other physical manifestations on the ground such as relics of footpaths, dust, and grassy vegetation after long periods of lying on the ground, and which indicates insufficient or non-existent care of the public space or a certain indifference or even resignation in this respect. The environmental waste mixture often consists of broken glass bottles of alcoholic beverages - especially hip flasks - diapers, fast food packaging, other beverage containers, gnawed-off bones, vomit, and in some places an excessive amount of empty condom packaging along the bushes of less visible green areas. In addition, there are no public waste bins to be found, such as those in downtown or the LSU campus, which could be used by passers-by and emptied by a city cleaning service. The mummified skeleton of a medium-sized dog corpse in the drainage channel of Plank

Fig. 6.40 Absence of care in public spaces: skeletonized dog body in the gutter, dilapidated footpaths, signage; own photographs 2019

Road, which forms an amalgam of decomposing organic materials, street dust, marginal vegetation, and rubbish that has been reduced to small particles, also indicates a lack of care and a low level of responsibility with regard to public space. Judging by the state of decay, the dead dog's body has probably been at this location for about three years already (cf. Fig. 6.40). Another striking feature is the significant frequency of religious institutions whose physical representations are in a similar state of need of renovation as the buildings surrounding them.

The 'Smellscape' of Plank Road is dominated by seized brake pads and the dirty exhaust gases of vehicles lacking the filtering of catalytic converters. No olfactory conclusions can be drawn about the preparation of food, neither regarding commercial food supply nor private households. While the humming of high horsepower piston engines can be heard again and again, the chirping of crickets can be heard, on those occasions when the traffic is temporarily stopped by crossing trains.

In the heat of the day, there are hardly any expressions of shade manifesting along the way, for example through trees that could provide shelter and coolness, while in the few covered passages of former business buildings, dogs panting heavily from the heat seek cooling, of which it is hardly possible to judge whether they are stray animals or whether possible owners can be found.

The frequently encountered walking or bicycling population is predominantly composed of African-Americans of all ages. This accumulation of pedestrians and bicyclists is occasionally countered by suddenly protruding concrete slabs or holes that appear in the pavement, which are aging footpaths with a high risk of injury. While pedestrian paths - albeit in poor condition - are available, designated bicycle paths are completely missing, with large, unused, and forgotten asphalted areas along Plank Road, which are overgrown with a particulate mixture of waste, grass, and concrete (see exemplary Fig. 6.40).

The buildings that pedestrians pass along Plank Road show serious signs of structural decay, which is often - if at all - only rudimentarily countered with inadequate materials and methods instead of comprehensive restoration measures. Thus - if tackled - the upper quarter of the windows are omitted, missing or broken windowpanes are replaced by lower-priced pressboard or foil and adhesive tape are applied as a temporary solution. The physical remains of bullet holes in windows are also more likely to be covered with foil and duct tape instead of a comprehensive repair and replacement of the windows. When walking along Plank Road, the nonlocal pedestrian passes orphaned shops, where those unfamiliar with the area cannot tell at first glance whether they are still in use, because the condition of the building seems so disastrous that no sure signs of use/non-use can be seen. In retrospect, it becomes clear that the properties with a higher state of maintenance, which were located around the Memorial Stadium, particularly in the paths leading to Plank Road, often had a fence around the property with chest-high (at OK, at C.J. eye level) wire mesh fences on stable, embedded metal posts that are not visible along Plank Road or even with a view of the adjacent access roads. Conversely, the walls of the orphaned buildings are often the subject of murals in memory of killed rap musicians, also drugs or HIV/STD-related as well as sexualized murals can also be viewed on the walls along Plank Road (see Fig. 6.41).

In general, Plank Road is lined with a considerable amount of abandoned commercial buildings and business properties, often with reminiscences of physical manifestos of 1950s/1960s billboards. The remaining locations offer a range of services and goods in

Fig. 6.41 Memorials for killed rappers as well as drugs, HIV/STD-related, sexualized murals; own recordings 2019

Fig. 6.42 Signs reduced to function along Plank Road

the lowest price segment with clear signs of use, such as used mattresses, tires, previously owned dishes, etc. whereas many shelves of the only local department stores are empty, which in the middle of an industrialized country leaves the viewer with the unreal impression of being caught in the film set of a disaster movie, only to realize that a neglected supply chain and goods management of the discounters has established itself in an economically and socially underprivileged area and is a further indication of limited (life) possibilities. In the most discounted section are also used car rentals, which offer 20 to 30 year-old, run-down limousines for 10 dollars a day. The areas of the few remaining used car dealerships are framed with small flags, whose original glory was long ago burned away by the sun and ruffled by the winds. What is striking is the high presence of nail and hair studios, which instead of professional billboards draw attention to hand-painted signs with ethnic-specific advertising. The signs are often not focused on their aesthetic form, but rather reduced to their function: often without design, without a specific name or elaborate play on words, they announce purely functional, e.g. nail design or barber (see Fig. 6.42).

The physical manifestations of decay and violence form a strong contrast to the overall relaxed, friendly and even peaceful atmosphere. In this friendly aura of resignation there is no hecticness, one is as a pedestrian - completely in contrast to Florida Street/ Blvd or Government Street - greeted by car drivers when crossing the street, permitted to cross the street, and perceived as a road user. The clear majority of people in public places apparently go about their daily business, wearing work clothes that identify them as either construction worker, restaurant or supermarket employee, thus referring to jobs at the lower to bottom wage levels. Also, the high frequency of people waiting at bus stops that are hardly recognizable as such, repeatedly highlights the importance of local public transportation for this part of the city.

Overall, Plank Road presents itself to those unfamiliar with the place as a space without any recognizable spatial coherence. For outsiders, the physical arrangement and inventory seems to be reduced to pure function without any significant creative will, certainly not beyond the individual object. In addition, as a whole, the public space is also characterized by a low to only rudimentary level of care and design. However, due to the large number of underused, unused, and derelict areas and buildings in the relative proximity of the city center, the diagonally intersecting access road in the north of the city also harbors an extraordinarily high development potential.

6.7.4 Florida Street/Blvd. and Government Street as Complementary Inverse Urban Landscapes, Plank Road – The 'Mostly-Out-of-Order-But-With-High-Potential' Road

In the synopsis of the Phenomenological Walks, the fragmentation of the city and its unsynchronized differentiated development becomes particularly clear: Florida Street/Blvd. seems lethargically resigned to the modernity of decades past and seems to have come to terms with it. In its spatial expression, it appears like the physical precipitation of a spatial texture which, after reaching its maximum, no longer developed any further vision. Government Street, on the other hand, exhibits a series of postmodern approaches within which emerging ruptures are not ignored, but rather transformed into cyclical spatial utilizations and woven into new structures. Here, the inverse landscape - i.e. that which does not or cannot manifest itself spatially (cf. also Kühne 2013a, b) - where one is the physical manifestation of the other: along Florida Street/Blvd. currently hardly any owner-managed restaurants or bars, artistic groups, or - with the exception of the car-related trade - an upscale range of goods and services can be found, whereas these are clearly more common on Government Street, for example, when antique dealers instead of pawn shops and owner-managed restaurants instead of a monopoly of fast food chains can be found. In addition, comprehensive revitalization measures in Government Street can also be marked by the integration of pedestrian and bicycle paths, which indicate a will to design beyond the individual object. In Florida Street/Blvd., numerous car-related businesses such as gas stations, garages, and tire-dealers line up side by side, but no will can be seen to design beyond the individual object. The physical arrangement of Plank Road in its phenomenological assessment, on the other hand, leaves a broken impression on those not familiar with the place: the clear manifestations of decay as well as violence are contrasted during the time of the Phenomenological Walk by a friendly, neighborly, and downright peaceful atmosphere, as is the clear potential of Plank Road, which seems to lie in an atmosphere of rather friendly resignation within a slumbering Sleeping Beauty.

6.8 Evolutionary Compartments in the Baton Rouge Spatial Pastiche - An Interim Conclusion

In this chapter, one thing became very clear: the fragmentation of the Baton Rouge spatial pastiche. Baton Rouge presents itself from partitions that can be understood as 'spatial-islands', which can be connected almost exclusively through the use of the automobile. Footpaths are not very continuous, when they do exist at all. This is particularly problematic for residents of the 'triangle of diminished life chances', an area in which access to healthy food is as limited as access to higher education (already purely physical-spatial), which is characterized by poverty and particularly high ecological burdens. And by the dominant share of the underprivileged inhabitants being people of black skin.

These 'compartments' are subject to an uneven development and available strategies in dealing with decay and the structure in public space. While Government Street has a strategy of cyclical space utilization and integration of new structures, which aims at a design beyond individual objects and provides space-integrating measures for pedestrians/cycle paths; Florida Street/Blvd. is currently characterized by a leaving behind of decayed structures without 'Plan B' and the aesthetics of the purely car-oriented spatial structure concentrates only on the design of individual objects. In Plank Road, on the other hand, the decay and structure of public space seems to meet with a resignation that lacks the means to leave or overcome these structures. Instead, rudimentary repair measures - if at all - take the form of temporary provisional measures that have hardly any design elements but are reduced to their pure function.

After having devoted ourselves to the experience of Baton Rouge (or some parts of it) in the past, we will now look at Baton Rouge from a more distanced perspective, namely the investigation of mass media presences, focusing on a medium that has been linked to a new emergence of social life in the past two decades: the Internet.

7 Baton Rouge and Mass Media: Of Descriptions, Interpretations and Stagings

The influence of the mass media and, since the end of the twentieth century in its expansion to include the world wide web, in which technical progress and increasingly user-friendly (mobile) end devices allow content to be produced, uploaded, multiplied, and distributed with increasing ease, on the mediation and construction of the world has remained unbroken to this day. This has been and is, in part, critically assessed, particularly by sociologists, linguists, cultural and media scientists (Curran 1997; Dill 2009; Luhmann 1996, 2007 [1995]; McCombs and Shaw 1972). Already at the end of the nineteenth century (Speed 1893, p. 711) stated in one of the first media analyses ever that "if the New York newspapers ever recorded history accurately and with any appreciation of the significance of the events occurring, they do it less now than heretofore, for now everything is so covered with the millinery of sensationalism that none but the wisest can detect the truth beneath". Mass media, which in the present understanding are understood to include in particular content accessible via the world wide web, which can be accessed by users via keywords in search engines, play a central role in the search for information. For example, the global market leader Google alone processed an estimated 3.8 million search queries per minute in 2016, which is an estimated 5.5 billion a day (Sullivan 2016). These modalities of information generation also affect places and spaces, some of which serve as the basis for synthesis to landscape.

Accordingly, the following will focus on which mass media representations Baton Rouge is subject to on the identified 'homepages' on the basis of a Google keyword search (Sect. 7.1), the visualizations in Internet images and videos (Sect. 7.2), the tourist representations of Baton Rouge at TripAdvisor (Sect. 7.3) as well as the supra-regional representation of Baton Rouge using the example of selected years (1978, 1998, 2018) of the New York Times (Sect. 7.6) in order to be able to determine temporal changes.

O. Kühne and C. Jenal, *The Multivillage-Metropolis Baton Rouge*, RaumFragen: Stadt – Region – Landschaft, https://doi.org/10.1007/978-3-658-30716-5_7

7.1 Descriptions, Interpretations, and Staging on Internet Homepages Based on a Google Keyword Search

If information on a question, topic or even a place is searched for, this is often done via a keyword search of a search engine. The provider Google has by far the highest market share among the most-used search engines worldwide. In the years 2015 to 2019, for example, it fluctuated between about 85 and 90 percent (Statista 2019a). A more detailed analysis of the respective click rates shows that more than 99% of search queries only consider the search results of the first hit page (Beus 2015; Petrescu 2014), and the search is aborted at the latest after visiting the third hit page (Beus 2015). Accordingly, in the present analysis, the results of the first three hit pages of a Google keyword search for the term 'Baton Rouge' are subjected to a more detailed analysis. The corpus was compiled on 30.04.2019 from a German IP address "(see Annex, Tab. 11.1)".

If the data corpus thus collected is structured according to the inductively determined categories (cf. Mayring 2003), then, in relation to the total number of homepages determined, the first thing to be considered are tourism-related web offers that provide information and tips on Baton Rouge (e.g. go-05, go-06, go-16), followed by websites of restaurants (see Fig. 7.1), which do not, however, refer to restaurants in Baton Rouge itself, but to bars and restaurants with the name 'Baton Rouge' outside the USA (go-07, go-09, go-10). Behind them, the other places are ranked by homepages of educational

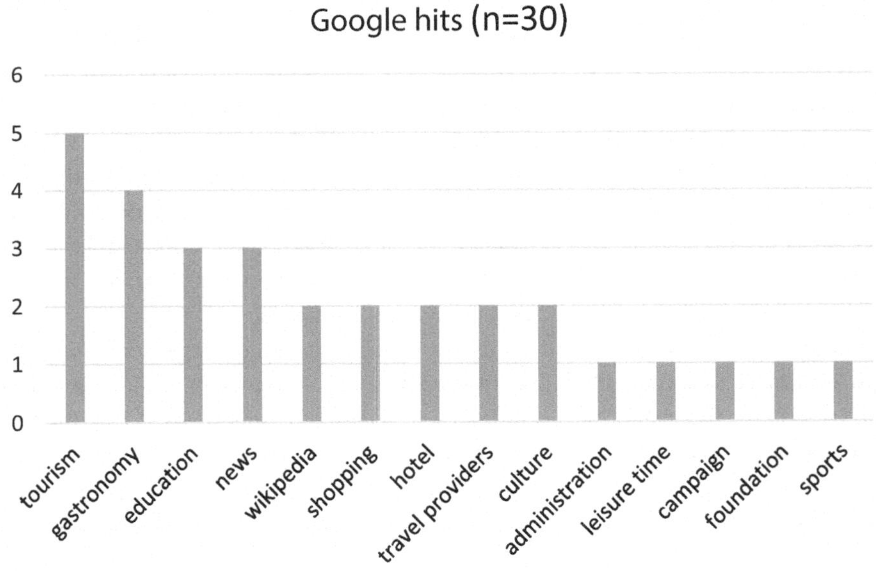

Fig. 7.1 Categorization of the first three pages of a Google keyword search for the term 'Baton Rouge' with German IP address; retrieved on 30.04.2019. (Given in absolute numbers; own survey and presentation)

offers and news portals and on a low level by a number of goods and service offers such as online shops for the 'Baton Rouge' guitar, hotels, travel agencies, and leisure offers etc. (see Fig. 6.42).

If the data are sorted according to their appearance on the respective result pages, the first hit pages mainly contain tourism-related offers such as the official tourism website of the city (go-03) or the United States (go-05), which refer to sights and attractions of the city, followed by the Wikipedia entries in German and English language on Baton Rouge, but also e-shopping offers on the Baton Rouge guitar (see Fig. 7.2). On the results of the second hit page, the results pages differentiate between further tourist offers and websites of a Canadian restaurant in English and French, as well as e-shopping offers for the Baton Rouge guitar and tips on leisure activities. Only on the third hit pages do offers of cultural or educational institutions and events appear, as well as a hotel or bar in Baton Rouge (see Fig. 7.2).

Even if tourist offers dominate the result lists of the google search, it is the contributions of the online platform Wikipedia that prominently lead the hit lists, and thus also assign dominant interpretations, associations, and attributions to the researched terms – such as Baton Rouge. A closer look at the online entries (as of June 2019) reveals that the German-language Wikipedia article on Baton Rouge, for example, is very much like a woodcut and provides only rudimentary key data on history, population, and economic sectors. Sections or even references to culture or places of interest are completely missing, whereas references to high cancer rates are cited, which in turn – as well as high

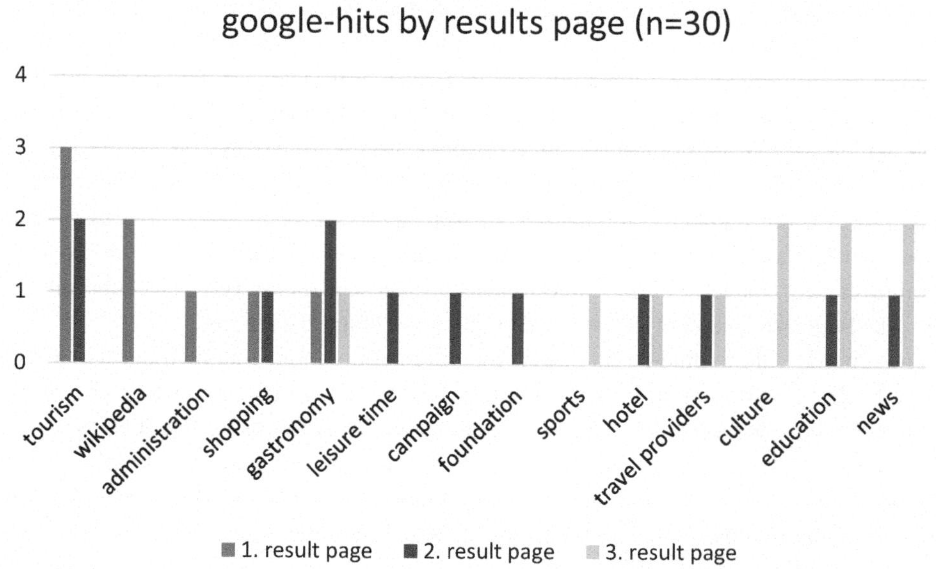

Fig. 7.2 Display of results of the Google keyword search 'Baton Rouge' by result pages (n=30; own survey and display)

crime and HIV rates – are not a topic in the English contribution. If the comparison is made with the German-language Wikipedia entry on New Orleans (as of June 2019), it becomes clear that registered Wikipedia users and authors pay considerably more attention and care to this entry and accordingly focus more strongly on it than, for example, Baton Rouge and thus also refer to a subordinate interest in the state capital. A comparable picture emerges regarding the English-language Wikipedia article on Baton Rouge (as of June 2019). Here, the representations on history, geography, economy, culture, education, etc., about Baton Rouge are much less comprehensive and there is a much more intensive engagement with New Orleans. In addition, the presentation covers fewer social issues but aims to focus on the cultural elements that have emerged in the course of the revitalization of the downtown area.

When considering the various tourism strategies for Baton Rouge (e.g. go-03, go-05), they seem to attempt to do justice to stereotypical expectations of destinations that are interpreted as attractive for tourists. Thus, the (external) representation of Baton Rouge is based on existing stereotypical expectations of (mass) tourist destinations considered valuable by the masses, such as plantation houses, historical buildings, etc., and endeavors to meet these in its tourism-related communication to the outside world. In the (internal) observation of people living in Baton Rouge, however, the community is attested little agreement regarding stereotypical visual expectations of urban picturesqueness. The attractions of the city are rather selective on micro-levels and often require knowledge of the place or the local area in order to discover them (see Chapter 8 in detail).

The listed gastronomic offers of the first two hit pages (go-07, go-09, go-10) refer to restaurants and bars outside Baton Rouge. Accordingly, there is no reference to 'hip' local clubs, bars, or restaurants on site, which initially suggests the absence of a 'nightlife culture' or 'scene' and appears unusual in the context of state capitals and is more a feature of rural regions and therefore does not necessarily increase the appeal of a visit to the city for urban 'scene goers' and 'creative people'. Instead, the Baton Rouge guitars and ukuleles are strongly present online in connection with the keyword 'Baton Rouge' and are also dominantly represented among the Google suggestions for Baton Rouge (see Fig. 7.3; the number combinations such as ar21c/ace or ur 4-s denote instrument types and series), whereas these references are less obvious locally. Also, with regard to the universities, they are either - like Louisiana State University (go-11) - subordinated or - like Southern University - not listed at all on the first three hit pages.

In summary, the online-based search for information about Baton Rouge via a simple Google keyword search initially provides selected attractions and places of interest, which can, however, often be found in this or a similar way in other cities such as plantation houses, historic government buildings, zoo, museum, etc. and thus initially appear arbitrary. Accordingly, attributions and interpretations of Baton Rouge seem to be constituted in an interchangeable and outward-looking way rather than from an internal perspective. The lack of references to a capital city nightlife or 'scene life' in the Google search also makes it clear that people with preferences for urban lifestyles may be less likely to be addressed here.

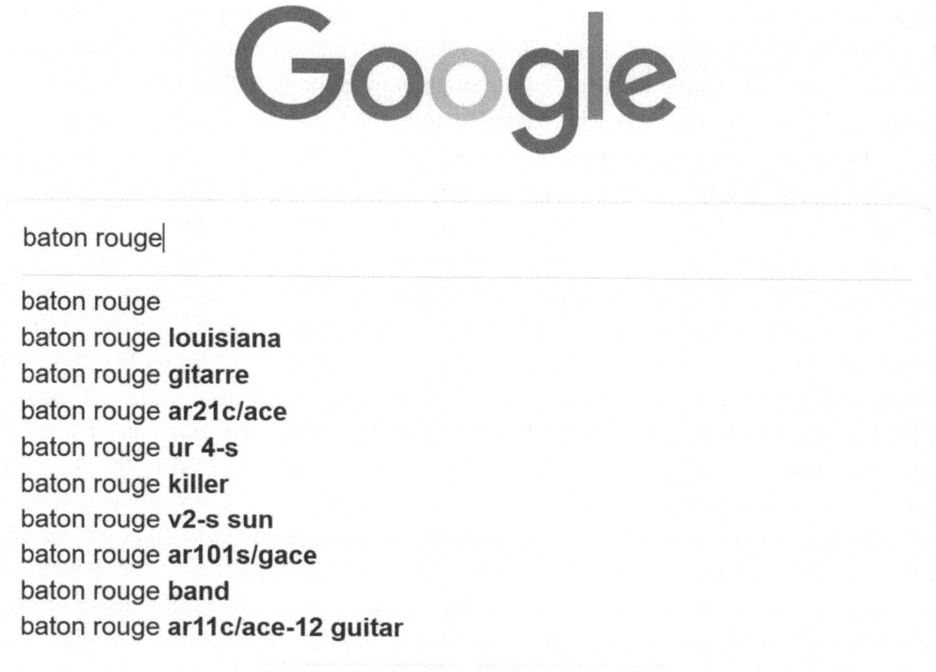

Fig. 7.3 Screenshot of the completion of the keyword search for 'Baton Rouge' on google.com (retrieved on 06/14/2019)

7.2 The Visualization of Baton Rouge: Internet Images and Videos

As mentioned above, the Internet is currently the main source of information and impressions. The information is very often not (solely) textual, but rather visualized, especially photos and audio/video. These visualizations are particularly suitable for synthesizing, decomplexing, and, last but not least, focusing a multitude of information. Accordingly, they are particularly suitable for updating and consolidating spatial (in general) and landscape stereotypes (in particular; Bittner 2014; Glasze 2014; Kühne 2008b). In this respect, they represent an essential access to research on stereotypical interpretations and knowledge (Kühne 2012a; Kühne and Schönwald 2015b; Linke 2017).

7.2.1 Baton Rouge in Internet Images

With the aim of outlining the pictorial representation of Baton Rouge on the worldwide web, the Google hits were determined using the image search. This search was carried out by different computers, both in Germany and in the United States. The result of this implementation showed only minor deviations. The first 100 suitable photos were subjected to a more detailed examination with regard to the content presented. Not evaluated were hits for the Baton Rouge guitar (twelve hits), a band called Baton Rouge (one hit), a street sign indicating the name of the street in question with the place name 'Baton Rouge' and interiors (six hits). The (round) number 100 resulted by chance from content considerations, since the following hits did not offer any more content that could be evaluated meaningfully. Compared to other studies on Internet images of spaces (such as California, Los Angeles, San Diego, Lower Bavaria, but also landscape in general; see Kühne 2012a; Kühne and Schönwald 2015b; Kühne and Weber 2019a; Linke 2019), the proportion of oblique aerial images is very high: 41 of the 100 photos evaluated fall into this category, while 58 can be categorized as photographs with the camera location close to the ground, one of which is a map showing the location of Baton Rouge in the United States. Eleven of the oblique aerial shots have identified the location of the top of the Capitol, which is open to the public (and may have triggered some visual convention with respect to the Baton Rouge overview). This is also reflected in the thematic focus of the evaluated Internet images (Fig. 7.4): in 25 of the 27 cases, the category of the 'overview' of the city consists of oblique aerial photographs. This type of recording is also found in the second most common category, the depiction of the downtown area, although there are also photographs that deal with individual buildings or streets from a pedestrian perspective. The thematic category 'Mississippi and Horace Wilkinson Bridge' follows at a clear distance, although this distance should be put into perspective

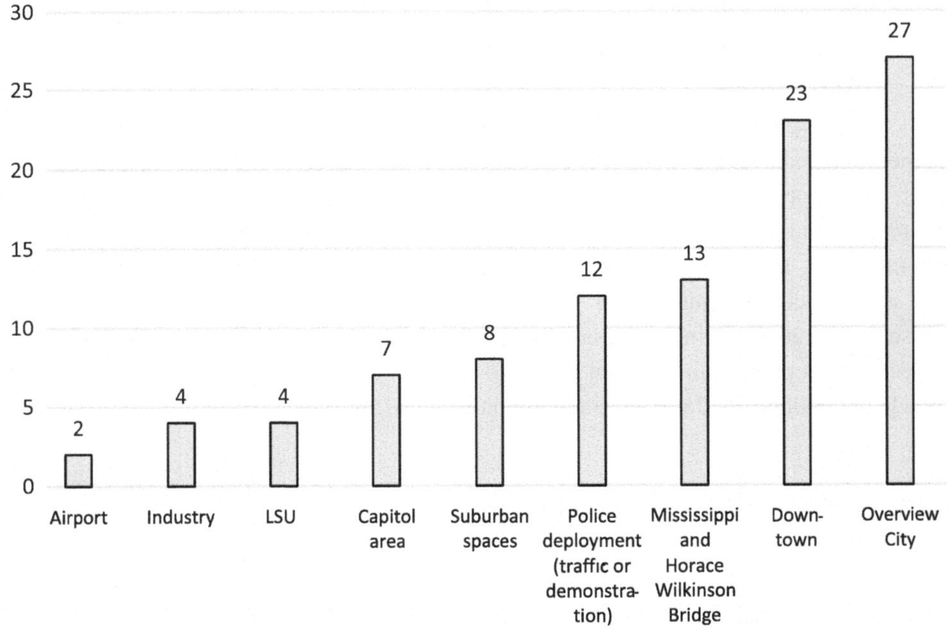

Fig. 7.4 The thematic foci of the 100 examined illustrations on Baton Rouge (own survey and representation)

in that both are also frequently seen in the overview photographs, although the focus is not on them, which also applies to the downtown category. In this respect, it can be stated that the vast majority of pictorial representations of Baton Rouge on the Internet are thematically focused on the downtown area and the Horace Wilkinson Bridge, which spans the Mississippi River. This focus is complemented by depictions of the Capitol and its surroundings, which adjoins the city center to the north. Suburban areas also have a certain importance in the pictorial representation on the Internet, whereby these are often images from sales ads of suburban buildings. Other thematic foci, such as the airport, LSU, and industry (Exxon) are dealt with in relatively little intensity. There is a certain presence of police operations in relation to traffic or demonstrations, with Baton Rouge being only the scene and not a specific location (such as the Capitol or downtown).

The dominance of downtown, Mississippi River, Horace Wilkinson Bridge and Capitol, which has been worked out with regard to the thematic focus, is also found when dealing with occupied objects and object constellations that are particularly symbolic in Baton Rouge (Fig. 7.5). Whereby the significance of the Capitol does not only arise from its

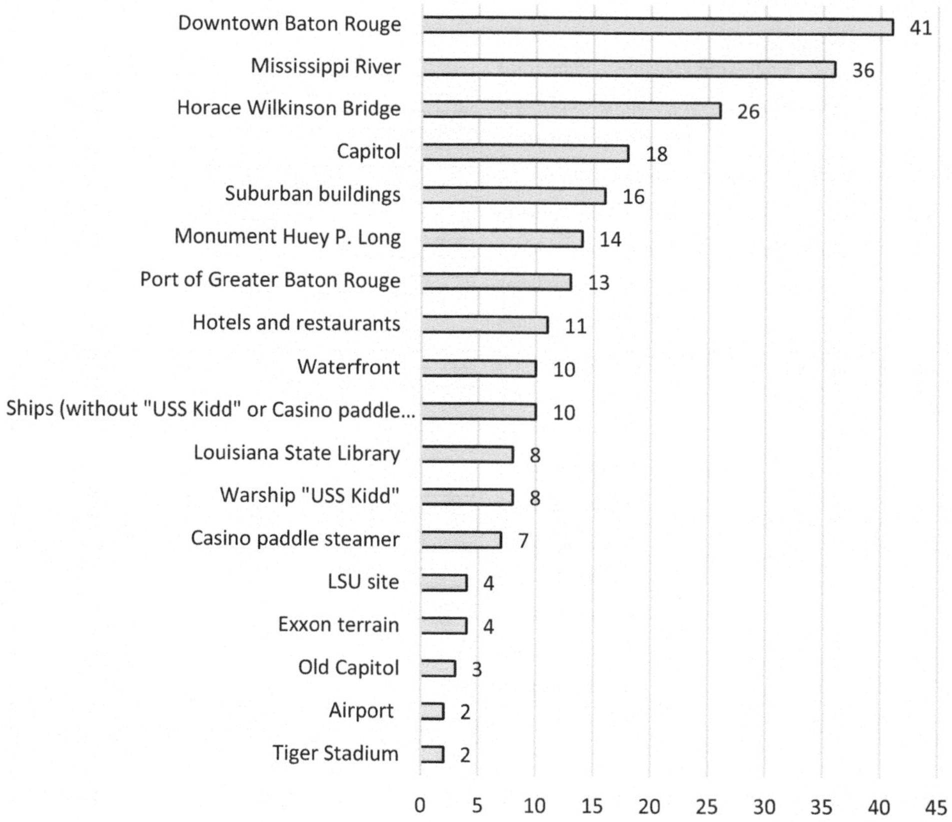

Fig. 7.5 Special symbolically occupied objects in the 100 illustrations examined. (Own survey and representation)

characteristic as a photo motif, but additionally (as mentioned above) as a location from which photos of Baton Rouge are taken. This dominance of the area in and around downtown is complemented by the monument of Huey P., located in the Capitol Park. Long, the Port of Greater Baton Rouge on the western bank of the Mississippi, the advertising shots of hotels and restaurants, mostly located in or around downtown, the depiction of the Waterfront on the bank of the Mississippi near downtown, often in connection with the former warship 'USS Kidd' moored there, today a museum, also the two casino paddle-steamers, located southwest and northwest of downtown on the eastern bank of the Mississippi, can be added to this context, which also applies to the Louisiana State Library and the Old Capitol. In contrast to this focus on the downtown area and its Mississippi-related surroundings, the (in terms of area dominant) suburban parts of the Baton Rouge metropolitan areas are poorly represented. The same applies to LSU, but also to the petrochemical industrial complex around the Exxon Refinery and even the LSU Tigers' stadium.

If the focus is directed towards the investigation of individual material objects and object constellations (Fig. 7.6), a certain dominance of those objects and object constellations can be found to which the attribute 'green' can be attributed: among the five most frequently represented categories there are single standing trees (rank 1), lawns (rank 3)

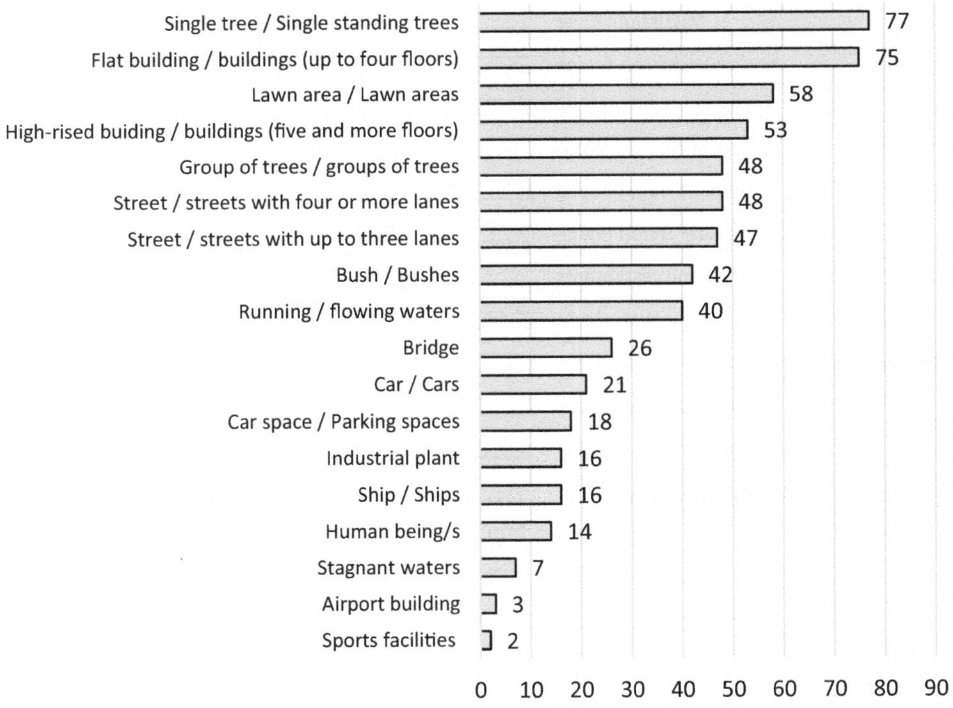

Fig. 7.6 Material objects represented in the 100 evaluated illustrations. (Own survey and representation)

and groups of trees (rank 5), a supplement can be found with the category 'bushes' ranking 8. Also due to the manageable number of buildings with five or more stories (especially in the downtown area, rank 4), lower buildings (rank 2) dominate in the evaluated figures regarding the buildings shown. Transport infrastructures (both in terms of roads and, to a lesser extent, parking spaces) also have a clear overall visual representation. The clear presence of parking spaces in the downtown area is also evident in the photos. Especially because of the Mississippi River, but also because of stagnant waters, such as the University or Capitol Lake, Baton Rouge is represented as a city rich in water. Compared to the clearly pronounced representation of transport infrastructures, the low presence of cars is striking. On the one hand, this is due to the high frequency of oblique aerial photographs, which mostly do not show individual cars, and on the other hand, to the tendency of the photographers to focus on objects with a high spatial persistence, which is not only true for cars, but also for human bodies. Even photographs of the downtown area and its districts are taken with extensive rejection of (individually recognizable) vehicles and people. Industrial plants, on the other hand, are frequently represented, albeit mostly as a backdrop in the depiction of the Mississippi and Horace Wilkinson Bridge, in the form of the industrial plants around the harbor on the western bank of the river. Both the airport and sports facilities are of secondary importance as photo motifs.

All in all, most of the pictures show an effort to portray Baton Rouge as a green urban center on the Mississippi River, which is spanned by a bridge of remarkable dimensions. What is remarkable is that this production is carried out with the representation of human bodies largely omitted, whose representation clearly recedes in comparison to that of human artefacts. But not only people can be identified as part of the inverse cityscape of Baton Rouge (in relation to the images examined), the parts of the community outside the downtown area (even Beauregard or Spanish Town) are not taken into account; this also applies (except for one photo in the context of crime) to the parts of the community inhabited by the black majority of the population, which are still not addressed in the exhibition, vacant and dilapidated buildings. If LSU is already poorly represented in the photos, Southern University is not mentioned at all. Equally unrepresented is the Huey P. Long Bridge, the second bridge from Baton Rouge across the Mississippi, in the north of the city. There is no mention of the railway infrastructure, which, despite the discontinuation of passenger train operations, is present both through track systems and also through freight traffic - including in the city center. But the inverse landscape also includes fences, which are remarkably rare (not only in terms of media representation, but also in the physical space of Baton Rouge).

7.2.2 Baton Rouge in Internet videos

With the increase in transmission speed and the reduction in the cost of transmitting large amounts of data for users, the foundations for the mass distribution of Internet video were laid. With the fall in the price of recording devices and the spread of less complex video editing programs, the production of Internet videos has become possible for a wider range of people. This creates opportunities for low-threshold communication for a manageable circle of addressees, but is associated with the danger of communication bubbles with the effect of recursively consolidating one's own patterns of interpretation and evaluation of the world while at the same time reducing the acceptance of alternative patterns of interpretation and evaluation (see inter alia Nagle 2017; Paris 2011; Schach and Lommatzsch 2018; Weber et al. 2018). In comparison to images in which ontological, moral, and aesthetic norms of photographic and publication values coagulate in a statically depicted motif, videos not only dynamize images, but also allow for a decomplexation of the interpretation process, since the author of videos can present ontological, moral, and aesthetic ideas in an explicit form.

On March 7, 2019, a total of 55 evaluable videos about the city of Baton Rouge were found on YouTube, the most important platform for Internet videos, among the first 100 hits, the rest were about the Baton Rouge guitar or the rock band Baton Rouge "(see Annex, Tab. 11.2)". These videos had a total length of 5 h, 15 min, and 47 s, and on the day of the study had just over 14.4 thousand views, with an arithmetic mean of 261,968 (median 75,540), minimum 139, maximum 4,677,913; the shortest video had a length of 0:00:38, the longest one of 0:28:37 (arithmetic mean 0:05:44, median 0:05:40). Dichotomously, the videos can be divided into two categories, those that deal with violence and crime and those that do not (Fig. 7.7). These are often self-portrayals of conflict and violence, such as videos of brawls or threats to rivals communicated through the medium of video (see serial numbers 8 and 35). These videos are generally characterized by a low degree of formal professionalism in terms of camera work, sound, and editing techniques (Fig. 7.9). However, there are also reports – appropriately professional in form - from television stations, such as the BBC (serial number 15), on violence and crime in Baton Rouge, but also reports on police violence against blacks (numbers 57 and 51; see also Fig. 7.8). Almost a quarter of the music videos have different formal professionalism, ranging from a picture of the band (number 42) to professionally produced videos (number 23). One-fifth of the videos are dedicated to Baton Rouge as a collection of physical exhibits, mostly – more professionally produced – promotional videos of tour operators (serial number 2) or – less appropriately professional – travel documentaries of tourists (number 55) as well as of car rides through

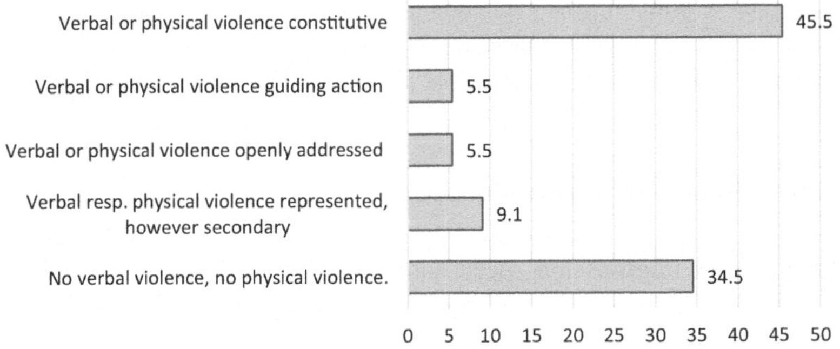

Fig. 7.7 Degree of depiction of violence in the examined videos in percent, n = 55. (Own survey and depiction)

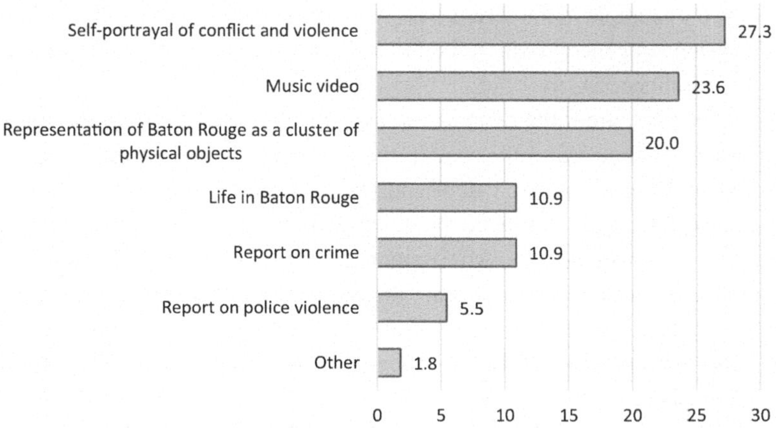

Fig. 7.8 Inductive developed thematic categorization of the examined videos in percent, n = 55. (Own survey and depiction)

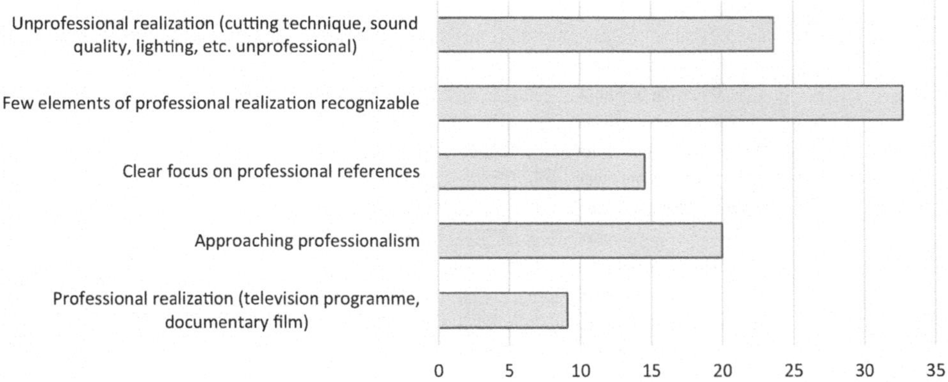

Fig. 7.9 Degree of formal professionalism, categorization of the examined videos in percent, n = 55. (Own survey and depiction)

parts of Baton Rouge (number 49). Quite diverse are the videos in the category 'Living in Baton Rouge', they range from short videos about a Mardi Gras parade in Spanish Town (number 19) to the car tuning scene (number 9) to introductions to slang expressions (videos 30 and 31), these videos generally show a low level of formal professionalism. One video could not be classified into the above categories: the video with the serial number 54 also deals technically with an accident at the Exxon Refinery. Compared to similar video studies (Kühne 2012a; Kühne and Schönwald 2015b; Kühne and Weber 2015, 2019a), a rather rustic degree of technical implementation of the videos can be observed. Most of the videos examined (38) show no signs of polyvalence, irony, or plagiarism, but instead the effort to present 'facts' or authenticity prevails, only 17 of the videos show (involuntary) signs of irony, for example in the seriousness with which the video with the serial number 7 criticizes Louisiana in general (with reference to Baton Rouge) or stages the downtown and adjacent areas (number 55).

Thematically, there are three thematic areas, which are dealt with in the following (in ascending order of numerical importance): the representation of downtown and its adjacent areas, Baton Rouge in music videos and violence and crime in Baton Rouge.

Travel-related references dominate the videos depicting the downtown and its adjacent areas. Apart from downtown, the Mississippi River, the Horace Wilkinson Bridge, the Capitol, but also hotels and restaurants are the main topics of discussion here (Fig. 7.10). However, videos that do not show significant spatial references, e.g. videos shot indoors, dominate. The classification of Baton Rouge (in its entirety) is usually based on its location in Louisiana on the one hand, and on its relations to other settlements on the other hand (Fig. 7.11). New Orleans alone serves as a qualitative benchmark (not only in this category; in terms of urbanity), while the other communities, mostly neighboring districts of lesser functional importance and population, and are mentioned more for information. A further spatial contextualization is made by (thoroughly romanticizing)

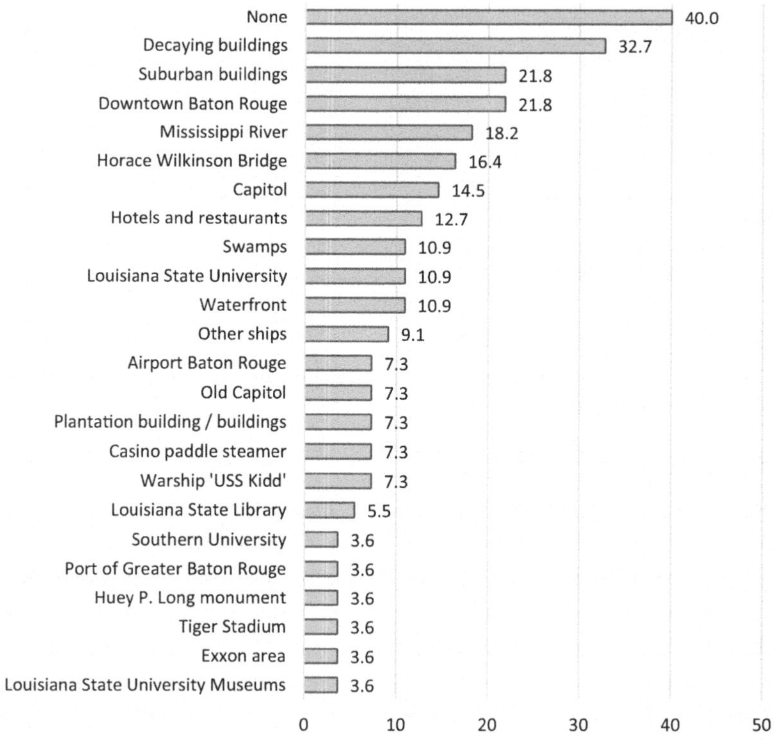

Fig. 7.10 Objects and object constellations shown in the videos with special symbolic charge, in percent, n = 55, several objects in one video possible. (Own survey and depiction)

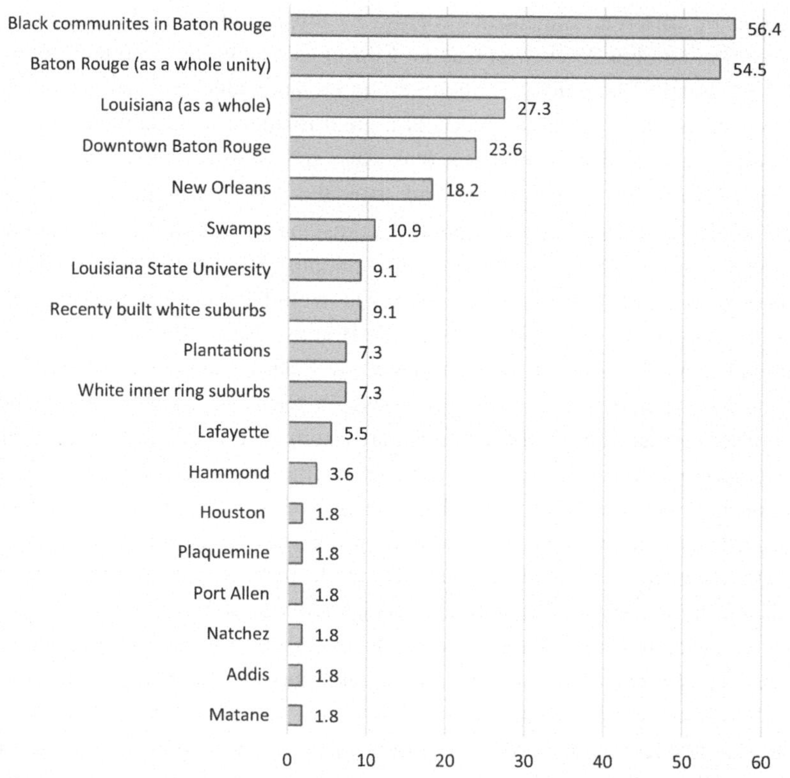

Fig. 7.11 Themed locations in the videos examined in percent, n=55. (Own survey and depiction)

references to the (past) plantation culture around Baton Rouge as well as the swamps to the south and west. The presentation of Baton Rouge downtown and adjacent areas is specifically aimed at not only presenting the historical character of Baton Rouge, but also at staging its uniqueness (especially in videos dealing with urban development, such as serial numbers 40 and 46) and at marketing it to tourists in the form of the definition of 'sights', such as Old Capitol or museums (such as number 34). The staging of the Waterfront and Downtown does not lack a certain irony in view of their physical form and social significance, for example in relation to the density of shopping or gastronomy facilities or accessibility (Waterfront over railway tracks). The people appearing in these videos are white in color and – as far as can be seen – have a middle-class habitus (busy, culturally active, committed, correctly dressed etc.; see Bourdieu 2016). In these videos, Baton Rouge is summarized as an economically and politically significant river crossing town with distinct urban qualities and a largely white population, shaped by its history.

The function of Baton Rouge in music videos, on the other hand, is relatively undifferentiated. Partly, this is due to the fact that the country song "Callin' Baton Rouge" is interpreted by different singers in 7 of the 13 music videos evaluated, in the video with the running number 25 even in Gaelic. Conversely, as in the song mentioned above, Baton Rouge mostly functions as a relatively interchangeable physical-spatial location. For the song to work, all that is needed is a great distance to be covered on the highway to justify the intensity of affection that was expressed in hourly telephone calls to the woman with whom the lyrical ego had spent the previous night, whereby a certain familiarity of the town (here as the capital of Louisiana) is helpful to the target audience (in this respect, in the USA, for example, Sacramento, Atlanta, or Jackson, Mississippi, in Germany also Saarbrücken or Gelsenkirchen would have been possible; the same applies to the song sung in French by Issabelle Boulay "Entre Matane et Baton Rouge", serial number 24). The song of the same name by Palatine (numbers 10 and 53) shows a more intensive connection with Baton Rouge in terms of content, on the one hand because it refers to the Cajun culture and is partly sung in French, and on the other hand, because the video with the consecutive number 10 (the one with the number 53 shows a studio recording) is strongly influenced by generally known South Louisiana motifs in a graphic-abstract manner, such as buildings on the waterfront, a boat ride on a bayou, swamp vegetation up to the call of death in female form in the swamp. Here, too, the protagonists in the videos are - if depicted - white skin colored.

This is completely inverse compared to the category 'violence and crime in Baton Rouge'. People of white skin color (as individuals) appear here as a person living in the black neighborhood and designated as a "white nigga" (consecutive number 48), with the exception of a resident in a murder case, only in the form of police officers. These, in turn, appear in the analyzed videos in an inappropriately aggressive to violent manner, for example in the form of the arrest of a clearly unarmed black nurse during a demonstration or the firing of a shot at a black man lying on the ground (serial numbers 51 and 47). The content of these videos is also strongly related to local black neighborhoods and their conflicts. The conflicts among rappers in Baton Rouge are strongly present, which are carried out by means of insulting statements on the net and rap songs as well as with firearms, which has led to the death of several participants documented and commented on in numerous videos as well as the arrest and conviction of the perpetrators (for example in the videos with the serial numbers 33, 34, and 38). In general, the videos in this category give the impression that violence is recognized as a general means of social hierarchization; this is documented not only by verbal statements, but also by the presence of (sometimes apparently brutally executed) brawls (such as numbers 8, 11, 13 and 14). A function of clarifying one's own position in the reference group (Dahrendorf 1961; Riesman 1950) can also be seen in posing with firearms, money, gold chains, and expensive or converted cars (including consecutive numbers 14 and 48). In general, it can be said that protagonists of videos apparently show traces of the influence of drugs (about number 37). The problem of violence as well as the helplessness in dealing with it is also evident in the videos with the serial numbers 43 and 44: both videos form a commentary framework to another (at least partially recorded) Internet video. This recorded video shows a young black woman shooting a live video of herself on Instagram, while apparently another woman is being raped in close proximity (she screams "get out off me" several times; serial number 43) and shows no reaction in the form of an intervention, but instead continues to shoot the video. The commentator of the recorded video in turn presents morally accepted options for action in response to this: either calling the police or helping the woman being raped himself. In the video with the serial number 44, the further development is described, the internet video had triggered police investigations, which, however, had not led to a result, the screaming woman on the video had assured that it had not been a rape.

All in all, the videos in the category 'Violence and Crime in Baton Rouge' show a low level of trust in state institutions, especially the police, and subsequently, a tendency to carry out conflicts - rapidly escalating - with physical violence. Conflicts are understood in a personalized way, and more abstract social references are not established (for example, regarding the still unequal educational opportunities). Baton Rouge, which is socially constructed in this category of videos, has no intersection with Baton Rouge from the other categories: it is characterized by mostly young violent male blacks in a poor environment, in which whites only appear as assaulting policemen (or judges who do not understand the customs of the neighborhood), older people only marginally, and women as inferior victims. An exception to this is the video with the serial number 51, which deals with the arrest of a confident black woman and her interpretation of the situation on CBS, including the dilemma of explaining her arrest and conviction to her six-year-old son, after all she tried to make him understand his entire life, only "bad people" would go to prison.

If the result of the image analysis (Sect. 7.2.1) corresponded in large parts to the category "downtown and its adjacent areas" in the Internet videos, with the exception of the people more strongly present in the videos, black living environments were not present there – unequally as shown – in the evaluated videos. Overall, the videos are dominated by the depiction of people in different spatial contexts (Fig. 7.12). In contrast, road traffic infrastructures are even more prominent than in the photos, here also with a higher presence of vehicles. Baton Rouge also appears in the videos as a green city.

If the parts of the urban landscape inhabited by blacks formed the inverse landscape in the Internet images, they are strongly present in the videos. Here the 'white suburbs' in the south of the city form an essential part of the inverted landscape. Even though the universities (especially LSU) have a stronger presence in the videos, they are mostly mentioned only marginally in the videos. As shown, the videos around Baton Rouge focus on the presentation, commentary, and discussion of (personal) conflicts, which in turn are presented without significant spatial symbolism (Fig. 7.10).

The evaluation of internet communication about Baton Rouge shows the communicative split along the city's division as outlined in the previous sections in terms of ethnicity, income, education, social references (i.e. symbolic capital), and overall life chances between North and South in communicative terms: the 'other' Baton Rouge is socially so far away that there is hardly any direct involvement with it, the 'other' enters into one's

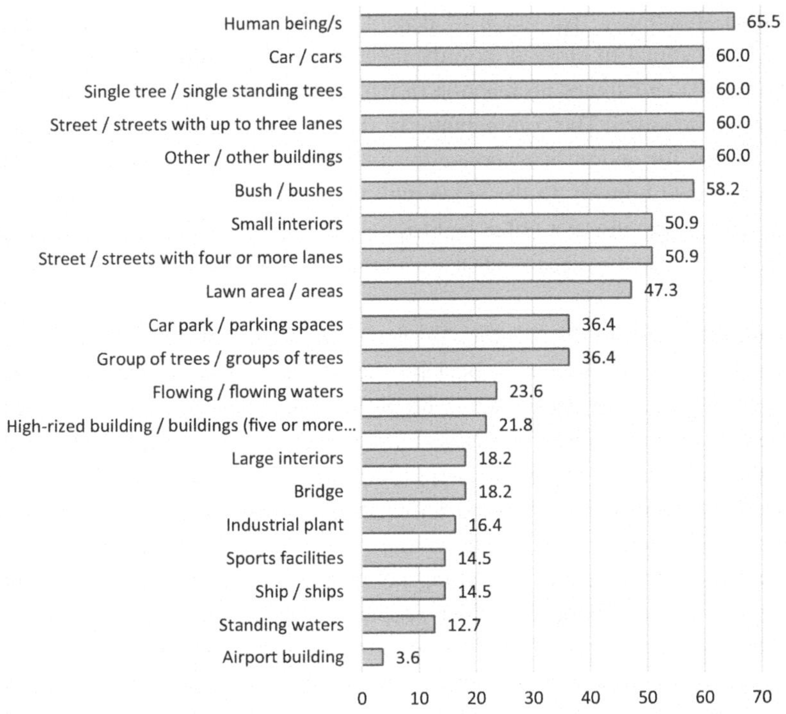

Fig. 7.12 Material objects and object groups represented in the videos examined in percent, n=55. (Own survey and depiction)

own (communicative) life in particular in the form of generalizations, which is solidified into a stereotype according to the logic of communication bubbles (see on this process for Louisiana in general: Hochschild 2016).

7.3 The Tourist Representation of Baton Rouge: Tripadvisor

The internet platform TripAdvisor offers the possibility to publish individual travel experience reports and evaluations and can be used on this basis for personal travel planning. According to TripAdvisor (as of 26 April 2019), it has 730 million reviews and opinions, 8.1 million ratings of accommodations, airlines, experiences and restaurants and has 490 million monthly average unique visitors (TripAdvisor 2019). In the following, the 15 sights and landmarks in Baton Rouge will be briefly presented in terms of their dominant ratings as retrieved on April 26, 2019, in order to preserve further aspects of the social construction of Baton Rouge, here under an explicitly tourist mode of perception and interpretation. The presentation is based on the (descending) order of the listings on the reporting date:

1. Alex Box Stadium/Skip Bertman Field (138 ratings): The view of the field, the atmosphere at matches, the service, the low-ticket prices and the parking facilities were rated positively. JohnHough3 describes superlatively: "This is probably the best college baseball fan experience in America", a rating that is mentioned more often in a similar way.
2. LSU Tiger Stadium (851 ratings): Again, the good view of the field is praised, as well as the overall program by the LSU Marching Band, the atmosphere (and the volume) during the match and the friendliness and communicative orientation of those present. Numerous comments positively emphasize the symbiosis between LSU Tigers and the city, which is expressed in extreme enthusiasm at home games in the stadium, its surroundings and throughout the city.
3. Louisiana State University (243 ratings): LSU is present in the reports primarily in three aspects, firstly as a basis for sporting events (Tigers) or cultural events (concerts), secondly, as a campus designed according to general landscape preferences and oriented towards pedestrians, thirdly (far less intensive), as an educational institution, whereby in this context, too, reference is made again and again to celebratory activities that formed a part of campus life at LSU. The strong presence of restaurant chains on the campus and its surroundings has been criticized.
4. Louisiana State Capitol (659 ratings): Here, the architecture, both inside and out, the murals and especially the view of Baton Rouge and the surrounding area from the viewing platform are highlighted. The surrounding Capitol Park is also frequently mentioned, as is the possibility of finding parking spaces near the Capitol. Remarkable (compared to the LSU) is the presence of the Capitol as an educational institution in which the history of Louisiana is presented (although the presence of Huey P. Long in the presentations is sometimes criticized, which indicates that his assessment is still controversial today). Criticism is primarily directed at the failure of the elevator to the viewing platform.
5. Magnolia Mound Plantation (138 assessments): Here the 'realistic' depictions of life on a pre-war plantation and the good accessibility are praised. The facility provides an opportunity to experience life in the plantation era, especially because of the 'authentic' design of the buildings, although a more critical view of slavery is sometimes called for.
6. The Old Governor's Mansion (117 ratings): Generally, the possibility to gain insight into the history of Louisiana as well as the knowledgeable and friendly guides are rated positively. The object of criticism here (like the representations in the State Capitol) is a strong presence of Huey Long in the statements.
7. Atchafalaya Basin Bridge (99 Ratings): In general, the view over the bayous and swamps, which is connected with the drive over the Interstate 10 west of Baton Rouge, which runs partly on stilts, as well as the length of the bridge (about 18 miles) is rated very positively. On the other hand, it is regretted that the bridge has narrow lanes on the one hand and is often heavily trafficked on the other, so that the focus of the driver's attention must be on the traffic; in this context, a lack of viewpoints is also formulated.

8. Downtown (164 ratings): The users' evaluation of the downtown area is quite divergent, the State Capitol and the Mississippi are positively emphasized (often with reference to the 'USS Kidd'), the design of the waterfront area is less clearly assessed, while the new design is emphasized here, there exists also its low public resonance. This interpretation is also given to the entire downtown area, the majority describe it as quiet, which is combined with the attributes family-friendly, clean and safe, as well as with boring. The pedestrian friendliness is also emphasized here. If a chronological comparison is made, the development is assessed positively, the number of restaurants has increased, and the Waterfront has been developed.
9. Louisiana Art & Science Museum (116 reviews): The emphasis of positive reviews is on the scientific part of the museum, especially the planetarium, the part where art is shown is much less emphasized in the reviews. In general, the small size of the museum is mentioned, which is either positively interpreted as clear and focusing on the essential or - with regard to the thematic combination – as arbitrariness or critically related to the admission fees. The child-friendliness of the exhibition, if addressed, is either evaluated positively (by parents with children) or criticized as not being adequate for their own interests. If mentioned, the accommodation in the former ice rink depot is praised.
10. Pete Maravich Assembly Center (28 ratings): The third LSU Tiger sports facility, primarily used for basketball, is praised (like the others) for its visibility of the court, as well as for the professional and helpful staff. However, the sports facility is sometimes criticized as structurally outdated and parking nearby is described as difficult.

Of the other places, follow St. Joseph Cathedral (28 ratings, commonly described as affirmative and beautiful), the Pentagon Barracks (17 ratings, the historical significance is valued, the accessibility is criticized), the Louisiana Memorial Plaza (6 ratings, the historical significance is also the focus of the ratings, partly with patriotic connotations), Saint Alban's Chapel (1 rating, which emphasizes the peacefulness of the place in the context of student life) and Greenoaks Funeral Home and Memorial Park (2 ratings, which emphasize the adequacy of care).

In the top 15 evaluations of the sights and landmarks in Baton Rouge, the presence of LSU (first three places) and 10th place is conspicuous, although the focus is less on its function as an educational institution than as a sports institution or ensemble of buildings and parks. Vitality is primarily associated with sport, while the other sights and landmarks shown are said to be rather quiet. In this context, it is somewhat ironic that a funeral parlor and cemetery is ranked among the first 15 sights and landmarks (albeit in 15th place) of TripAdvisor. The question of accessibility to the respective objects and object constellations runs right across the evaluations of the sights and landmarks of Baton Rouge; this is only in the case of the LSU campus or the downtown area (also) in relation to pedestrian traffic. The question of whether sufficient parking spaces are available generally dominates. Notably, the sights and landmarks with references to Louisiana's history show how strongly Huey P. Long, who was shot more than eight decades ago, polarizes.

The spatial distribution of the presented sights and landmarks shows a similar set of objects and object constellations as that of the Internet images, although the weighting is different: in TripAdvisor, LSU is more present (specifically from a landscape perspective or in relation to sports facilities and events), both have in common the focus on the downtown area, including the Mississippi River and the Capitol area. TripAdvisor also emphasizes the inverse landscape of Baton Rouge in the north of the city: in general, the parts of the city inhabited by African-Americans are ignored, but especially – and this becomes particularly clear when comparing LSU and its sports facilities – Southern University, where college sports are also played in appropriately equipped sports facilities, supported by enthusiastic fans.

7.4 Artistic Approach to Baton Rouge - Baton Rouge in Song

Art can be understood as a human form of practice that stands in a "continuity to other human practices" (Bertram 2014, p. 12), since it can only gain its potential through reference to other human practices, and thus these are fundamental to it (Bertram 2014). Accordingly, art as a human form of practice is not delimited from other human practices, but only differs from them in the way it is related (Bertram 2014). The constitution of the everyday world takes place reflexively for the individual and results from a complex interpretation procedure, "in which current perceptions are related to earlier perceptions" (Schütz 1971 [1962], p. 123 f.). It is also important for the individual, under the omens of constant social change, to take a position on himself again and again in a reflexive recourse (see Bertram 2014). In this context of the continuous redefinition of the individual, art "represents a specifically reflective form of practice [...] by means of which people take a stand on themselves within the framework of a cultural practice" (Bertram 2014, p. 13). Art not only liberates the subject by breaking through its self-control and self-determination in the sense of Nietzsche and Schopenhauer, but, according to Bertram (2014, p. 14), it rather makes "a contribution to becoming a subject". Accordingly, an analysis of artistic approaches to Baton Rouge was chosen in the present context as well, in order to take a look at the reflexive recourse of individuals to the respective constitution of 'Baton Rouge worlds' and to see whether, and if so in what form, changes occur over the years with regard to a continuous redefinition and artistic reference to Baton Rouge. Musical forms of expression were used, as these also contain linguistic elements and can therefore be evaluated quantitatively and qualitatively. In July 2019, 45 song titles were compiled on the music platform Amazon Music as one of the most rapidly growing music streaming services (Nicolaou 2019) and, in addition, on the music and video streaming platform YouTube as a platform with currently over one billion users (YouTube 2019) under the keyword search 'Baton Rouge'. However, there is no claim to completeness – rather, the aim should be to find out which songs are primarily offered to users in the context of a song search for the keyword 'Baton Rouge' "(see Annex, Tab. 11.3)". These were further differentiated and quantified on the basis of inductively

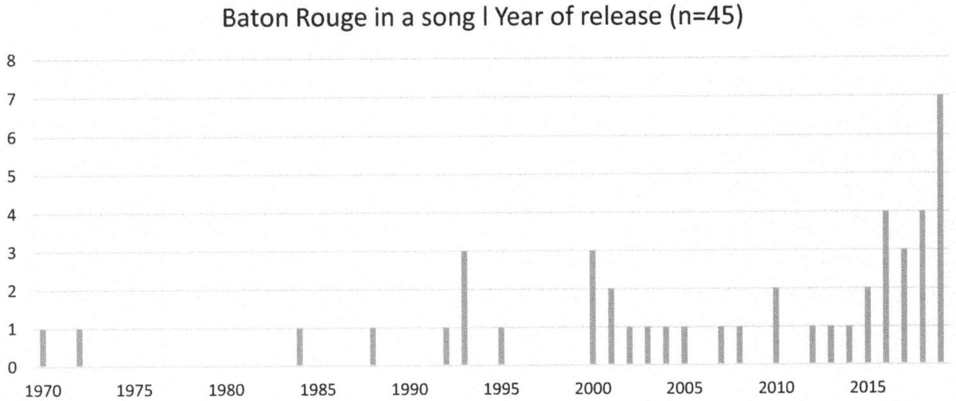

Fig. 7.13 Representation of the songs determined for Baton Rouge according to release years; own survey and representation

formed categories in the songs of the topics dealt with, and then qualitatively evaluated. In this context, a significant increase in Baton Rouge-related songs in the last five years (see Fig. 7.13) is fundamentally evident at first, as is a genre change in the music styles from blues, country, and Zydeco to songs dominated by rap music, as will be shown later.

If the corpus compiled in this way is classified into the categories 'Songs up to the year 2000′ and 'Songs from the year 2001 onwards' and evaluated according to the different thematic categories, it becomes apparent that emotional references expressed in the songs up to the year 2000 with hope, happiness, and a sense of home, tend to be dominated by emotions with positive connotations (see Fig. 7.14). Also, in terms of recourse to human relationships, the focus is primarily on lovers, followed by family and friends, although the relationship to the gang is still relatively undefined. With a view to life-worldly, leisure-related aspects, the songs up to the year 2000 still update especially bars or festivals as well as culinary delights – drugs still experience a rather marginal treatment here. Also, the city of Baton Rouge itself is the subject of much attention, whereas crime and violence in the songs until 2000 is not the subject of any attention.

In the songs from 2001 onwards, on the other hand, an increase in the number of feelings with negative connotations can be observed, within which the emotions of hate and anger in particular increase, as does a feeling of sadness, and the emotion of being lost is also added (Fig. 7.15). In the context of interpersonal relationships, gang and friends now take center stage, and are given equal importance, even if the references to lovers and family are still pronounced. From a life-world perspective, a decline can be observed regarding bars and festivals, while references to drugs are increasing. A 'new' subject area is the area of crime and violence, which could not be identified in the songs before 2000 and, especially in the songs of the newer and newest generation, concentrates primarily on violence between gang members and crime in general (Fig. 7.15).

In a qualitative approach, the songs up to 2000 focus in particular on the processing of individual experiences such as a lost (SG42; SG05; SG30) or newly won love

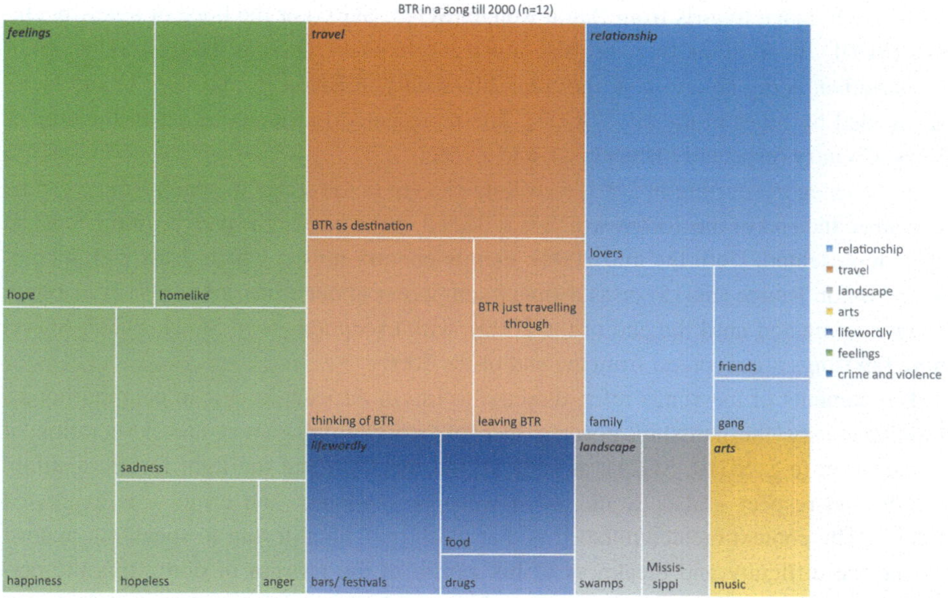

Fig. 7.14 Treemap of the topics covered in the songs on Baton Rouge up to the year 2000, n = 12; own survey and presentation

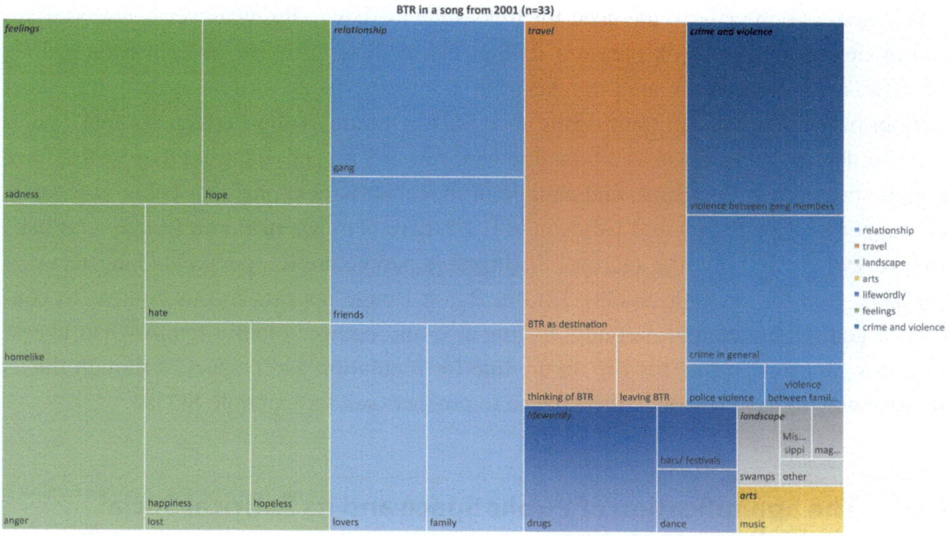

Fig. 7.15 Treemap of the topics covered in the songs on Baton Rouge from 2001, n = 33; own survey and presentation

(SG08), a homage to girls from Baton Rouge (SG43; SG03) or the hope of a new beginning (SG01). In addition, there are also musical references, such as the (re)production of the narrative, being able to hear the 'real' blues only in Baton Rouge (SG11) and being surrounded by blues in the city (SG09). The represented music styles are in the area of Blues, Country (especially Bluegrass), and Zydeco.

In the category starting in 2001, a striking change in the songs to Baton Rouge can be seen since the end of the 2000s, which is reflected in the content as well as in the musical styles represented. Until the mid-2000s, mainly country music, which deals with memories of Baton Rouge (SG15) or as a singer-songwriter about a lost love (SG12) in Baton Rouge, dominated until the end of the 2000s, with exceptions (e.g. SG24, SG27, SG41) more rap songs are recorded from the end of the 2000s.

The contents of the songs refer, directly, to life in African-American neighborhoods. Simultaneously, the hardship, poverty, lack of opportunities, anger and desperation of the situation (e.g. SG10, SG21), as well as, the rebellion and the fight for recognition, wealth, and respect – also by means of weapons, violence, and crime – is expressed (SG37). The context created/pursued is that of an individual trying to speak out against the intense difficulty and bleakness of his world. In the process of doing this the performers are demanding respect, showing their toughness and fearlessness while trying to generate fear in those opposed to them and their friends/gang members. Also is an underlying hope that the recognition and respect they seek will bring financial reward and the chance to leave this environment (see exemplary SG14, SG39). Baton Rouge City and life there get the usual Gangsta/Rapper description, hardship, hostility, hate, and hopelessness giving rise to the need for folks such as they to do whatever is necessary to find money, stay alive, seek pleasure such as can be had, stave off or fight back against the efforts of others in the area to take what is already his as well as to deny him what he is in pursuit of with all their desire (see SG14). Some of the Performers make great effort to declare themselves as a bonafide gangster and rapper, claiming to go hard, have women, be fearless, do drugs, and engage in illegal activities, plus that being a gangster and rapper and tough guy and lady's man is all there is to life in Baton Rouge – perhaps life everywhere and that the successful/admirable people are Rappers and Gangsters that 'go hard' (see for example SG39). Life in Baton Rouge appears in this context to consists of getting stoned and partying, having sex, and being violently tough when necessary as a gangster putting his life on the line for reputation and respect as though this is the only way to have respect and a chance to prosper (see for example SG39).

7.5 The Spamification of Public Space and its Satirical Social Media Representation

The approach to Baton Rouge as a physical space is confronted with an unusually intense confrontation with large-format billboards along busy thoroughfares as well as entrance and exit roads. This type of advertising is not unique in the United States and other parts of the world, yet the density of these billboards in Baton Rouge is remarkable. People – usually drivers – can hardly escape the messages because of their physical presence. From a landscape point of view, these arrangements can be described as 'spamscape' (Kühne 2012a), following unsolicited advertising sent by e-mail.

The billboards in Baton Rouge are neither physically nor thematically evenly distributed (Fig. 7.16): as mentioned above, they are piled up along busy roads, but especially where the passage of more affluent drivers is suspected. Following Gerhard Hard's (2008) call to include not only what can be seen, but also what cannot be seen and why it cannot be seen, an 'inverse spamscape' can also be constructed (Kühne 2013a, b). It is missing not only

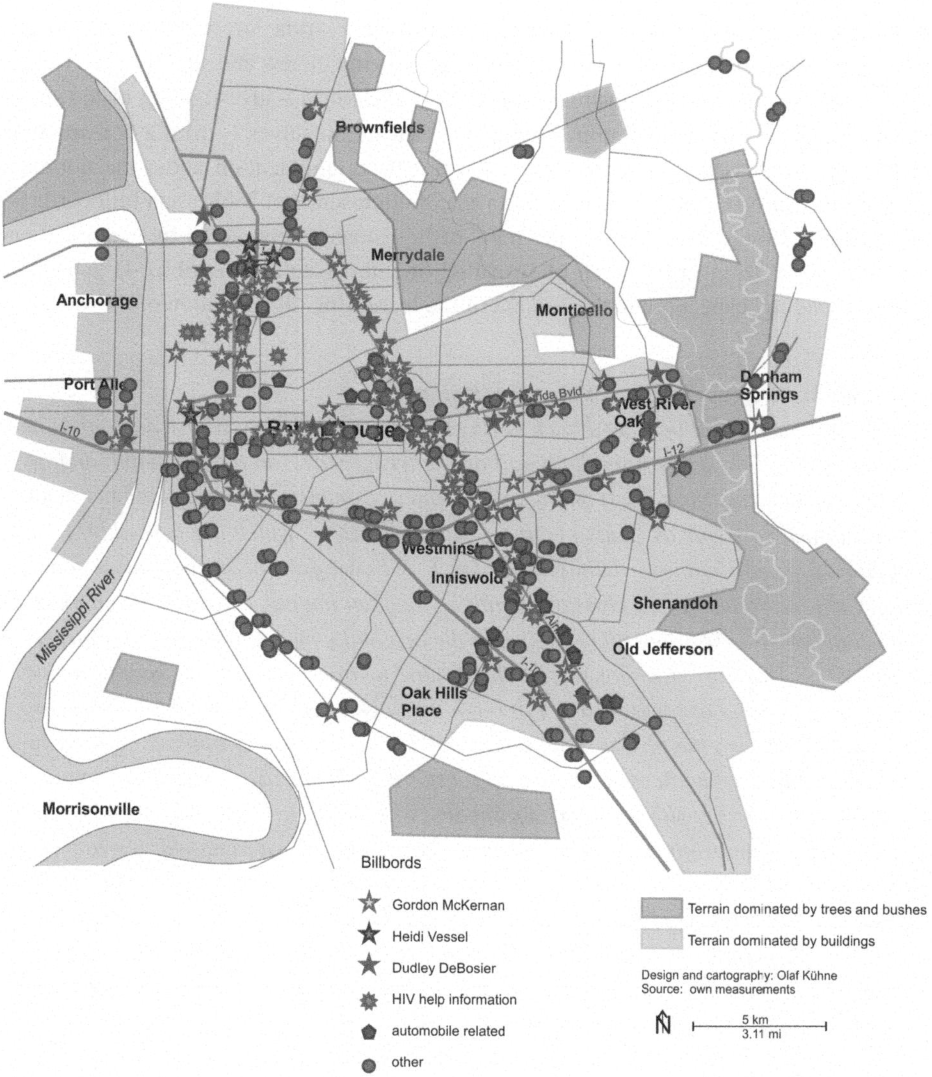

Fig. 7.16 Spamscapes. The distribution of large-format billboards and their contents in Baton Rouge, September 2019 (own surveys)

where there is less traffic, for example, in parts of the city primarily used for residential purposes, but also (largely) along streets that are primarily frequented by people with a low level of symbolic capital (i.e. specifically in the north of the city). Thematically, a clear presence of advertising by traffic lawyers can be demonstrated, whereby – the numerically dominant – Gordon McKernan is present in particular on the Airline Highway, Florida Street/Boulevard as well as Interstates 12 and especially 110, whereas the law firm Dudley DeBosier is present in particular in the south, Heidi Vessel exclusively in the north with web boards. Except for the colored Heidi Vessel, all law firms (including those not separately mapped here) advertise with white, mostly middle-aged males. In addition, there are numerous HIV information boards in northern Baton Rouge, as well as car advertisements on the Airline Highway south of the intersection with Florida Street/Boulevard, an area where there are also numerous car dealerships specializing in new and slightly-used cars.

After this – following the positivist approach to the world – investigation of the spatial distribution of billboards in Baton Rouge, we dedicate ourselves – from a constructivist perspective – to media content analysis of satirical communication about the dominant motif of billboards: Gordon McKernan. For the omnipresence of McKernan billboards not only in Baton Rouge, but also in other parts of Louisiana, is countered by a satirical and humorous negotiation in the form of so-called 'memes' (see also Textbox 4), which are rapidly spreading, especially in the context of the increasing digitalization of the world.

Textbox 4: Excursus - Outline of the term 'Memes'.

The term 'meme' can be traced back to Richard Dawkins (1978), whose neologism in the pre-digital era was originally based on the term 'genes' on the one hand, and on the Greek mímēsis (imitation) on the other hand. He defines all non-genetic behavior and cultural concepts as 'units of conceptual replication', which are passed on from person to person within a culture and which, following an idea of 'natural selection', can reproduce themselves differentially, i.e. can mutate, change or even extinguish in the process of passing on. Even if the original concept of memes was largely rejected in Dawkins' reading, it found a return to Internet-based discourse in the context of Internet use and its users through the back door: "[i]n the vernacular discourse of netizens, the phrase 'Internet meme' [...] is commonly applied to describe the propagation of content items such as jokes, rumors, videos, or websites from one person to others via the Internet" (Shifman 2013, p. 362). These may well be disseminated in the original version, but they also produce "user-created derivatives" (Shifman 2013, p. 362), which achieve enormous digital ranges (Davison 2012). Central is "the speed of their transmission and the fidelity of their form" (Davison 2012, p. 122). Internet memes often have the same structures in place: usually text elements are integrated into image macros, although some image macros may contain the same text to different images and vice versa, the same image macro with different text (Davison 2012). The punch line often arises from the fact that two or more originally separate areas are correlated with each other, both textually and visually.

This also applies to the digital exploitation of the omnipresent billboards, especially those of the law firm McKernan, in the context of which the spamscapes are decontextualized and placed in new contexts in a humorous and pointed manner, thus transforming the permanent and unsolicited exposure to the advertising messages into a creative outlet such as memes, or in this case McMemes. These are communicated via social media such as Facebook and Instagram and are thus brought to a public hearing, which now makes it possible to create a forum in digital niches (here individual profiles in social media) for the - in this case - humorous processing of jointly experienced spamscapes, essentially by means of likeness, advertising slogan and billboards of one of the main advertisers are related to popular TV formats and app applications, current events and happenings or simply existing memes from other contexts are modified and applied to the McKernan case (see, for example, the collage of GordonMcKernanMcMemes in Fig. 7.17) A certain reinterpretation and restructuring of the visual mode and patterns of interpretation regarding billboards and spamscapes takes place in physical space: from the often more or less informative, painfully or unconsciously 'experienced' omnipresent advertising message to the running gag of the highway, which instead of a passive 'letting pass' stimulates the recipient to a possible creative use of jointly experienced 'spamscapes'.

This exposed position of the spamscapes - here in terms of creative exploitation and thus their emphasis among a whole series of other physical manifestations in the space along the infrastructure veins - shows that they are not 'stereotypically' included in the social construction or in the learned visual patterns of 'urban spaces', but that they are physically present in such a way that they can no longer be ignored in the context of the social construction of the normal native landscape. In the form of their satirical reinterpretation as memes, however, they are reintegrated into the show of normal native landscape and thus reappropriated in a modified form.

The receiving internet community makes this clear in the form of comparatively high subscription figures for the respective profiles in the social networks (Instagram: around 3,000 subscribers, Facebook around 14,000 subscribers; as of November 2019) and their ratings or comments on the profile page: the McMemes are not only a '[g]reat laugh' (fb02), but rather 'nothing has ever felt so necessary as this' (fb08). Because 'if you drive around Baton Rouge, this is a must follow' (fb21) and 'if you live here you need this' (fb26). However, there is also criticism that this is a 'Subpar page for a subpar person and predatory lawyer firm', which '[l]iterally only exists as an extra marketing gimmick. 0/10.' (fb13; similarly also fb04). However, subscribers take this reproach with composure: 'he certainly did not get it done' (fb16).

After the depiction of the spamification of public space and its creative negotiation in 'new' forms of mass media representation such as social media, the following section will turn to a more 'classic' mass medium, the national daily newspaper – here the New York Times as one of the highest circulation national newspapers in the USA – and in particular look at the contexts, classifications, and interpretative contexts in which Baton Rouge is (re)produced from an external perspective.

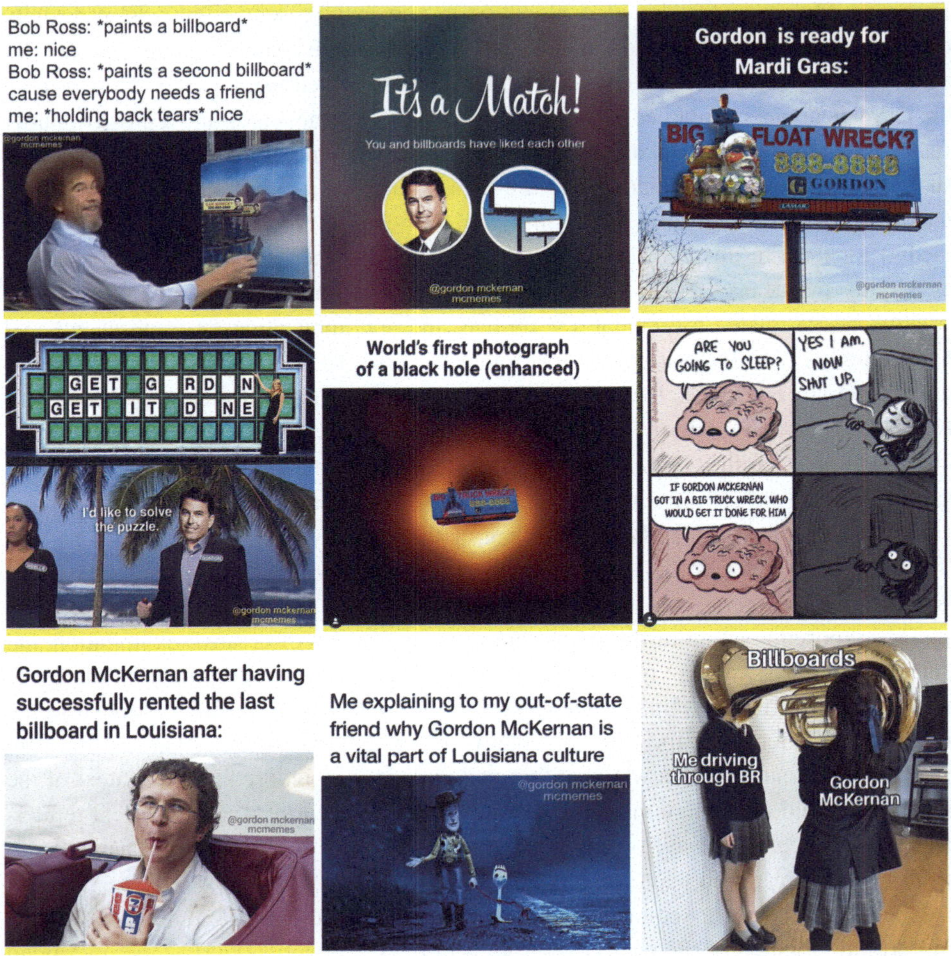

Fig. 7.17 Satirical communication on McKernan billboards in Louisiana in the social networks made in Baton Rouge from January to October 2019 in chronological order (selection); with the kind permission of the account operator of the Gordon McKernan McMemes profiles on Instagram (https://www.instagram.com/gordon.mckernan.mcmemes/?hl=de) and Facebook (https://www.facebook.com/gordonmckernanmcmemes/) All rights by ©Gordon McKernan McMemes

7.6 The National Representation of Baton Rouge Using the New York Times as an Example

With regard to Baton Rouge's representation in the national media, the New York Times (NYT) has been used as a reference, as it has become one of the highest-circulation nationwide newspapers in the United States with about 500,000 copies (Monday–Friday) since its foundation in 1851 and has recorded more than three million digital subscriptions in the growing digital market in the fourth quarter of 2018 (Statista 2019b). Accordingly, in May 2019, the online archive was compiled using the keyword search 'Baton Rouge' to find all articles on Baton Rouge published in the NYT and made available through the online archive from 1978, 1998, and 2018, whereby the data set was adjusted for family and bereavement announcements in order to be able to analyze only news-related announcements on Baton Rouge. By highlighting selected vintages, the aim was to find out whether and how so, which thematic focuses and related shifts in the national reporting on Baton Rouge can be identified "(see Annex, Tab. 11.4)". At the same time, various phases of the city's development should be taken into account, such as the boom caused by the petrochemical industry in the 1970s, the years after the depression and before the restrained turnaround at the end of the 1990s, and current trends at the end of the 2010s. Accordingly, a short quantitative and qualitative analysis of the articles of the respective vintages followed.

If the articles are categorized and quantified in absolute numbers according to their thematic focus, the greatest emphasis is on sports-related reporting, followed by the category 'crime', which is ranked second, along with 'arts, literature, music' (cf. Fig. 7.18). This is followed by society, politics, and economic issues. Travel and climate/floods and science/education are in the bottom third, with health, transport and infrastructure, and culinary issues bringing up the rear. The areas of transportation/infrastructure and food still play a central role in the context of the 10 things you hate about Baton Rouge (see Sect. 5.4).

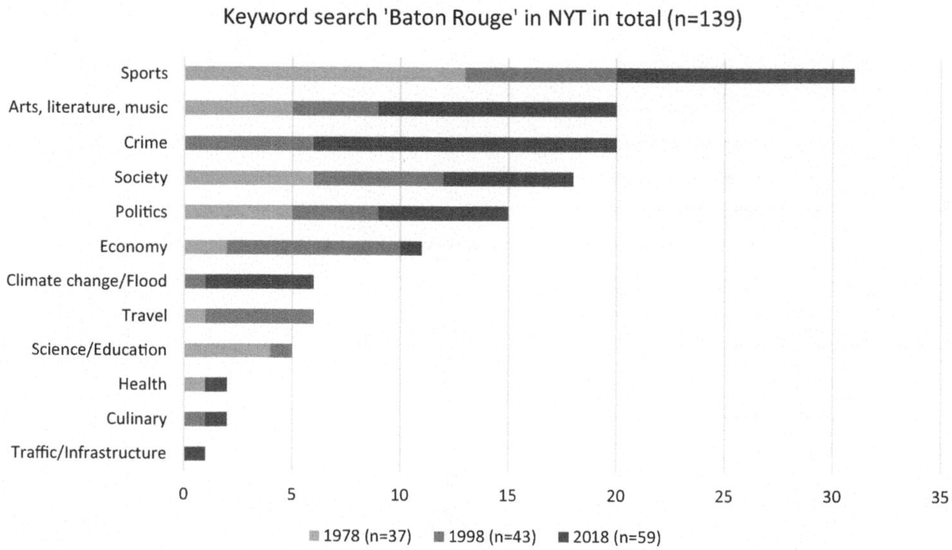

Fig. 7.18 Keyword search for 'Baton Rouge' in the online archive of the NYT for the years 1978, 1998, and 2018 by categories TOTAL in absolute numbers; accessed in May 2019

If the results are presented according to grouped columns, it becomes very clear at first that the reporting on crime in connection with Baton Rouge more than doubled between the years used and has thus risen to the topic most frequently associated with Baton Rouge in 2018 (Fig. 7.19). Sports reporting with references to Baton Rouge is in second place in 2018 and has thus been displaced by the issue of crime from the sports theme that still dominated in 1978. There is also an increase in the topics of climate change/floods in the ascending years, which were not yet a topic in the issues of the 1978 volumes. But also, the coverage of art, literature, and music in connection with Baton Rouge has doubled in 2018 compared to the previous year, whereas society and politics are on a similar quantitative level. On the other hand, travel-related reporting in connection with Baton Rouge, which was still in the top third at the end of the 1990s, has declined. By contrast, science and education-related topics, which were particularly prevalent in the 1978 vintage, are no longer discernible in the 2018 vintage. Even economy-related topics, which were still the dominant category in 1998, are hardly to be found in more recent issues (see Fig. 7.19). Accordingly, the respective focuses of the reports change, with references to Baton Rouge by sport in 1978, to economic topics in 1998, and to crime-related reports in 2018 (see Fig. 7.19).

When the contributions are subjected to a brief qualitative analysis, it becomes clear that in 1978 'sport' was the strongest category, but that it was essentially short reports, which in particular contained results from various sports such as basketball, baseball, athletics, etc. (NYT 1978–05, NYT 1978–10 or NYT 1978–11), but also include reports about selected people in the sports scene, such as a sports scout who discovered pitcher

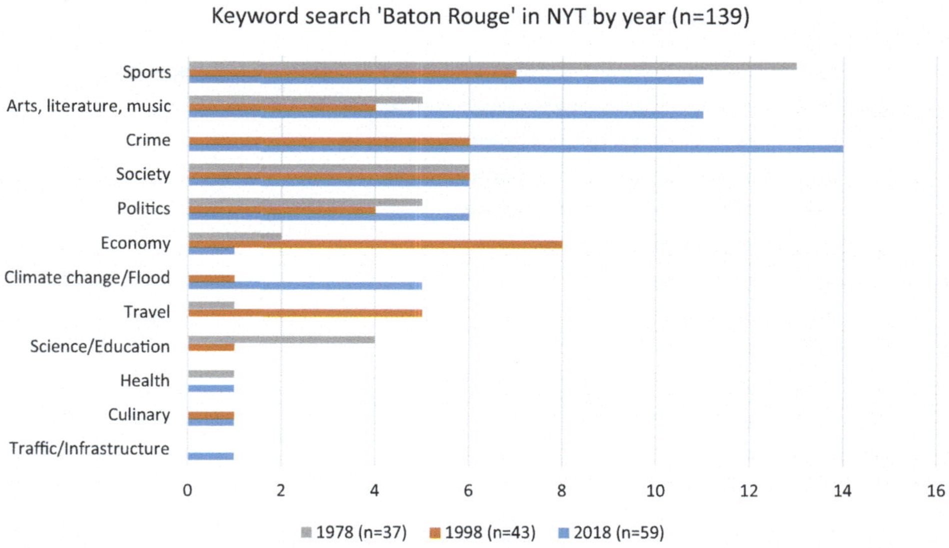

Fig. 7.19 Keyword search for 'Baton Rouge' in the online archive of the NYT for the years 1978, 1998, and 2018 by category and year in absolute numbers; accessed in May 2019

Ron Guidry, who played for the University of Southwestern Louisiana/Lafayette, and brought him to the New York Yankees (NYT 1978–17). What they all have in common is that Baton Rouge is either mentioned here only casually or focuses on the successes of athletes who came from the region but only made their sporting breakthrough in other clubs or towns.

In 1978, the Society section focuses primarily on events such as oil tanks caught in flames, serious traffic accidents, and house fires (NYT 1978–03; NYT 1978–07; NYT 1978–25). However, socio-critical issues are also addressed, such as the review of a study of municipal employment patterns in Southern cities, including Baton Rouge, which found that blacks and women continue to be employed in jobs of low social status and earn correspondingly low wages, while predominantly white men tend to hold the prestigious jobs with responsibility and above-average salaries – which, according to the research team's forecast, will take at least another 50 years before parity is achieved (NYT 1978–20).

Within the category of art/literature/music there are especially book reviews of stories set in the Southern States, such as the collection of short stories by Barry Hannah (NYT 1978–21) or Ernest Gaine's post-Civil Rights novella 'In My Father's House', which is based on a combination of literature and literary sources deals with the persistent conditions for the black population despite the Civil Rights Act of 1964, which are also reflected in the socio-spatial separation of the city of Baton Rouge into a north and a south Baton Rouge.

In the context of the political complex of issues, the focus is particularly on the implementation of the Civil Rights Act of 1964 and complaints by white workers from factories along the Mississippi River between Baton Rouge and New Orleans, in the context of which the white workers now fear discrimination in the filling of jobs and are taking legal action against it (NYT 1978–06, NYT 1978–08). But also the tightening of the abortion law into one of the most restrictive in the United States, which was hardly noticed at that time and accordingly nobody on the ground opposed, which was assessed by the then director of the Louisiana Bureau for Women in Baton Rouge, Pat Evans, as meaning that there is "a general feeling in Louisiana to simply avoid the abortion problem because of the political ambience of this state, its folklore and culture, the influence of the Catholic Church" (quoted after NYT 1978–14).

Correspondingly, 'Baton Rouge' appears in the national coverage in 1978 as rather marginal, which is already indicated by the low number of hits for the keyword search 'Baton Rouge' (n = 37) compared to other capitals of the Southern States such as Atlanta (n = 1,660), Nashville (n = 313), Austin (n = 236), Raleigh (n = 155) or Montgomery (n = 104) in the online archive of the NYT (as of July 2019). In addition, the city is only mentioned in passing in subordinate clauses – for example, as the origin of successful sportsmen, artists, etc., who came from other places. In addition, Baton Rouge appears in the reports as rather conservative, in that social progress and processes tend to be approached restrictively and corresponding social changes tend to be delayed – the reports give the impression that in and around Baton Rouge, the preservation of what is known or the status quo is a priority goal.

In 1998, the economics category is the one to which most of the articles in this year's volume could be assigned under the keyword 'Baton Rouge'. These reports relate to news about individual companies located in Baton Rouge or their subsidiaries in various industries, such as finance, petroleum, and marketing, who are seeking new partners, downsizing, or expanding their businesses, without Baton Rouge as a city playing a special role in this context (NYT 1998–10, NYT 1998–09, NYT 1998–23, NYT 1998–23). In addition, there are overarching economic issues within which Baton Rouge is also mentioned only marginally, such as the development of gas prices or the handling of the accusation that the environmental policy of the Environmental Protection Agency (EPA) represents an unfair burden on companies from the perspective of the states (NYT 1998–33, NYT 1998–38).

Even though the category of sports is comparatively strongly represented and ranks second, the focus here is on player transfers (NYT 1998–15, NYT 1998–41) but also on proceedings involving gambling fraud in a sports-related context (NYT 1998–17). As in the comparative year 1978, athletes who had stopped off in Baton Rouge at an early stage of their successful career (e.g. NYT 1998–42) are also discussed.

With regard to the category Society, the reports refer, for example, to the closure of a former leprosy hospital or the exhumation of the alleged assassin Carl Austin Weiss (NYT 1998–28, NYT 1998–30), whereby Baton Rouge is otherwise only mentioned in passing in a half-sentence as a place of reference or transit (NYT 1998–27, NYT 1998–34). On the other hand, a detailed reportage can be found in Laura Mansnerus' text 'Market Puts Price Tags on the Priceless', in which the US-wide shortage of white babies given up for adoption is addressed. This would create a veritable market, which in some states is subject to strong legal requirements and high fees charged by licensed adoption agencies. The State of Louisiana, with less restrictive and controlling adoption legislation, is a place where wealthy white couples can be placed with white babies much more quickly, often by drug-addicted mothers, and a Baton Rouge-based placement agency has become a 'marketplace' of supply and demand for white couples and mothers from across the United States (NYT 1998–11).

In addition to the treatise on new provisions of a drinking law in Louisiana, which aims in particular to reduce mortality rates from traffic accidents, since slightly more than half of all road deaths in one year were attributable to alcohol (NYT 1998–36), the Baton Rouge-related crime category focuses particularly on the debate about the devastating conditions of brutality, nepotism, and neglect in the juvenile justice system in Tallulah in the impoverished Mississippi Delta, which many legal experts claim "it is the worst in the nation" (NYT 1998–26; also NYT 1998–06). There are also reports of a general sharp decline in crime in American cities and its causes, which are difficult to identify, especially since crime has also increased in some cities, including Baton Rouge (NYT 1998–35).

In the travel category, Baton Rouge is described in 1998 as a capital city, "today, the state capital thrives, a city well-endowed with politicians, professors and petrochemical magnates" (NYT 1998–20), whose environs are seen especially as a cultural melting pot of cultures that have settled here (see also NYT 1998–43). Otherwise, Baton Rouge is often a place of transit, where you can land by plane to see the surrounding area as the actual destination of the trip (see e.g. NYT 1998–19).

Thus, in 1998, it can be summarized that Baton Rouge, at least in the most frequently mentioned categories in the 1998 articles, often occurs in the context of problematic social circumstances: poverty, the lack of regulation of emerging social changes, but also rising crime rates, while in many other American cities these rates are just falling. In addition, Baton Rouge is often named merely as a spatial reference point to other places or the place in transit with supra-local functions, without the city itself being the subject of discussion.

In the 2018 issues of the NYT's keyword search, Baton Rouge appears most frequently in connection with crime-related reports, such as the killing of police officers on duty in various American cities by firearms, including Baton Rouge (NYT 2018–19, NYT 2018–19). However, it is in particular the case of the killed African-American Alton Sterling as well as the trial of the police officers involved and the debates about urban gun violence, police violence, and the discriminatory treatment of blacks by police officers that culminate in this case and which also occupy the public debate two years after the crime. Although the incidents are framed as a general problem that is not solely related to Baton Rouge, the problem areas involved are implicitly centrally linked to Baton Rouge in the reporting (see, among others, NYT 2018–41, NYT 2018–42, NYT 2018–43, NYT 2018–44, NYT 2018–45, NYT 2018–49, NYT 2018–50).

In the context of the category sport, however, the picture is already the same in previous years, with sports results and reports on new entrants, changes of coaches, or deceased sports stars (see for example NYT 2018–08, NYT 2018–09, NYT 2018–15, NYT 2018–34), in which Baton Rouge is mentioned by name only in a half-sentence, possibly also as a stopover. The category of arts, literature, and music includes in particular reviews of the eventful lives of the greats who died in 2018 in the respective fields (including NYT 2018–25, NYT 2018–31). But also – analogous to sports reporting – news about up-and-coming artists who come from Baton Rouge but are successful elsewhere (e.g. NYT 2018–22, NYT 2018–23).

Within the category of politics, by making Louisiana a top failed state under Governor Bobby Jindal in particular, and by his tax policy, in which income tax cuts were made while at the same time doubling corporate tax breaks, and currently seeking ways out of the self-inflicted misery (NYT 2018–54). Although Baton Rouge is generally more of a marginal note in the articles, its past and present also means that the city is constantly being updated as an arena in the struggle for racial equality (e.g. NYT 2018–29). The ethnic segregation of the city is also implicit in a contribution to Sarah Smarsh's article "Liberal Blind Spots Are Hiding the Truth About 'Trump-Country'", compiled from a variety of reader reactions, when, for example, a resident of Baton Rouge states: "To be honest, most of the white people I know don't have many African-American neighbors … sometimes co-workers, yes, but many times not even that" (quoted from NYT 2018–26). Accordingly, he is not sure whether "there's much opportunity to be, in a personal sense, mad at co-workers or neighbors of color" (quoted from NYT 2018–26).

Little or no role in the compiled articles by a keyword search on Baton Rouge in 2018 occupies the city in the context of travel, science or business. Likewise, the low total number of hits of n = 59, adjusted for family and bereavement reports, indicates a low relevance compared to other capitals of the Southern States such as Atlanta (n = 1,513),

Austin (n = 599), Montgomery (n = 207), Raleigh (n = 171) or Nashville (n = 135), which have a significantly higher hit rate in the online archive of the NYT (as of July 2019).

If all years are considered together, Baton Rouge is updated in the national reporting in the context of poverty, social grievances, mismanagement, and especially ethnic-social segregation, as well as violence and deadly shootings, which in part originates from the street, but also from the police itself. The omnipresent petroleum industry, on the other hand, which is rather treated in all the articles viewed under the heading "other", is hardly discussed and hardly appears in connection with Baton Rouge, as if all the petroleum-related industries did not exist there. There are also few references to the Mississippi River, which is not only one of the most symbolic and important rivers in the United States, but also dominates the city economically and physically.

The references to important artists, particularly musicians, are often within the framework of established genres such as jazz or blues, whereby musical innovations or further developments are less in focus. Even if their roots or interim stays are in Baton Rouge, the city is only a stopover on their way – although success often only comes elsewhere.

Thus, in the syntax of national reporting, Baton Rouge is mostly relegated to the subordinate clause as a casual mention; even when shocking events occur or serious social grievances are pointed out, they are treated in an abstract and generalized manner on a meta-level and are discussed less as a concrete problem of the city itself. In addition, a shift in the thematic focus within which Baton Rouge is updated in national reporting can be seen between the years, when, for example, sports reporting in 1978, then business-related news in 1998 and, correspondingly, crime in 2018 are among the most frequently reproduced contexts, and in a certain way – one might think – also trace lines of development in the city. Especially the purely quantitative comparison to other Southern capitals shows how clearly Baton Rouge flies under the 'radar' of national reporting.

7.7 Baton Rouge and Mass Media: of Descriptions, Interpretations, and Stagings - an Interim Conclusion

In accordance with the underlying neopragmatic approach, the present chapter aimed at recording and analyzing World 3 phenomena, in this case mass media representations, in relation to Baton Rouge (see Chapter 2). In accordance with the increasing importance of digitized information processing and distribution, online freely accessible content in textual and visual form was investigated as an essential access to research on stereotypical interpretations and knowledge of Baton Rouge. A simple Google search query shows that Baton Rouge is a tourist destination whose listed sights – with the exception of the State Capitol, which is (re)produced in many ways as a landmark in the mass media – have a low degree of unique selling points. In contrast to the communicated inscriptions and interpretations of Baton Rouge in the context of official channels of communication such as tourism, education, etc., the contents are placed in the context of the unofficial video presentation of Baton Rouge, which – in addition to the State Capitol, Downtown or the Horace-Wilkinson Bridge – refers in particular to African-American neighborhoods and paints a picture of brutality, crime and police violence and shows a city of de-industrialization and urban decline. Depending on the mass media approach chosen, very different spatial and social pastiches to Baton Rouge and the associated attributions and interpretations are (re)produced and updated (North/Old South Baton Rouge, Downtown, LSU etc.), which, however, have hardly any points of contact with each other and strongly correlate with the segregation of the city. The spatial pastiche of Baton Rouge is represented in the investigated mass media by spatially, thematically, socially, and aesthetically distinct compartments. The focus is on North Baton Rouge (decay, crime), the downtown area with the State Capitol, Mississippi River, and bridges (touristic), LSU (primarily connoted with sports and (landscape architecture) as well as scattered spatially reminiscences of Southern tradition (plantations). Especially, vernacular parts of the Baton Rouge spatial pastiche, which are far less suitable for a media production, are excluded. In this sense, these form the inverse landscape of media representations.

In the following Chapter 8, we will take a closer look at how effective World 3 is in the form of attributes and interpretations (re)produced and updated in the mass media in the process of social construction on World 2, the individual living world, in which individual ideas and references to Baton Rouge and its compartments are only partially gained through our own experiences in direct confrontation with the physical-spatial initial substructure.

8 Baton Rouge as a Living Environment

The explanations given so far have been based on structures and processes, especially scientific literature, statistical data, mapping, analysis of media content, and our own inspections. In this chapter, we will now let the people of Baton Rouge have their say on how they experience and appreciate Baton Rouge, i.e. what are the perspectives of residents (and to a lesser extent outsiders) on Baton Rouge. Thus, in addition to the previous Baton Rouge surveys and analyses of World 1 and 3, the analysis of World 2, which includes in particular the contents of consciousness, individual thoughts and feelings, now follows with reference to the theoretical framework, an approach to the social construction of the life-worlds experienced in Baton Rouge. This is mediated by World 3 through World 2, also having an effect on World 1, since people inscribe their ideas in physical spaces by placing, shaping, and removing animate and inanimate material objects. As an empirical social study, it thus aims to observe and analyze this particular section of 'World' (see in more detail the theoretical framework from a basic social constructivist perspective in Chap. 2), which requires an investigation of individual and subsystemic approaches to and perspectives on Baton Rouge and allows for a detail-oriented consideration of qualitative approaches (see fundamentally and many others Flick 2007; Gläser and Laudel 2010; Helfferich 2011, 2014; Misoch 2015). Guideline-based qualitative interviews are particularly suitable for this purpose, since this form of

---

The original version of this chapter was revised: Video 1 has been updated. An correction to this chapter can be found at https://doi.org/10.1007/978-3-658-30716-5_12

Electronic supplementary material The online version of this chapter (https://doi.org/10.1007/978-3-658-30716-5_8) contains supplementary material, which is available to authorized users. The videos can be accessed by scanning the related images with the SN More Media App.

O. Kühne and C. Jenal, *The Multivillage-Metropolis Baton Rouge*, RaumFragen: Stadt – Region – Landschaft, https://doi.org/10.1007/978-3-658-30716-5_8

methodological procedure allows for openness in the conduct of the interview, which allows interview partners to set their own priorities with regard to individual life-worlds, while still maintaining the focus on the relevant topics and questions through the interview guide (Helfferich 2014). Special focus in the preparation of the guide was initially derived from the cognitive interest of the processes of construction of living environments in Baton Rouge such as personal relationship, wishes for the future of 'their' Baton Rouge, etc., but also from research and analysis of Baton Rouge, such as the segregation of the city plus existing challenges and conflicts. The transcribed interviews and anonymous manuscripts were structured with the aid of software using inductively determined categories and evaluated in terms of content analysis according to Mayring (2003).

In order to ensure a differentiated determination of the social construction of the living environments of Baton Rouge, guideline-based interviews were conducted with a total of eighteen Baton Rouge residents and two one-day visitors in the period from February 2019 to September 2019. The selection of the interview partners consisted essentially of two focus groups: persons residing in Baton Rouge without a professional or volunteer connection to Baton Rouge (BR-01 to BR-13), but with different socio-demographic characteristics such as ethnicity, age (approx. 20 to 75 years), gender (14 male, 6 female) as well as experts on Baton Rouge with different professional or volunteer activities, who are particularly responsible for the city due to their fields of activity (administration/education; BR-14 to BR-20). While people with a professional or volunteer background in Baton Rouge were specifically contacted and asked for an interview, the acquisition of interview partners, especially for people living in Baton Rouge, was done in an open process. As an example, if in the context of interviews already conducted, additional interview partners could be found whose perspectives for the research process, and in regard to the subject matter of knowledge were assessed as broadening and meaningful, in addition, randomly encountered people on the street to overcome possible 'communication chambers' that could arise within circles of acquaintances. While the interviewed experts often preferred the respective offices or meeting rooms as interview locations, people living in Baton Rouge often chose their own homes or cafés or, in the case of randomly selected people, the interviews were conducted in the open street.

Derived from the results obtained so far in the previous section, the evaluation of the collected data will focus on the construction of everyday life in Baton Rouge in general (Sect. 8.1) and on the segregation of the city in particular (Sect. 8.2). In addition, the focus is also on potentials and restrictions for the further future development of Baton Rouge (Sect. 8.3) and finally, a summary of the results (Sect. 8.4).

8.1 Everyday Life in Baton Rouge

In connection with the social construction of Baton Rouge, several areas were focused on, which emerged in the course of the preliminary studies and analysis, such as the processes of constructing and interpreting a division of the city. These are also connected to the epistemological interest of the analysis of social negotiation processes concerning the meaning, interpretation, and assignment of individually experienced environments.

Thus, in Sect. 8.1, everyday worlds of Baton Rouge are of central interest, addressing existing personal references, the attribution of stereotypes as well as challenges, conflicts, as well as social, environmental, and economic issues and more. Section 8.2 explicitly refers again to the social, ethnic, and spatial segregation of the city and then looks at the future prospects and desires associated with Baton Rouge aspects of the Baton Rouge residents interviewed (Sect. 8.3) and draws a preliminary interim conclusion with a summary of the results (Sect. 8.4).

8.1.1 Baton Rouge as the Center of Life – the 'Big City Small Town'

When people first talk about the reasons why they live in Baton Rouge, of all places, or even why they have returned to Baton Rouge from other locations to make the city their center of life again, one thing becomes particularly clear in the conversations: there are neither specific (natural) spatial areas that are stereotyped as 'beautiful' nor 'attractive' neighborhoods that offer a wide range of leisure activities, such as those that attract people from outside the city, often as part of one-day excursions, but rather strong emotional ties and references to Baton Rouge. For example, in the form of social relationships with life partners, family, friends or neighbors, familiar work colleagues combined with a job (offer) or as positive experiences, which makes Baton Rouge a center of life for many of the interviewees and will remain so for the foreseeable future (see in detail Textbox 5, a-f), because "otherwise there's nothing special about Baton Rouge" (BR-01; Textbox 5, e).

Textbox 5: Why Baton Rouge? Work and social/family

a) "It was just **different**, I guess. [...] **I had made some friends here who are still here**. And so, when I came back I was looking for a new job, role within my company. I was looking at many different places. So, this kind of had [...] **the job that I wanted, in an area that I already knew** and **had some friends and trusted co-workers**. So, I guess **there's more of that support system was already here**." BR-02

b) "Two reasons for coming back, one was **I had in-state tuition. I knew the area. One of my best friends was still living in the area.**" BR-12

c) "It was **family reasons and I had just sort of burned myself out in a particular job** in [City]. And **I had an opportunity here** that seemed kind of interesting. So, you know I wasn't ready to just strike out in another city where I had no connections. **I had some pretty strong connections in Baton Rouge. So that's why I came back.**" BR-18

d) "I **used to come up here once when I was a little boy**, about a teenager. And **since then I fell in love with** and I wind up came here, moved here." BR-07

e) "And then **I was still want to leave Baton Rouge** but then when [my wife] and I started dating she took me to Mardi Gras parades and St. Patrick's parades. So,

it's just these events that kind of made me realize Baton Rouge wasn't so bad. [...] I don't know, probably **just having somebody to do fun stuff with** helped. Yeah, because **otherwise there's nothing special about Baton Rouge**. To me there's nothing that Baton Rouge has that other cities don't have. I mean it's got a university crowd, other places have university. It's got a river, everybody else has a river. I think that's it." BR-01

f) "I grew up [...] west of here [...] and I went to college [state east of Louisiana]. So, **for two and a half decades I drove through Baton Rouge** and I never stopped. I don't even remember stopping for gas. I remember crossing the Mississippi River was always kind of a big deal because I was a little bit closer to home **but never occurred to me... to stop in Baton Rouge**. So, **I knew nothing about it when I moved here**. The **only time I've been here was a couple of times prior to visit at LSU** because that was where I wanted [...] to grad school and eventually I made it here and **I had no intention of staying here, at all**. I lived in an apartment in a kind of [...] not the best part of town, [...] I was going to try and graduate and get a job teaching and just get out of Baton Rouge. And now, **I live here because I met [my wife] and got married. And I love Baton Rouge. I'm happy. Like I have no reason to want to leave here**. So, I like it." BR-01

Similarly, Louisiana State University and Southern University educational institutions promise independent student life and adulthood beyond the parental home and even more rural areas from the perspective of the younger adults surveyed, many of whom also come from the surrounding smaller urban towns. So, it was "the first time I got to move away from home and you know kind of go and do my own thing, so naturally it was all very exciting, you know. And I was ready for, you know, I was mature enough to, like, leave home and be on my own" (BR-11; Textbox 6, b).

That Baton Rouge has less of a reputation as a dazzling tourist destination is illustrated by the statement of a day visitor who originally came to New Orleans for Mardi Gras and wanted to take the opportunity to look around the area a little in the remaining time: "we actually came to New Orleans for Mardi Gras. And then we wanted to discover some surrounding places which is Baton Rouge" (BR-09; Textbox 6, c). The overall appeal of Baton Rouge is thus accorded less the role of a main tourist attraction and more that of a chance encounter or one option among many, which is only visited when the destination already in mind has been reached.

Textbox 6: Academic education as a pull factor, Baton Rouge in terms of tourism rather the add-on to New Orleans

a) "I'm not from Baton Rouge [...] **I've chosen to come here for school.** " BR-06

b) "Yeah. Well, when **I moved here I was 19 years old, 20 years old** or so. And it was new [...] I grew up in [City north of Baton Rouge]. So, I was ready to get out of there. I wanted a change, you know, and it was like the **first time I got to move away from home and you know kind of go and do my own thing,** so naturally it was all very exciting, you know. And I was ready for, you know, I was mature enough to, like, leave home and be on my own [...] or as much as **being in college** is on your own. It was exciting, I suppose." BR-11

c) "So, **we actually came to New Orleans for Mardi Gras**. And then **we wanted to discover some surrounding places** which is Baton Rouge." BR-09

The people living in Baton Rouge, on the other hand, appreciate the cultural mix, which is reflected primarily in traditional cultural assets such as customs and traditions, food and architecture (see Textbox 7, but this mix is less reflected – as will be shown in more detail in Sect. 8.2). The French heritage in particular is referred to, which evolves over time with other influences such as African-American/Creolean culture and is thus constructed in (re)construction as a sui generis way of life. This also serves to provide an identity in contrast to others, if the local customs and food cultures are perceived as 'different' and considered 'special' compared to other locales where the interviewees have already lived: "So, total culture is... little different, food is different, food tastes is different." (BR-04; Textbox 7, b). So, it is "nice though – different cuisines, styles of food, as well it's a whole different culture to where I'm from. So, there is a lot more black people so it's a lot bigger influence in the culture and things like that [...]. So, those are things that I enjoy" (BR-06; Textbox 7, c; see also Textbox 8).

Textbox 7: Appreciation of traditions and cuisine

a) "Then **like a Mardi Gras.** So, [...] **the customs what they have started that's still good continuing** it. So, **a lot of French things are continuing.**" BR-04

b) "[...] So, **total culture is little different, food is different,** food tastes is different. [...] no matter where you go **you still find the French dishes to eat** then. Their way of living was there like how the French Quarter Design building you still happen to see those things here." BR-04

c) "So, they have **a lot of good tasting food,** where I'm from you either they get hamburgers or hot dogs and maybe some pizza. But it's nice though – **different cuisines,** styles of food, as well **it's a whole different culture to where I'm from.** So, there is **a lot more black people so it's a lot bigger influence in the culture** and things like that [...]. So, **those are things that I enjoy.**" BR-06

In addition, the general friendliness and helpfulness in Baton Rouge, but also in Louisiana as a whole, is experienced as particularly pronounced, especially when the construction of world is processed on the basis of perceptions of difference, in which the encounter with (guest) friendliness and helpfulness can be related to previous experiences in other states. Thus "[f]rom there [the former place of residence; author's note] as well as in compare with a lot of States in the United States, people are friendly here" (BR-04; Textbox 8, b), one has "just friendly, really just different vibes. How we approach people, [...] we're just different" (BR-03; Textbox 8, c).

Textbox 8: Friendliness and helpfulness

a) "From there [the former place of residence; remark of authors] as well as in compare with a lot of States in the United States, **people are friendly here**." BR-04

b) "Yeah, **Baton Rouge usually in Louisiana, people help each other** [...]." BR-04

c) "To me, just **the southern hospitality**, the style, the music, the blues, jazz, food – food definitely, most definitely food, Louisiana style food. The people. [...] I mean, **they're real friendly**. You can talk to others. I had, don't get me wrong, I have been out of Louisiana so I have met different other people but [...] I just mean [...] **we have just friendly, really just different vibes. How we approach people, [...] we're just different**." BR-03

Respondents also attribute a more small-town and rural character to Baton Rouge, despite its capital city function: "This is a big-city small-town" (BR-14; see Textbox 9, a-b). This evolving of urban and rural elements into hybrid forms gives rise to physical and social spatial pastiches that confound traditional dichotomies between 'city' and 'country' and the patterns of vision and interpretation they are subject to – both in spatial and social terms. As an example, one of the interviewees reported an encounter with a hunter who, half a mile as the crow flies from the seat of government in Louisiana, trained his dog to hunt hares on an abandoned railway site and found it to be an excellent location for this activity, since it is near both the residential areas and unpopulated open country (BR-14), and thus the middle of the capital of a U.S. State – an assumed bulwark of urbanity – an activity can be engaged that would primarily be spatially classified more in rural contexts. Also, with regard to the concentration on and orientation towards neighborhoods and rather smaller districts or developing street lines instead of a dominant center, and the rather restrictive attitude towards the big city (see Textbox 9, b-c), shows the evolution of the urban and the rural in an urban–rural hybrid on a physical-spatial as well as social-community level (see also Kühne 2012a).

One of the interviewees reports that a certain degree of ruralism is also attributed to the city from the outside: "I know that people in New Orleans think that we are, you know, Philistines, uncultured, and that Baton Rouge is just a backwater, just provincial"

(BR-18) and thus also seen as an antithesis to New Orleans as an urban metropolis (cf. Textbox 9, c).

Textbox 9: Urban–rural hybridity not only physical-spatial, but also social-community

a) "This is **a big-city small-town.**" BR-14

b) **"I like the small-town quality to it.** Because, like, it's big enough but... because I know, like, there's places even smaller than Baton Rouge but it's still not like as big as some other cities we would have like outside of the Louisiana." BR-08

c) **"People in New Orleans, Natives in particular, barely know that Baton Rouge exists,** you know, I mean really... because everybody in Baton Rouge knows about New Orleans and, you know, might go there for... well, whether it's perhaps for Mardi Gras, for dining, for just the culture. [...] **I know that people in New Orleans think that we are, you know, Philistines, uncultured,** and that **Baton Rouge is just a backwater, just provincial,** you know, 'What's in Baton Rouge???'. That would be the attitude of most people in New Orleans. [...] And **people in Baton Rouge would say that New Orleans is dangerous** and, you know, if **Baton Rouge is sort of culturally conservative they would certainly see New Orleans as being sort of decadent,** right. And **New Orleans kind of sells a little bit of its decadence as part of its appeal.** And that **is just the antithesis of the way Baton Rouge thinks about itself as a place for families.** People don't think of New Orleans as a place for families. So [...] just culturally the **mindset of the two is is very very different.**" BR-18

8.1.2 The View from Outside – 'Baton... What?'

The view from outside and the associated attributions and interpretations are followed by the question of what reactions the respondents receive when they tell third parties that they live in Baton Rouge. Despite the city's function as a capital city, a weak or no association at all is often associated with Baton Rouge, possibly even its relative geographical proximity to the southern metropolis of New Orleans: this is often the case "even though it's the capital of Louisiana" (BR-05; Textbox 10, a). Even newcomers to the city could not preconceive any associations with Baton Rouge before their move, they said they were "not aware of too much of the Baton Rouge" (BR-04; see also Textbox 10, b-c). Even if people have heard of Baton Rouge, "all often the next question was 'how far is that from New Orleans?'" (BR-19; Textbox 10, d). Even if Baton Rouge is "a blank slate in people's minds" (BR-19), this would affect a number of other "mid-sized cities" (BR-19), so that the low level of perception and identity associations for Baton Rouge is not at all similar to other state capitals, but rather to other urban cities of similar size, and thus relativized. The perceived identity of the city is not only low outside Louisiana, but possibly "even within Louisiana" (BR-19; Textbox 10, d).

Textbox 10: Baton... what? Baton Rouge as "a blank slate in people's minds"

a) "**Sometimes they probably don't know the city even though it's the capital of Louisiana.** Most of people when I go back to [US States] [...] when I say I'm from Baton Rouge, **they don't know what kind of city it is.**" BR-05
b) "No I don't know that I knew anybody from Baton Rouge. I didn't know anything about... **I just don't think I had any preconceptions about Baton Rouge.**" BR-01
c) "When **I came I was not aware of too much of the Baton Rouge.**" BR-04
d) "People had heard of it, **all too often the next question was 'how far is that from New Orleans?'**, right. So, I think **Baton Rouge is a blank slate in people's minds.** And I would say that **that's the case for most mid-sized cities,** you know. If you said you were from Toledo, Ohio, people would be like, you know, there. **It does not have a strong identity outside of Louisiana, I would say even within Louisiana,** I don't know how strong our identity is really, I really don't." BR-19

Even if a low level of identity is perceived regarding the city, this is all the more pronounced with regard to the sports teams of LSU, the LSU Tigers. The latter is so dominant that other aspects of Baton Rouge are overlaid by it, which is also viewed critically by the interviewees (see Textbox 11, a-c) and Baton Rouge is thus, in the associative external perspective, primarily attributed a sports town instead of a capital city or science town.

Textbox 11: LSU as dominant to overpowering in its association with Baton Rouge

a) "I really didn't have a perception, **I didn't really know what Baton Rouge was about, you mostly hear about LSU and things like that.**" BR-06
b) "They usually say 'oh Baton Rouge. How far is that from LSU?' They always think of LSU, especially where I'm from, they might think of LSU a lot. Like I'm about 15–20 min from LSU. My school has nothing to do LSU. People would be you would be mad if I told you my school was LSU and I took you over here. So yeah. But **mainly I feel like the perception is about better or more about LSU and they don't really know about really what goes on in Baton Rouge other than LSU.**" BR-06
c) "**LSU football!!!** That's one thing, **LSU sports... beyond I'm not sure people have much of an image of Baton Rouge.** Maybe, they maybe have seen a picture of our State Capitol, possibly, beyond that it would be very interesting to

ask people what their perception… perceptual regional kind of… their image…. 'stock image' of Baton Rouge." BR-13

8.1.3 Searching for the Center

In the context of the question of the city's identity, there is also the question of a city center that brings the all the neighboring communities together. In the case of Baton Rouge, however, it is an eccentrically located downtown, which experienced a boom, especially in the 1950s and 1960s, and has 'classic' downtown functions such as shopping and promenades, restaurants, bars, etc. In the following years with the recession from the 1970s onwards, leisure facilities developed a reduction in function and significance with the corresponding consequences such as vacancies and increasing physical decay of the building and the associated social interpretations and (evaluations) assessments (see Sect. 6.1 for more on downtown development). Thus, especially in the context of moving to Baton Rouge, many of the interviewees – here the focus is primarily on those who have already lived in Baton Rouge for a decade – first oriented themselves towards the downtown area as a presumed spatial-physical as well as social-community city center, in order to then create a "necropolis" (BR-13; cf. Textbox 12, a-c), in which neither cars nor people were on the move. "When I moved here […] for 15 years 16 years and we went to the downtown, I mean, that's, you know, you go to a city you go to the downtown, […] there was not a car, there was not a person. All the buildings were vacant, not all of them, but many of them were vacant" (BR-17; Textbox 12, d). For example, at the time of their move to Baton Rouge ten years and more ago, the interviewees described the majority of buildings in the downtown area as abandoned and in a state of ongoing decay, which in the view of some interviewees was exacerbated by the relocation of state agencies and administrations to other parts of the city", "So, there was no nobody even coming in for that" (BR-14; Textbox 12, c.; see Textbox 13 for development). While the downtown area could not come anywhere near meeting the stereotypical expectations placed on a CBD of a state capital, the campus of Louisiana State University" with the live oaks… the Spanish Moss" (BR-13), on the other hand, forms the epicenter of stereotypical visual expectations of a section of the city that is conceived of as a 'campus' (BR-13; cf. Textbox 12, a).

Textbox 12: Downtown as 'necropolis' and antithesis to the LSU campus.

a) "So, of course, everyone remarks, which is true, **the campus is a very pretty campus with the live oaks the Spanish Moss. You know, it fits into kind of an image. But one of the things that I felt immediately going down to the CBD because it was dead. There was nothing, it was like a necropolis, nobody there.**" BR-13

b) "When I moved here in 2010, like, we've said**, downtown wasn't a place where you would think to go out.** If you were in college, you go to the college bars south of LSU, if you few were older, our age, you probably had kids and you didn't go out. And I think that was pretty much it." BR-01

c) "**Twenty years ago, downtown was empty. Literally empty.** Where the Hilton is now, [...] all of that was empty. The windows were broken out. You could smell mold coming out when the wind blew. **Third Street which is now, they're trying to make it hip and happening, it was all abandoned.** All those buildings were empty. One or two had businesses. **The State Government** had moved their offices out because the population had gone East and South. And so they **removed their office from the Capitol grounds.So there was no nobody even coming in for that.**" BR-14

d) "When I moved here [...] for 15 years 16 years and we went to the downtown, I mean, that's, you know, you go to a city you go to the downtown, [...] **there was not a car, there was not a person.** All the buildings were vacant, not all of them, **but many of them were vacant.** And that's because of this phenomena throughout the United States [...] where people left the downtown." BR-17

e) "**There was nothing downtown** when Kip Holden took office. And that was one of his things. [My husband] used to work downtown. And I would take him to work and on the way back **I'd see these poor tourists who had come on a riverboat and they get up and there was absolutely nothing to do downtown.** They would just walk around. [...] They could go see the State Capitol." BR-16

The state of almost complete inner-city wasteland at the end of the "twentieth and beginning" of the twenty-first century is still unimaginable. The twenty-first century is used by the majority of respondents as a point of reference for interpreting and classifying current downtown developments (see Textbox 12 and Textbox 13), in the light of which progress already achieved is seen as "tremendous change" (BR-18), which "makes us feel like we're kind of a city, you know, and not this overgrown small town" (BR-18; Textbox 13, b), so that even the most minimal and marginal efforts to improve the status quo constitute progress, and the comparison with other, more comprehensive urban development transformations has not even begun. This can also be seen as a reference to the rather hybrid self-image of the Baton Rouge residents between the stereotypical self-image of a 'village', on one hand, and the claim of a 'metropolis', on the other.

Textbox 13: Small steps for downtown development, but a giant leap for Baton Rouge

a) "And then about 20 years ago, they started a plan to redevelop downtown. They moved the state offices back. They've spent a lot of money and incentivized a lot of redevelopment. So, what it looks like now at least from the river to Third Street there is some activity. When you get above that there's not much anything. Because it's still undeveloped, still surface parking lots. So, Baton Rouge tore down all the old buildings stock that in other cities when they redevelop it's there and it's cheap because it's already there. You don't have to build new. And then by moving out people had no reason to come downtown. There is no central part of town. It's a move with the population and there's no center. They built a Mall in the '60s, just outside of downtown.“ BR-14

b) “I am somebody who can **say I have seen such tremendous change**, I mean, downtown is a classic example. I mean, 15 years ago, even… I mean, downtown was… was… was dead. So, **I've seen the energy there which is very positive, makes us feel like we're kind of a city, you know, and not this overgrown small town**. And this and seeing what's happening on Government Street, I mean, that's… those are things that are very encouraging and positive.“ BR-18

Going out in Baton Rouge is seen as a certain challenge due to the lack of a central downtown area as a focal point of social and community life: thus, it was not so much the "pristine shiny countertops" (BR-08; Textbox 14, a) that imposed themselves on people seeking pleasure, perhaps even those in transit, but rather the hidden pubs and restaurants, often known only to locals, that shaped the nightlife of Baton Rouge (see Textbox 14, a). The conflict, which is difficult to resolve between the desire for prosperity and development and the attempt to maintain places such as an "old-fashioned diner" (BR-08; Textbox 14, a) as "individual" and "special" without being overshadowed by the big corporate chains, becomes clear. In addition, the unconnected supply structure and the further distances to bars, restaurants, and even grocery stores are experienced as 'disappointing' and the desire for shorter distances and centrality on a smaller scale is evident, as nearby locations in particular can represent a high reference and social value within their own radius of movement (cf. Textbox 14, b-c).

Textbox 14: Nightlife BTR – going out in Baton Rouge as a challenge

a) “But, you know, **I just love those root beer floats there** and, like, this is one of those things where, like, **I wish more people could come to see this old fashioned diner,** like, it's really old fashioned, […] **there's no pristine shiny countertops but that's like the quality of life,** you know. And you can tell it's been there for awhile. But I […] **wish some people that would come here, you know, would know about it** but not many don't because it's sort of like tucked, like, right next to like a street light. **So it's not really easy to find** unless you really drive around downtown […]. And but at the same time **I don't want too**

many, like, big chains to come in to the point where shops like that actually close.“ BR-08

b) “**It's disappointing how it's not as many things to do right off campus or right on campus.** [...] Even going to the grocery store you have to go 10, 15 min away. And if you want to go to a nice restaurant, same with that as well. So that's a little bit more disappointing.” BR-06
c) “One thing that I tend to do when I go to new places, I don't tend to do it I always do, is I'll go look for a good bar. And **there was one by campus and it just ended up being my bar** [...]. So, it was like a place where I could go be a solitary person that I am sometimes.” BR-01

Shopping as a part of everyday practice was less intensively addressed by the respondents, but – as already apparent in the context of restaurants and bars – there was a preference for short distances and a more fragmented supply structure, especially with regard to goods and services to cover daily needs (Textbox 15, a), The provision of these services could currently be an initial spark for further investments at local level (Textbox 15, a), but also against the background of the rapidly replacing shopping infrastructures of the past (from the downtown shopping mile to the ever larger mall complexes to the simulation of a city with a series of specialist retail stores; see 6.1.1), and in some cases can also be viewed with skepticism. Although the current trend is moving away from the large, roofed mall centers and towards open-air strip malls – thereby returning to the shopping structures of the past – increased online shopping is a central factor in the failure of mall concepts (see Textbox 15, d) and their future is therefore uncertain. The main focus of the criticism is the observed sequences of developing and leaving without a viable plan B for the shopping infrastructures of the past, which were then ‘left to themselves’, in which existing physical potentials fell into disuse and, resultingly, high new investments had to be made, which had set in motion an unfortunate cycle (BR-14; Textbox 15, d) with regard to the development of shopping infrastructures (see Textbox 15 in detail).

Textbox 15: Desire for supply structures to meet everyday needs in the immediate vicinity; develop-abandon as an ‘unfortunate cycle’ of the supply infrastructure

a) “So, like, if you all have a fast food or if you want to have a grocery store or a Walmart or something, **should be in a close vicinity. So you don't have to go far**.“ BR-04
b) “The White Star Market was a pretty significant investment there on Government Street. So, that was a real milestone.“ BR-18
c) “All of these dumb little shopping centers that are popping up in places that are trying to be Perkins Road or something like that. I think those are just stupid

and some idiot sold that idea to other idiots who said, yeah, let's go ahead and do that […].“ BR-01

d) "Perkins Road […] which is developing and developed the town center Perkins and Bluebonnet. […] **It seems like the shopping malls are dying**. And now that's multiple reasons for example online shopping. The thing is **people are now getting back into strip malls**.“ BR-12

e) "They built it [shopping mall; remark of authors] and then they abandoned it and moved out; and they built it, and they abandoned it and moved out. Now they can't go any further. So now **malls are failing** because of larger issues. The Internet is one and **they destroyed their traditional business base of individual stores**. And now they've done this develop-abandon and develop-abandon and develop-abandon and now everything has changed around that. So, **they've wiped out everything they could use**. Now they have to build new which is very expensive, so the rents are higher. **It's an unfortunate cycle**.“ BR-14

8.1.4 Petrochemical-Industrial Imprinting as 'Pact with the Devil'

The oil and petrochemical industry, despite recent attempts to attract other industries to Baton Rouge, especially IT companies, remains the central industry that has shaped the city and its environs for more than a century, both economically and spatially, since its establishment in 1909, and has made a decisive contribution to population growth and the modernization of the infrastructure by providing thousands of jobs (see also 6.2). However, the relationship of the interviewees living in Baton Rouge to Standard Oil's or ExxonMobil's production plant located in the north of the city is seen as a 'pact with the devil', partly because 'Baton Rouge as a modern city would not have existed if Standard Oil had not come to […] Baton Rouge' (BR-18; Textbox 16, a), and also because it is recognized that the growth of the city is largely due to the company's location (see Textbox 16, a-b). In contrast, the location of the company and the associated production facility is questionable from both an ecological and a human health point of view, as the feeling arises "that what's coming out of those chemical plants is probably going to kill me, people that live in Baton Rouge" (BR-12; Textbox 16, b), but the petrochemical companies are (re)produced by the interviewees as "too integrated into our economy to advocate for a retrenchment of the petrochemical industry" (BR-18) and updated as a "fact of life" (BR-18; cf. plus Textbox 16, a and b). The job creation effects, which particularly ExxonMobil has on the city and which the proponents of the petrochemical industry argue about, are much smaller than often assumed: instead of creating more jobs, services would be outsourced to subcontractors and the company would also endeavor towards "hiring fewer and fewer people with greater skills" (BR-14; Textbox 16, c). However, because of the dilapidated public school system, they are not trained locally and have to be recruited from abroad first (BR-14). Although ExxonMobil does not currently rank among the 100 largest employers in Louisiana (see also Kolmar 2019), the narrative of the most effective and economically most important employer in the region is so powerful that, despite the ecological and health-related side effects, "nobody's going to tell them to go away, even if it is killing us, you know, it's stupid. (BR-14; Textbox 16, c). ExxonMobil and the other petrochemical companies are often so fundamentally and significantly integrated into the social construction of Baton Rouge that, despite the (life-threatening) side effects associated with them, they are constructed as the more 'pragmatic' and probably more 'familiar' evil in the face of the fear of a possible economic vacuum. Accordingly, the inhabitants are developing a more-or-less pronounced "fatalistic approach" (BR-12; Textbox 16, b) instead of a revolutionary dynamic towards the presence of petrochemical companies (see Textbox 16).

Textbox 16: 'Fatalistic approach' to the prevalence of the petroleum industry in Baton RougeExxonMobil and the establishment of petrochemical companies as a 'pact with the devil'

a) "It'd be devastating for Baton Rouge ["When Exxon leaves the city: remark by the authors"]. But **we made a deal with the devil**. 75 years ago, I mean, [...] **we've benefited, greatly,** from the petrochemical industry here in Baton Rouge **but there have been consequences for that relationship**. [...] **Baton Rouge as a modern city would not have existed if Standard Oil had not come to**, which is now ExxonMobil, had not come to Baton Rouge [...]. So, **we owe our growth in many ways to that**. [...] **But [...] there's been downsides to that marriage**, as well. So, [...] even with all of that, [...] it would not be a good thing for Baton Rouge [If Exxon Mobile were to leave the city; note from the authors.]. It's just, **it's too important** and **it's too integrated into our economy to advocate for a retrenchment of the petrochemical industry**. I mean, that's the case for the whole state of Louisiana. So, **it's just a fact of life**, I think." BR-18

b) "I realize **it's an integral part of the city** and [...] [a]nd I realize that there are chemical plants, I mean on either side of the river. [...] But **for an ecological standpoint, yes I would like to see it** [Exxon leave the city; authors' note]. [...] **I have a fatalistic approach, I feel like that what's coming out of those chemical plants is probably going to kill me, people that live in Baton Rouge**. But so I would like to see them somewhere else. It'd be **nice if you could put them out in a marsh somewhere**. Have the infrastructure and **they'd be far away from people**. But that's not gonna happen. **But economically I understand that they're very important to the city and they're necessary**." BR-12

c) "So they're having fewer and fewer people they actually have to have on the plant. **They have hired outside contractors to do a lot of work. So they are not hiring those people. Those jobs are still attached to the corporation.**

But they're not of the corporation. So at this point, Exxon is not even in the top 50 employers in the state. [...] So **Exxon is not the big employer everybody thinks they are**. Which is a decision that the corporation made to price their profit. They're not concerned about hiring more and more people. **They're concerned about hiring fewer and fewer people with greater skills. Which the schools don't provide here**. So the people they're hiring to work **now they're hiring from somewhere else**, to come in here. That said, there's still a very large economic impact from that Industry. So **if they just left, it would be catastrophic**. It would take a lot of money out of a local economy. Not only through their taxes. As low as they are. But because of all the downstream effects of people having jobs. So **if there was a way to replace the jobs, yes.** [...] **[T]hey don't call this cancer alley for nothing**. So that's **there are a lot of problems with that plant**. But the economic side of it is so great that **nobody's going to tell them to go away, even if it is killing us**, you know, it's stupid.“ BR-14

The fact that ExxonMobil, located in the north of the city, is constitutively integrated into the social construction of Baton Rouge is shown by the recent discussion about tax payments of $290,000 to the city (see Textbox 17, a-f), in the context of which the company had announced a relocation of the industrial site in the event of implementation – which was ultimately not implemented, instead an outcome of an expansion of production occurred (see Karlin 2019a). However, "the perception [...] that the corporation has such a large economic impact that we can't do anything to upset them because they would go away which is like... utterly insane" (BR-14; see also Textbox 17, b). However, even if the majority of the interviewees accept the establishment of the petrochemical industry as a 'fact of life', there persists a demand for compensation regarding the irreversible negative consequences of the companies and the assumption of responsibility in the form of financial participation in the community through higher taxes, which should be invested in the ailing public school system in particular. Even the most recent success would remain well below the possibilities, because Exxon "spend[s] more money on toilet paper and donuts every year" (BR-14 Textbox 17, b; see also a. also BR-13, BR-01) than is paid to the city as taxes. There was also criticism that industry did not want to contribute to the training costs although it was dependent on highly trained employees (BR-19). A responsible use of resources such as drinking water is also required (e.g. BR-14, EI-2, BR-16), so that the residents of Baton Rouge definitely recognize ExxonMobil as an industrial enterprise in the heart of their city in different respects (economic, social) and different gradual gradations (higher, lower), and – despite all accompanying circumstances – wish to keep the enterprise, so that the central demand of the interviewed Baton Rouge residents consists only in financial participation with regard to the maintenance of central structures of the city such as infrastructure, groundwater supply, and educational system, but no general exodus is desired (cf. plus Textbox 17, e). ExxonMobil's company with its large production facility in the city area is thus constructed as an integral part of the city, which would not, however, fulfill its obligation in terms of its social, economic, and ecological responsibility towards the city.

Textbox 17: Taking responsibility – paying taxes, responsible use of resources: Corporate strategies for tax avoidance as 'the boy that cries wolf sort of thing'.

a) "So yeah, [...] **the company was here and it provided a lot of jobs and stability in a lot of money and that had effects throughout the community** because there was more money in work and all that and that could be spread around. **Part of the reason that the company stayed here is that the state gave them just gave away everything,** they're not charging much in the way of tax they have lots of programs to give tax back so that the **industry is making more and more profit without having to pay taxes. They do pay taxes, but not very much compared to what they're making.** They got the legislature to declare water a mineral. So we get our water from an aquifer not the river. So we're pulling ancient water out of the ground. When the industry got the legislature to declare water a mineral had meant that they could take water from the aquifer because it's on their property. So **now the industry takes twice as much water every day as the city takes.**" BR-14

b) "**The perception is that the corporation has such a large economic impact that we can't do anything to upset them because they would go away which is like... utterly insane**. And so that played out in a thing recently when they changed the industrial tax exemption program (ITEP) where in the past the state controlled that and **recently the state gave control to the local communities who were the ones who benefit from the tax andfor the first time the school board here said, no you have to pay the tax**. And Exxon was like well never mind then. Then they withdrew their application. It was a minuscule amount of money compared to what they make. **But it was the first time they were told they couldn't do whatever they wanted to do. Andit scared the hell out of people here in town**. The first thing they said is they're going to go away. We're going to be like Detroit. We're just going to die. And my point was **any decision that the corporation is making they're making for the corporation they're not making it for Baton Rouge**. In any decision they make, if it's a viable corporation, is not going to be based on a one-off tax problem that amounted to two hundred and ninety thousand dollars ($290,000) [...] **They spend more money on toilet paper and donuts every year**. So that wasn't a thing the money wasn't the issue." BR-14

c) "**The gas and oil lobby and power in this country is just ferocious,** the hydrocarbon – in fact, of course, that's what the whole world is going to have to fight, as if it were an alien from outer space invader. **It's going to take a whole global movement to destroy the monster that is the gas and oil**

hydrocarbon industry. And here we are. I mean, **we're a colony of it. They don't tax it. Exxon was just absolutely outraged at the idea that they might have to pay just a little bit of taxes for the schools.**" BR-13

d) "But their workers wouldn't have to pay as much tax if they paid a bit more. So, I mean but the industry and the **industry lobbying groups** always say – **well the industry brings jobs, they bring payroll and that is what's supposed to pay for education and all other public services.** We have a terrible road system we have terrible education compared to the other national standard averages. **So we have a lot of work to do and all that's tied to taxes.**" BR-19

e) "The legislature has now allowed municipalities to decide if big industry... big industry right now for 60 years when they when they build a new plant in Louisiana they can get a 10 year exemption on property tax which pays... and property tax is used to fund education. So, the big industries... the big Exxons have been getting for almost perpetual... property tax exemptions, so there's **less funding than there should be for education. So, the industry that needs educated people is unwilling to help share the cost** but the legislature has decided that now localities can make that decision not before the state made that decision. Now the localities and our school board just said we want the tax money give us the tax money. And they're going to make a ton of money and the industry is not moving. I mean the argument against it was, **well [...] if we tax the industries they'll leave, they have too much invested. They're not gonna leave.**" BR-19

f) "So, that's why I think... and, you know, there was very recently this controversy about property tax exemptions that it just sort of been rubber stamped for years. And finally, because now there's some local say in that matter for the first time ever they were denied an exemption from some taxes and for schools. And they threatened they acted like 'if Baton Rouge is going to continue to behave in this way we'll have to look at other opportunities for future investment'. Well, **within a month** of that they just **announced now a 500 million to one billion dollar expansion of the large plant here.** So, you know, **the boy that cries wolf** sort of thing. So, it's **an interesting dynamic, interesting relationship.**" BR-18

g) "No, they give a lot of people jobs." BR-17; "I'd like them to do things correctly." BR-16

However, employees in the petrochemical industry are also noticing a change of thinking in their companies: for example, in the company of one of the respondents, the filtered and purified water of the Mississippi River would be used for production processes instead of groundwater, and the chimneys would only emit water vapor instead of smoke (BR-02; Textbox 18, a). The company has also set itself the goal of CO_2-neutral growth and is also endeavoring to communicate this progress more effectively: "they're working very hard to maintain a good image" (BR-02; Textbox 18, a). At the same time, by supporting local events and festivities, they have also tried to show "that they're part of the community and they support it" (BR-02; Textbox 18, a). Moreover, industrial enterprises are not the only emitters of pollutants, as the omnipresent cars of the inhabitants also contribute to the impact on air quality (BR-01; Textbox 18, b). Furthermore, it is primarily the socially disadvantaged who are particularly exposed to the direct emissions of industries through the locations of the low-cost housing, and who are thus exposed to social inequality in terms of exposure to industrial pollutants, thus further reducing the life chances of underprivileged sections of the population (BR-01; Textbox 18, b; see also 2.4 in this book and Blodgett 2006; Bullard 1997, 2005; Wright 2005).

Textbox 18: Petrochemical industry in general – social inequality with exposure of sources of pollution.

a) "I mean, I don't have a negative impression of it **because I understand the misunderstandings that people who don't work with industry**. So, people drive by and see what they think is smoke coming out of a stack. Now, that's steam and it's clean and that's fine. And we use river water and we put it back in the river cleaner than when we got it. So, I don't. And just, like, yeah when you see a report, well Exxon had 50 reportable releases of chemicals. And you, like, take it with a grain of salt because OK, they're legally doing what they're supposed to reporting and then you see you later like a week later I got a comment… well forty of those were precautionary, like, they didn't know how much but they follow the rule and it was very very small amount. So […]. Yes, I think they try very hard to reject that image like a Louisiana Chemical Council or something like that. **All the leaders of all the companies are involved with the politics and legislation and really trying to show how we're getting ahead of it**, you know, we have company goals on, like, carbon neutral growth and expansion. That's all happening. And **there is a really big push on that**, you know, we're good places to work for like when you see this poster, you know, billboards on the highway. They all support various activities. So, there's supporters of Life After Five and the Crawfish Competition and the Jambalaya Festival and all the chemical people and all the local businesses are there supporting it with their name everywhere. So **they're working very hard to maintain a good image, that they're part of the community and they support it**." BR-02

b) "I'm not a chemical engineer. I don't know how this works I know that the water they put in the Mississippi River is cleaner than the water they take out. So, I've been told by an official. But when you say it's part of the community it's literally a part of the community because whatever does go into the air and it doesn't matter […] if it's an industrial petrochemical plant. But **their point sources for any pollution we have cars or transportation from all over the**

place as these non-point sources of pollution for the air that we breathe. **So everybody's guilty of everything**. But in terms of **that industry in particular, that's where the lower income people live who can't get away from there because that's where the cheapest housing is**. So, **any negative health consequences from that source of pollution is obviously social environmental inequalities going on there and that's an understatement**." BR-01

Baton Rouge as a whole is perceived by its residents as a life-threatening location because of the pollutants produced by the industries and the extensive use of individual automobile traffic. Yet, this does not lead to any of the respondents wanting to move from Baton Rouge to a less polluted community. Among the respondents, the progress achieved by individual companies is only updated for those that are employees in the respective companies.

Accordingly, the hope of those surveyed who are not employed in the petrochemical industry also lies in the development of other branches of industry, such as the IT industry and the creative or film industries (cf. Textbox 19, a-b). In the construction of the self-portrait of Baton Rouge as a business location, only a low level of competitiveness and innovative power is attributed to it, so that a concentration on niches such as trying "to create pockets of expertise about wetland restoration and hydrologic engineering" (BR-19: Textbox 19, a) is more promising and – it would appear – no thought is given to setting larger benchmarks for itself.

Textbox 19: New economic sectors sought: tech industry, creative industry, hydrologic engineering.

a) "**The city is trying hard to attract tech industry to bring in young professional adults**. You know the 30 year olds and millennials. And they haven't been entirely successful in other, you know, other areas with concentration of Universities like in North Carolina, you know, they developed these really big think tanks, sort of, and an intellectually oriented sort of businesses. **Baton Rouge and New Orleans have never really been able to kind of create a similar sort of synergy** between… with UNO, Tulane, and LSU. They haven't been able to align themselves… with complementary programs that really augment, although right now, there is a big push to emphasize all aspects of water management. **We're trying to copy the Deltares, the Dutch**, to try **to create pockets of expertise about wetland restoration and Hydrologic Engineering** and all things that deal with people in water so that… that is something that's going on in both Baton Rouge and New Orleans and that may pay dividends in the long run. So, that's one area. We'll never compete on in the movie arena, **we'll never compete in tech because we're just too small**. We can't compete with you know M.I.T. or North Carolina or Silicon Valley, or Seattle, **but we can maybe carve out a niche where we already have strengths**. And I think that's a sensible thing that they're beginning to do, **they're really trying hard to make that happen, develop this water expertise.**“ BR-17

b) “I remember. I remember a lot of people come to Baton Rouge […] **we had a smaller tax compared to other people that came for filmmaking. And I think they altered that and that's sort of like scared some people off**. I think some of these… if they were able to revert back to the original… whatever original value they had for that tax. That probably would help and just you know. Just certain things where… I don't know, because like I don't really get into that sort of side of stuff. But I know it's one of those things if that improved they probably would help. You know.“ BR-08

Thus, the question of the economic pillars on which the city is to base itself in the future is one that has so far hardly been answered, it is uncertain and unsettled. There is the strong dominance of and dependence on the petrochemical industry, which is, however, accompanied by heavy burdens on health and the environment and, as these are times of the emergence of research into alternative energy production and materials, may only be sustainable in the medium term; conversely, further opportunities such as the IT industry, hydrologic engineering, etc. are only seen at an early stage of development, which, however, are only half-heartedly assigned the ability – if at all – to be able to seriously compete with the dominance of the petrochemical industry at some point in the future.

8.1.5 The Public School System as a Forcing for and Manifestation of Social, Ethnic, and Spatial Segregation Processes

In order to be able to develop Baton Rouge, and also Louisiana, the interviewees see a need for a well-trained workforce, but there is a lack of political will to invest substantially in education (see Textbox 20, a). Not only is the underfunded and ailing public education system unable to produce the required labor force, but the disastrous state of the public school system also deters highly educated professionals and scientists – who, due to their acquired skills, are often in the privileged position of being able to choose between various job options – from settling in Baton Rouge, for example, and ultimately from being confronted with high costs for private schools for their children (cf. Textbox 20, b–c). Additionally, as in the case of the City of St. George's efforts to incorporate itself, this leads to the establishment of a school district by white, wealthy parents, since the situation of the public schools in East Baton Rouge, which are generally considered to be among the poorest in the state, would hardly improve in the foreseeable future (Harris 2019; The City of St. George 2019; cf. also Sect. 6.4.3 and 6.5).

Even if private schools were just as often below the required standards, they would still perform better than public schools, but private schools do not offer solutions for students with behavioral problems, for example, instead leaving these to public schools, which have to deal with discipline problems anyway (see Textbox 20, d–e).

Textbox 20: State of the public school system as disastrous – lack of appreciation and value of education as a social phenomenon Textbox 20

a) "**To really develop a really highly skilled workforce Louisiana in general**, and East Baton Rouge in particular, needs to have a stronger Elementary Education. And one of the reasons our educational system is weak is, we fought for 50 years, there was a lawsuit over integration in Baton Rouge which was just settle 10 years ago and we don't devote enough money. **The legislatures don't want to spend money on education. They would rather people send their kids to private schools.**" BR-19

b) "I mean, **maybe an improvement to the education so the public schools [...] to keep people here. So that they don't have to go out.**" BR-02

c) "I don't have children so it doesn't directly affect me but [...] **people have come in for interviews as full professors and have not come here because of the school system, the public school system**. They realized they would have to pay a lot of money to put their children in private school. So that... that's... that's a problem. And [...] you gonna to see when you just realizing that when you're dealing with the general population that **there's not a... there's not a value to education** and **not a value to higher education**." BR-12

d) "Under the previous governor or under Jindal [...] **they made it legal to take money from the public school budget to apply to vouchers for the [...] private schools**. Which brought a lot of private school companies here to establish schools. [...] **They haven't performed at all well because they're for profit, they're not for education**. So, they don't care, if there is a number of issues with it. **They're not performing up to a standard**. If a student has any kind of behavioral issue, they throw them out, back to the public school." BR-14

e) "I would say [...] that **private schools tend to have a better quality education**. They have more money. They have smaller classes. There's a better teacher to student ratio. They have better food. More opportunities outside of school. So [...] they may go to a museum, they may go to a competition of some sort. **And the public schools are underfunded**. They have way too big a class size. The student teacher ratio is really unbalanced. They're constrained to teach to a test because that's how the performance of the school is measured. So they're not teaching students to be 'critical thinkers'. They're teaching them to memorize things which is a very different way to educate. And **there are a lot more discipline problems in the public school**." BR-14

Correspondingly, the failures in the education sector are being followed with concern by the Baton Rouge parents and are seen as a failure of state bodies, which can, however, also arise in social contexts where education is not assigned a central value anyway (see Textbox 20, c). As already shown in the spatial statistical approximations, a large part of the underprivileged people of Baton Rouge live in an (higher) education desert, which at the same time means a reduction of their life chances compared to more privileged parts of the population. In particular, the low level of ambition for higher quality education for African-Americans, within which the majority are dependent on public schools because of financial limitations, not only manifested the segregation of the city, but rather exacerbated it (see Textbox 21, a–d). Thus, in accordance with the life chances approach, they are not given an option in the choice of a higher-quality educational institution, since they are neither entitled to participate in the education programs in other districts that are usually privately financed, nor are they offered the opportunity to select such programs in their district (for details on the ethnic, social, and spatial division of the city, see Sect. 8.2 and for details on the troubled desegregation processes in Louisiana, see Bankston and Caldas 2010).

Textbox 21: School education and segregation

a) "Drugs, poor education. **We have a history of [...] poor education for African-Americans**. One thing you know that such bigotry they… they didn't care about the schools for African-Americans." BR-16

b) "When I was a kid Black People did not eat in the same restaurant. They just didn't eat… they couldn't go to the same theater. Or if they did, there was usually a black seating section over the balcony and the Whites would sit down in the lower section. **So, we've come a long way in some respects, but school's still a problem and all that goes… still goes back to the fact of generations of inadequate education for blacks.**" BR-19

c) "Well I think, I mean, this kind of before I moved here was I think one of the major things that I've seen as a resident here now **is seeing the lasting impacts of it, was the desegregation of schools or rather the lack of**." BR-11

d) "I think for whatever reason Baton Rouge had a really hard time. Desegregating their schools or you know this obviously this is a very mixed culture in Baton Rouge. Over 50 percent African-American, you know, the rest… And then are Colored People and… or people of color rather than other Caucasians and Hispanics and Asians, I guess, but that's kind of the biggest define. One of the defining things I see about Baton Rouge in contrast to where I grew up because of, like, **I grew up with public schools where everybody felt very integrated.** There was a handful of private schools but for here **there's a lot of private schools and in turn that sort of… that sort of spills over into the politics of people and the culture of their mindset**, you know. And I think it's easy to see that there's two parts of Baton Rouge, you have at North Baton Rouge and, like, this area and then there's South Baton Rouge and I think they're kind of two different kind of almost two different cities you know within one." BR-11

8.1.6 'Crime is not a Baton Rouge-Specific Phenomenon, but a US-Wide One'

Already in the course of the evaluations of the YouTube videos on Baton Rouge, an intensified presentation of violence and crime with a focus on the African-American neighborhoods in the North and Old South Baton Rouge could be observed in comparison to the video evaluations of other metropolitan areas (cf. 7.2.2). The crime rate in Baton Rouge is significantly above the national average (neighborhoodscout.com 2020; Toohey et al. 2019), with the Baton Rouge locals surveyed seeing an increase here, particularly in the recent and most recent past. Because about 35 years ago "was the biggest news going on was an accident" (BR-12; Textbox 22, b) on the Interstate, "[t]hen for up until maybe 10 years ago there is maybe one shooting a month. Now, there's one every other day" (BR-12; Textbox 22, b). However, some respondents qualified the increase in violence and crime by pointing out that it is not limited to Baton Rouge, but "it's all over the country, too" (BR-12; Textbox 22, b). In the view of the interviewees, this is no reason to avoid Baton Rouge: "murder happens everywhere, things happen everywhere. So, […] I wouldn't tell anybody not to come visit because if you've never been to Baton Rouge, you know, you've never really got a Louisiana Culture experience" (BR-03, Textbox 22, d). Those who have moved to the city relate the topic of crime to experiences in other cities and tend to see it as an intensified one: "I don't think it's actually that much of a problem. It's not worse here than it is in Raleigh. It's just people here are more scared of it." (BR-20, Textbox 22, e).

Textbox 22: Increasing crime as a general phenomenon

a) "The violence is very high. […] It's above the national average. **It's like one of the most violent cities in this country […] in the United States.**" BR-15
b) "One of our reporters, who has since retired for health reasons, a lot of TV reporters talked **about thirty five years ago,** when she started out as a reporter, **they would send her out for a traffic accident on the Interstate because that was the biggest news in town.** That was the biggest news going on was an accident, 'there's been an accident on the Interstate!'. Then for up **until maybe 10 years ago there is maybe one shooting a month. Now there's one every other day.** […] And they're getting close to my house. But, so yeah, that… that **I see a rise in violence and I know it's all over the country, too.** " BR-12
c) "**You get used to it,** honestly, some day, everyday, most likely, **if you look at the news, somebody's got murdered everyday.**" BR-03

d) "To me, everywhere is going to be the same. Like, you can't stop people from being people. So, […] **I'm not gonna say this is a bad city, it's just people.** People in general. If I go… **if I go to somewhere… if I go to Dallas or I go to Atlanta. Anywhere. Murder happens.** Because, like I said, […] we're the number one in Louisiana. It's, you know, forty nine more States, you know, things of that sort. It's… murder happens everywhere, things happen everywhere. So, **[…] I wouldn't tell anybody not to come visit because if you've never been to Baton Rouge, you know, you've never really got a Louisiana Culture experience.**" BR-03

e) "My family comes from [City in South America]. And there everybody has, you know, fans with bars on the top and security systems because there's very little police infrastructure. But in the US, I've never really faced this problem before. Were here with all of the white people that they know have security systems with video cameras and gates leading into their driveway. And **I think it's very weird.** So, **you have the physical insecurity and sense of which I don't think it's actually that much of a problem.** It's not worse here than it is in Raleigh. **It's just people here are more scared of it.**" BR-20

While some of the respondents cited increased awareness of the accumulation of crime-related reporting in local news as a source, it is the young African-American adolescents interviewed who report crime in their immediate personal environment and are affected by it in their everyday lives and have to deal with it: "Three people, that I know personally, close to me, was murdered within two to three months apart from each other" BR-03 (BR-03; see Textbox 23, a). Rap music is framed as a catalyst for the criminalization of young African-Americans, especially those from difficult social backgrounds, by promoting a lifestyle of drugs, luxury goods, women, and fame/recognition through rap lyrics and videos, which young African-Americans from underprivileged contexts take as their role model, since they are hardly aware of alternative male role models. This is an outcome of the fact that in many cases there are hardly any family fathers who also have regular jobs, and therefore the young males orientate themselves towards the rappers who have attained recognition and fame and emulate them (BR-03; cf. Textbox 23, b–c).

Textbox 23: Rap music as a catalyst for the criminalization of young African-American adolescents, rappers as male role models

a) **"Three people, that I know personally, close to me, was murdered within two to three months** apart from each other." BR-03

b) "The murder, the crimes, the violence. […] 'cause I mean, the murder and this stuff is… it's really from… it's from.. I mean, it's older people but it's majority young generations, you know, […] So, it's from the music and… **it's really**

from the music. Honestly. You just… and, I mean, not Baton Rouge music **but,** I mean, **Baton Rouge music, celebrity music,** things of that sort […] Know that as I'm not **sure but I mean most rappers rap about, you know, the drugs… drugs and things of that sort. And most of these kids take the drugs can't handle the drugs, you know, they make you do stuff.** They make you do things. Or, you know, like I say we are… it's urban so**. It's urban, it's ghetto.** They're poor, so, you know, they need money, can't get a job. They'll rob somebody, kill somebody. Whatever they've got to do to get some money. So, you know, that's how it goes." BR-03

c) **"They don't have no role model. […] Their role models are the rappers.** They want… […] everybody wants to be a rapper. Everybody wants the […] drugs, cheers, girls, jewels […] everybody wants the cars, everybody wants the money. Those are their role models, they don't have real professional role models, no doctors, no lawyers or is rappers and drug dealers and gangsters is no professional role model. So, **they have no better to look up to. So, they look up to** that. Some, you know, […] some of them have moms, some of them, well all of them have mommas, and majority don't have daddies. So, there's no… **there's no father figure in their life.** So, like I say, there's nobody for them to look up to. So, **they look up to what they see on tv, which are the rappers. All they see out every day which are the gangsters, the drug dealers,** that's their role. If they had better role models then maybe they could be something. They can be something better." BR-03

However, this central attribution of significance to rap music by young African-Americans is contrasted by little or no knowledge of the success or even existence of the scene on the part of the white people interviewed, as will be explained in more detail in Sect. 8.2.

8.1.7 Disinvestment in Infrastructure, Existing Building Stock and Acceptance of the Housing Conditions of Less Privileged Sections of the Population

The infrastructural and building conditions of large parts of the city are classified by respondents as 'critical' and defined in many respects as in need of improvement: "so many things need to be developed in Baton Rouge" (BR-04; Textbox 24, a). In particular, the paradigm of a greenfield development, which triggers "successive waves of development and abandonment" (BR-14), is criticized "without any thought to what happens to the abandoned part" (BR-14; Textbox 24, b). This has led to a situation where a number of plots of land and buildings have been abandoned without subsequent use, in which there is either a lack of financial resources or a lack of interest to invest. Especially for newcomers to the city who have not been socialized into the urban landscape of Baton Rouge and for whom Baton Rouge is not the normal landscape of their home, "the degree to which there is disinvestment in the physical infrastructure" (BR-20; Textbox 24, c-d) is particularly surprising, whereas for longtime residents of Baton Rouge "there does not seem to be the kind of frustration with that" (BR-18; Textbox 24, e). So, there would be whole neighborhoods "within two blocks from here are housing conditions that are, you know, third world and we... that's acceptable" (BR-18; Textbox 24, e). Particularly when comparisons can be made through stays in other cities or states, the lack of understanding for the acceptance of the building conditions and the conditions under which people live is particularly pronounced: "And, you know, I've... I've often remarked along the lines of, you know, Baton Rouge does 'ugly' very well. We know how to do 'ugly. We do. And we're comfortable with ugly, you know. We're a dirtier city than a lot of places that I've been to. Trash! And it's just unbelievable to me" (BR-18; see incoming Textbox 24, e). Here, the construction of (urban) landscapes is based in particular on stereotypical-aesthetic visual expectations of (urban) landscapes, which do not serve large parts of the physical basis of Baton Rouge, but rather exhibit a postmodern aesthetic of a past and present hybridity (cf. also Sect. 6.7.1 on the hybridity of past and present in physical space). In addition to this, the interview partners criticize that the dilapidated state of existing buildings as well as the living and housing situation in less privileged neighborhoods is already implemented in the visual modes of many Baton Rouge residents in such a way that these are accepted as 'objectively given' and without special efforts to change them as 'unquestioned reality' (cf. including Sect. 6.7.1, 6.7.2 and 6.7.3 on the Phenomenological Walks Florida Blvd./Plank Road and the video analyses), and what makes them difficult to access for alternative interpretations, classifications, and symbolic inscriptions.

Textbox 24: Condition of the 'physical environment' as a limitation of quality of life

a) "I think there **are so many things need to be developed in Baton Rouge** but I think because of the funding they are lacking to do the development right. **You need a lot of developments in Baton Rouge**." BR-04

b) "And you can by going east on Florida see decade by decade. How they did these successful way **successive waves of development and abandonment**. Yeah, they just kept moving out that way. [...] **it's a very good indicator of of the way the city as an entity thinks about development** and clearly that is that you develop and abandon and develop and abandon and **development abandoned without any thought to what happens to the abandoned part**." BR-14

c) "There's many things that bother me about Baton Rouge [...] **The physical environment is the one thing that really affects our quality of life and they don't have an ability to change**, you know. Investment, at least." BR-20

d) “I **was very surprised by the degree to which there is disinvestment in the physical infrastructure**, I mean, I knew that the politics I was going to face, world views, and conservative views from people. I was not prepared and **I knew that they were skeptical of centralized government, but I wasn't prepared for how horrible like the roads are**.“ BR-20

e) “I mean, that **we have a tremendous number of blighted properties** and it's not been a priority from the elected officials to address those. **And there does not seem to be the kind of frustration with that, that there should be, because it has such a depressing effect, I mean, emotionally but economically on communities** and **we don't expect enough of our elected officials**. And they really should be held more accountable for things… for that and on that issue in particular. I mean, **within two blocks from here are housing conditions that are, you know, third world and we... that's acceptable**. And, you know, I've… I've often remarked along the lines of, you know, **Baton Rouge does ‘ugly’ very well. We know how to do ‘ugly'. We do. And we're comfortable with ugly, you know**. We're a dirtier city than a lot of places that I've been to. Trash! **And it's just unbelievable to** me. And it's sort of what… what is… what is in our DNA, what's in the water, what's in the air that accounts for that. You know, it's just… and **every time I am away** somewhere else, that just comes back to me and **it just makes me all the more frustrated with our lack of progress around some of those things**.“ BR-18

Respondents also view critically the central concentration of transport infrastructure on motorized individual transport without the development or provision of alternatives, e.g. by establishing a public transport system or not integrating designated cycle paths when renovating roads (Textbox 25, a). This lack of alternatives is particularly seen in the prevailing mobility paradigms with motorized private transport as the target (see 6.6 on the limited provision of alternative infrastructure such as accessible footpaths), when asked why cycle paths had not been taken into account in road renewal, the answer was “the people of Baton Rouge don't know what to do with the bike lane” (BR-14; Textbox 25, b). Even though transport systems focus on motorized individual transport, there is some irritation regarding the technical status of traffic lights, for example, but this can be regarded as a pars pro toto for the development of further issues in the field of traffic and transportation (Textbox 25, c).

In addition, the use of public transportation is constructed as an indicator of a low level of symbolic capital and classified accordingly, because “that's unspoken but this is a form of transportation for blacks. (BR-19; Textbox 25, d) and “if you take the bus, you are either a drug addict, prostitute or homeless person” (BR-01; Textbox 25, e), so that the use of public transport is subsequently rejected in the demarcation of oneself from less privileged sections of the population (Textbox 25, d–e).

Textbox 25: Concentration of infrastructure on motorized individual transport - from the mutual reinforcement of low-value infrastructures and socially inscribed interpretations of alternative forms of transport

a) "Those two things play together. [...] So, **if a city had a strong mass transit system and a dense footprint then that sprawling won't happen.** [...] You just keep moving out and out and out and because you have **a transportation system that is on the individual automobile.** Then you can just keep going. And now it's at the point where realtors tell people 'drive until you qualify'." BR-14
b) "And after a number of telephone calls, the man who was the manager of the program said **the people of Baton Rouge don't know what to do with the bike lane.**" BR-14
c) "And **it's almost impossible to ride a road bike here because of how many potholes and how bad the roads are.** And... it's amazing to me how even, you know, [...] I mean it creates an economic burden for them, which car insurance is higher. You have to replace things more often on your car [...], but even the traffic, **we are just now learning how to synchronize traffic lights to improve flow because we never allocated money to it.** I am like you sit in traffic for 45 min every day of your life, each way, why haven't we spent the money to synchronize their lights? I wasn't prepared for that. Which is incredible." BR-20
d) "People here don't want to ride a bus because (in a whisper...) 'because Black People ride the bus'... that's... **that's unspoken but this is a form of transportation for blacks.**" BR-19
e) **"If you take the bus,** you are either a **drug addict, prostitute or homeless person.**" BR-01

In the view of the interviewees, the insufficient or even lack of investment in the technical and social infrastructure particularly affects the African-American neighborhoods in North and Old South Baton Rouge (see also Chapter 6.6 on the limited and often impossible walkability of these areas), where not only do the residents often lack the financial means to repair or maintain the buildings, but the public authorities are also not making any efforts to rehabilitate the public space and thus there is also an imbalance of focus with regard to the spatial development of Baton Rouge (Textbox 26, a). These "nice areas when they were built" (BR-16; Textbox 26, b) with "healthy middle class blue collar workers" (BR-17) that were residents, and the current situation with "high crime, low education, [...] the fewest automobiles per thousand people [...], disinvestment" could simply be described as "sad" (BR-17; Textbox 26, c). However, hope is placed in a new transport project, so that "that place is going to change right now you go there and you think this is America (BR-17; Textbox 26, c).

Textbox 26: African-American neighborhoods – Old South Baton Rouge and North Baton Rouge

a) "**South Baton Rouge needs a lot of development. That's where the majority of the homeless is.** [...] I mean, it just needs to be [...] **remodeled**. Basically, South Baton Rouge, like I say, [...] when you get to certain points, let's put it like that and you get to a certain parts of something which for Baton Rouge is which is basically where LSU is which is the big school. The big big school. And, I mean, **where the school is everything's [...] nice. So, but when you get to the urban side of South Baton Rouge,** the ghetto... basically, if you want to put it like that, the ghetto, **it's... worn down, it's tore up.** So, you know, **they need to kind of focus more on the urban side instead of the college side.**" BR-03

b) "North Baton Rouge [...] those were nice areas when they were built. And then South Baton Rouge started being built and the white people moved out of there and they moved to South Baton Rouge. And so, then the African-Americans who had less money lived there. And then as they died. In those... those areas were rundown because they couldn't keep them up very well and as these people died without any wills. Then the deeds weren't in the children's names or the grandchildren's names and the land isn't worth very much and so they kind of gave up those and so the houses haven't been kept up. And so, as they get torn down then you have an empty lot. Nobody keeps up the empty lot." BR-16

c) "So, you know, Plank Road is another example. **Plank Road** [...] became the main commercial corridor and there was housing built on either side, small houses, small lots, middle class. Most of those people worked in the plants. It was not affluent **but it was healthy middle class blue collar workers. And now it's**... it's... it's, you know, it has the highest crime rate. **High crime, low education [...] they're below poverty level, the fewest automobiles per thousand people [...], disinvestment, the city has not invested anything, the roads are broken**, the sidewalks are, I mean, it's... it's... **it's sad, it's sad**. Well there is now a movement going on, an initiative to using as public transportation a high speed dedicated public transportation route, not to the airport, just south of the airport all the way to LSU going through downtown. There's a study being done right now. **If that happens that place is going to change right now you go there and you think this is America.**" BR-17

8.1.8 The Mississippi River as a 'Physical and Psychological Barrier'

The Mississippi River is another 'natural asset' that shapes the spatial structure of Baton Rouge. The relationship of the people of Baton Rouge to the river can be considered ambivalent: for one thing, the river is "really one of the main reasons the city is here"

(BR-19; Textbox 27, a), since, because of its orographic location on a flood-safe terrace, large parts of the downtown area, the universities, and the Exxon plant cannot be flooded because of their elevated position (see Chap. 5 in more detail), so that the city "has access to the river but the river doesn't threaten it" (BR-19; Textbox 27, a). The Mississippi is of central importance for industry, in particular, both in terms of production processes and the transport of goods and commodities, especially since large ocean-going ships can enter the Port of Baton Rouge via the Gulf of Mexico. However, New Orleans, compared with Baton Rouge, is so dominantly associated with the Mississippi River that New Orleans is even assigned – albeit erroneously – the geographical location at the estuary (Textbox 27, b).

Textbox 27: The river that makes the city possible

a) "**The river is really one of the main reasons the city is here.** It was the transportation corridor between the Gulf of Mexico, **big ocean going ships can come all the way to Baton Rouge from the Gulf of Mexico from anywhere in the world.** The city is up on high ground so it's not susceptible to flooding and never was, you know, in human history. So, it has access to the river but the river doesn't threaten it. So that makes this city really possible in many respects. We don't drink river water. Water is important for the industries for waste removal or for their processing. So, **it's really important for industry as well. But without the river, you know, who knows where the next big city and Louisiana would** be, you know after New Orleans." BR-19

b) "The river is an integral part of the city. The city life and everything, **I think, Baton Rouge is, maybe, not as associated with the Mississippi River as New Orleans is. Which is strange because you'll see people on TV talking about New Orleans is at the mouth of the Mississippi River and it's nowhere near the mouth of the Mississippi,** it's [...] hundred miles, but [...] I always consider Baton Rouge a river town or river city." BR-12

Despite its exposed location on one of the most culturally symbolic and geographically significant rivers (Schneider 2014), the Mississippi is perceived as a physical and psychological barrier in the everyday life of the Baton Rouge populace. Despite the efforts to upgrade the Waterfront, the interviewees see a number of physical objects such as the railway line running along the dikes or the River Road, which in their spatial orientation lay in the way of a more comprehensive development of a Waterfront: "[J]ust because of the nature of the way downtown is laid out and the way our levee acts as a barrier and then like I said with the railroad track and a fairly busy street that runs alongside the levee, I mean, they're just things that [...] will always make it a little less intimate than I think you might ideally want [...]" (BR-18; Textbox 28, a). There was a lack of cafés and restaurants overlooking the river, which could already convey an identification and attachment to the river through its visibility and physical presence. The bridge is also considered an "enormous choke point. So, people do not see it as a viable place to move or to invest in, simply because it's very difficult to get back and forth across the bridge at certain times of day" (BR-18; Textbox 28, b). The Baton Rouge citizenry, for example, interpret the crossing of the river primarily as an obstacle, so that the Mississippi is construed less as one of the most symbolic rivers with central functions and to which the residents are proud to return but more as "a river dividing places" (BR-01; Textbox 28, c).

Textbox 28: The Mississippi as 'a river dividing places'

a) "We've made improvements to our levee with a bike path. **The levee itself is a physical and psychological barrier in a lot of ways**. But it has improved. We definitely don't... still don't take advantage of our riverfront in ways that I think we could or as other cities have. You know, **there's a practical reality there's also a railroad track that runs right alongside the levee**. So, you know, there are just sort of landscape realities that will always be a challenge but it's improved. **It was nonexistent twenty-five years ago**, you could not go anywhere and have a drink or eat and actually look out over the Mississippi River. No place that was here. **Here we got the Mississippi River, right there, on our doorstep in downtown – not a single place where you could go and have a drink and enjoy the view of the river. That's changed**. So, it's better but it will always be... and **I don't know if it will ever be a complete marriage... just because of the nature of the way downtown is laid out and the way our levee acts as a barrier** and then like I said with the railroad track and a fairly busy street that runs alongside the levee, I mean, **they're just things that [...] will always make it a little less intimate than I think you might ideally want but it's better.**" BR-18

b) "Yeah, there's no almost no **that's the psychological barrier for people of going over the bridge, it's crazy**. I mean and there's also a real traffic issue. Our Bridge, **the main bridge**, is really obsolete... all of the traffic coming from I-10... on I-10 going east through Baton Rouge narrows down at the bridge to one lane. And it causes this **enormous choke point**. So, **people do not see it as a viable place to move or to invest in**, simply because it's very difficult to get back and forth across the bridge at certain times of day. I mean, you see that the cars are literally at a standstill across the bridge and into it. I mean, just stopped. **It's like a parking lot.**" BR-18

c) "**I have no personal connection to the river**. Like I said, when **I use to drive home from college, crossing the river was a big deal**. But I mean when I was a little kid, whether you crossed up Vicksburg, Mississippi on I-20 or here I-10 it was kind of an event because you never did that. And like I said **I used to**

drive back and forth every day. For me crossing the river it was really more of a barrier. Than, you know, in that sense of **a river dividing places**." BR-01

In connection with a rather weakly developed aesthetic-emotional reference to the Mississippi River, a merely rudimentary staging of the river scenically could also be possible. There are no regular excursion steamers or other activities on the river along the shore. There are also no cafés and restaurants from which to enjoy the view of the Mississippi River while having a drink, although this is also due to the construction of the dike (cf. Textbox 29, a), which must first be climbed in order to be able to see the Mississippi River (see also 6.7.2). However, the experience of the river by means of leisure activities is of central importance with regard to its aesthetic, but also especially its emotional appropriation: thus, in those places where this is possible, it was exciting "to live somewhere on the Mississippi" (BR-02), but in Baton Rouge these activities are not recognizable, so that "there isn't that connection now that I had with it. Those type of things don't occur' (BR-02; Textbox 29, b). How central the possibility of experiencing the river is for its appropriation is shown by the establishment of a cycle and walking path on the dam, from which the Baton Rouge locals can enjoy the view of the river and "really enjoy having the river there" (BR-11; Textbox 29, c). In the course of the low level of appropriation, the river is constructed as a "marginal place" (BR-20), as it lies within federal jurisdiction and is therefore not considered part of the city (BR-20; Textbox 29, d).

Textbox 29: Rudimentary staging of the river

a) "A long dam basically, so it's very high [the levee; note by the authors]. Right. Well comparatively high. [...] **So, we don't flood every year or every other year but now we can't see the [...] river anymore. And all [...] the business that used to be there with boats coming in and out** and the people being down there and the ferries back and forth when the businesses were right there at the river and you could see the river and people would be coming on and off and all this activity is gone." BR-14

b) "My dad grew up **in a very small town on the Mississippi River in Missouri,** two thousand people, and they were famous because they had a 'lock and dam' at that little town. **And so, when we would go visit we would always go to the locks see the ships coming through. They had the steamboats that go up and down.** So, whenever they would come through people would come out and throw candy at you, so, like, as a 10-year-old, **it was, like, the best thing ever.** Like, my grandma was friends with like a tugboat captain. So, like, one that I remember going out like my pajamas because he was coming through and I could go ride on it. So, I have all those things but yeah, **so it was exciting to me**

to live somewhere on the Mississippi. But, then, there isn't that connection now that I had with it. Those type of things don't occur.“ BR-02

c) “It doesn't really affect me personally other than, I mean, **I enjoy the view** because I ride my bike on the levee, you know, from LSU to downtown and so when I'm doing that I really enjoy having the river there. **I think that the river obviously is a huge asset to the industry here but it's an asset to people who live here as well** just, you know [...], I mean, [...] there's no mountains here so if there's any kind of scenic thing, you know, maybe you can find some beauty in that as well. But other than that I don't really have a lot of... **I don't work on the river, you know, I don't have to even cross it very often. I don't fish in it and I certainly don't swim in it but I do enjoy it when I'm riding my bike next to it.**“ BR-11

d) “I think **there's not a lot of connection** [...] And it's fairly common from a lot of cities because **the river is almost seen not as a resource and a part of the city, as it is a Federal thing**, you know, **we have very little control over it**. So, the city doesn't have much identity with it at all. It doesn't have much ability to interact with it. **It's one of these marginal spaces.”** BR-20

A central paradox is also the low visibility of the river in everyday life on the one hand and its enormous physical presence on the other (Fig. 8.1 and Video 1), so that in the social construction of Baton Rouge the Mississippi River is less strongly integrated, if at all, and thus the curious situation can arise that, contrary to its enormous land use, it is hardly noticed. Thus, parts of the respondents drive every day to work along the river “and it's literally next to me and of course I'm under it. So, I don't even think about it either” (BR-02; see also Textbox 30, a–b). Others report from an undergraduate student group where the teacher spoke of ‘the river’: “A student raised his hands, ‘what river are you talking about? I got... ‘the river... THE RIVER ... the big one over there! It's a real big one. You can't miss it. What do you mean ‘what river’ ?” (BR-12; see Textbox 30, c).

Textbox 30: "What river" – "The big one over there! It's a real big one. You can't miss it!"

a) “No. I couldn't... I **I could not even tell you the last time I saw the Mississippi River.** I think, I mean, yesterday we were right by the river. And **I don't know if I could even see the surface of the river because we were below the level,** might have been above it briefly, all I saw was a tugboat popping up above the level so I can't even see it.“ BR-01

b) “And I drive home on a road and **it's literally next to me and of course I'm under it. [...] So, I don't even think about it** [the Mississippi River; remark of authors] either“ BR-02

c) "I was teaching […] class to undergraduates that I was talking about the river. **A student raised his hands: 'what river are you talking about?' I got… 'the river… THE RIVER …the big one over there! It's a real big one. You can't miss it. What do you mean 'What river?' You could stand on the top of the building and see it'.** So, C'mon… what river?" BR-12

Video 1 Drone flight over the Mississippi/Baton Rouge. With special thanks to John McEwen for permission to include his video in the book © (https://doi.org/10.1007/978-3-658-30716-5_8)

Fig. 8.1 Mississippi River and Levee, Downtown. Photo: © John McEwen

8.1.9 Special Places of the Interviewees: Of Co-Existences of Stereotypical Visual Expectations and Postmodern Aesthetics of Decay

When the interviewees were asked about special places, a preference for landscape stereotypes is revealed, such as those found in the neighborhoods of University Lake and the Garden District in relation to the urban landscape. In addition, however, it is also waterbodies such as the Mississippi River and plantation landscapes that update landscape stereotypes and are named by the interviewees (cf. Textbox 31, a–e).

Textbox 31: Aesthetic references to landscape stereotypes - Mississippi, University Lake, Garden District, plantation landscapes

a) "The **neighborhoods around the LSU Lakes** [...] are beautiful. And if [...] someone come from out of town and I wanted to show them that Baton Rouge could be pretty, that there were pretty places. I would definitely bring them there. Where we call the 'Garden District' which is a neighborhood, it's in Mid-City, older homes, beautiful oak trees. **I love the Garden District**. I am just as a native, I'm proud of our downtown, I think it's no longer depressing. So **I would show off our downtown, now**. You know, with a lot of caveats that there's still a lot to do... I mean, a long ways to go but it's... it's yeah... **it's improved so radically that I do have a sense of pride for our community**. So I would say those. But **my favorite, I mean, yeah... I would say that area around the LSU Lakes is it's just beautiful**. And I would like to say you know this is where I live. So that would probably be it." BR-18

b) "I don't know that I have a single favorite place, mean I've really grown attached to walking in my neighborhood with all the big live oaks. [...] I always like to go out and walk on the levee. So I mean walking the levee above the river is always a favorite thing." BR-19

c) "I have two, three places, like, **I like to go and sit where I can see the water**. [...] We have our Mississippi River, so you how some places have a terrace so that you can sit, relax, have a drink, and then you watch the Mississippi River." BR-04

d) "**Down by the river** downtown." BR-07

e) "I do like the **plantation landscapes**. I mean, there's a huge problem of the 'exoticization' of the plantation. And now the plantation museums are just to fulfill a fantasy of what we think the plantations were like, but **there's still remnants of the physical infrastructure of landscapes that are fascinating to experience if you know the reality that's behind them.**" BR-20

However, a number of respondents also provide a functional reference in the context of the question about their favorite places in Baton Rouge, so that the focus is on the question of whether the place is 'suitable for me' – for example, as a backdrop for taking a walk or for the workplace (see Textbox 32, a–d).

Textbox 32: Functional references to space – dog walking, work, study

a) "There's a park on Highland Road [...] The park system here is really good. You **can take the dog loose**... but... let them run." BR-12
b) "I go for **walks with my little dog** every day. And we walk around the neighborhood." BR-16
c) "There is probably **LSU**. The college where I go." BR-04
d) "**My office** [...] I have all my books there." BR-13

However, with a large number of economically and functionally devalued buildings, Baton Rouge has a number of physical objects that are capable of connecting to postmodern aesthetics, but often only take effect within the framework of a practice of social distinction in conjunction with a certain spatial and living environment distance (Hauser 2001; Kane 2018; Kühne 2006). In many cases, a link is made to the aestheticization of the decay in the ruins as artistic expression (cf. inter alia Makarius 2004) by aestheticizing the "'natural' decay" (Hauser 2001, p. 181). For 'ruins' distinguishes "from the demolish-able building, from the useless remainder, from the rubbish [...] only the view that sees something in and on it that points beyond the only destroyed, only fragile, only chaotic" (Hauser 2001, p. 181; see also Edensor 2005). Thus, the same abandoned railroad warehouse district near the Capitol as a "spot of decay that's just fascinating" is also as one interviewee would "refer to it as the 'post-apocalyptic Baton Rouge'" (BR-14; Textbox 33, a).

Textbox 33: Postmodern aesthetics of decay

a) "Well there's a couple places that I really like. One is an **abandoned railroad warehouse district that's up behind the Capitol**. That's just gets spectacular... in the... in the weirdness of this abandoned decayed area that's a... clearly had been something else but has now become what it is **which is sadly changing and being improved**. But nonetheless it's kind of interesting to go up there, also because it's behind the Governor's Mansion and you can see the Capitol. **But it's this spot of decay that's just fascinating**. And I like to ride around at night **through different neighborhoods**. That most of the routes that I have have almost no traffic on them. And I refer to it as the '**post-apocalyptic Baton Rouge'. Because there's no one out. The lights are on but there's nobody around**." BR-14

Thus, in Baton Rouge, stereotypical visual expectations of landscape – such as the Garden District and plantation landscapes – are combined with postmodern aesthetics and reinterpretations of economically and functionally devalued building structures such as the abandoned railroad warehouse district in the immediate vicinity.

8.2 The Ethnic, Social, and Spatial Division of the City

As discussed in 6.4, Baton Rouge has a strong, spatially impacting ethnic and social segregation, which will be discussed below in terms of conceptions, anxieties, and traditional narratives (Sect. 8.2.1), the persistence of racial-based inclusion and exclusion (Sect. 8.2.2) as well as the territorial culmination point in the process of incorporating the City of St. George (Sect. 8.2.3), in the context of which the predominantly white, wealthier suburbs aim to establish their own school district.

Accordingly, the spatially inscribed ethnic, social, economic, and cultural division of the city is not only reflected in official statistics, but also shapes everyday life in a way that involves less open-ended exchanges and more subtle passive, as well as active, avoidance strategies. Consequently, these lead to the various forms of segregation, which subsequently also become tolerated and accepted in a certain way. Thus, it is "pretty clear that it's not a hard line it is a liminal space that divides the poor black north part of the city to the richer white part of the city" (BR-14; Textbox 34, a). Many neighborhoods are isolated due to "either economics or race or both" (BR-17), so that in the city "[i] solation tends to be the norm" (BR-16; Textbox 34, b).

Textbox 34: Segregation of the city

a) "It's also pretty clear that it's not a hard line **it is a liminal space that divides the poor black north part of the city to the richer white part of the city.**" BR-14
b) **"Many neighborhoods are isolated because of racial reasons.** It's either economics or race or both. The only one or two ways to get in and out of the neighborhood in the rest of it's fenced off." BR-17
c) **"Isolation tends to be the norm."** BR-16

The only places where the Baton Rouge citizens surveyed experience a mixture of different mixes are grocery stores or gas stations, so that the different ethnic groups are only mixed in functional areas of social life, whereas on the social level, for example in the composition of sports clubs etc., a separation of the different ethnic groups is still effective (Textbox 35, a–c). Accordingly, whites and blacks live "side by side, but they don't live together, if that makes sense" (BR-01; Textbox 35, d), although it is also pointed out that Baton Rouge "cannot be dinged as unique for still being a segregated city, you know, that's the same in almost every major American city" even if "it seems to me it's less obvious in other communities" (BR-19; Textbox 35, e). However, there were also areas in the city where social and spatial dividing lines began to dissolve, such as in the area of development processes around Government Street in Mid-City, because "the power of Mid-City is that this is truly kind of a meeting place of North Baton Rouge and South Baton Rouge in downtown which I think is very special and wonderful" (BR-19; Textbox 35, e), so that even the smallest beginnings of a change in the status quo of a segregated city meet with the approval of the interviewees and refer to the pronounced degrees of the segregated structures construed as 'objectively given'.

Textbox 35: Social division and places where it is overcome

a) "It's accurate. And then even people like... like people stay with... if you're black... if you're friends with black people... so, like, I remember when I was playing tennis. **There was a tennis team that had black people on it. And then there was all the other teams that were white people. And there is no mixing of that it**. Whatever reason if that's just the people you know and you play tennis. But I remember just being struck by that – I was, like, huh!" BR-02

b) "I can't think of anything like that either **I mean everything we see and do is white**. The **only place [...] where I see mixing is at the grocery store or the gas station**. I don't see mixing it like some of the restaurants around." BR-01

c) "And you have the **normal mixing at places like grocery stores and convenience stores**. But say in the bar, the bar landscape – **you've got black bars and you have white bars**. And I never went to any of the black bars because I'm white." BR-01

d) "But you don't see white, **you see white people next to blacks. [...] They live side by side, but they don't live together**, if that makes sense. So yeah. So **there is a geographical racial divide**." BR-01

e) "You know, **we cannot be dinged as unique for still being a segregated city**, you know, **that's the same in almost every major American city**. Where there is a **North–South divide in Baton Rouge, there's north of Government Street, north of Florida**. And what I think the power or at least what we say the power of **Mid-City is that this is truly kind of a meeting place of North Baton Rouge and South Baton Rouge in downtown** which I think is **very special and wonderful**. But, yeah, there's no question, **we remain a segregated, not by law, but obviously community** and we're not unique in that. It seems to me it's less obvious in other communities, I mean, in Washington and Dallas even but it's still, you know, that's an issue no question." BR-19

The educational system is cited as a model for the division of the city and its promotion, through which a segregation of the inhabitants of Baton Rouge is already brought about and manifested in its structure, in that the tendency among whites to move to districts with better schools is due to the striving for the best possible school education for the new generation, while a large number of African-American families, because of restrictions, especially of economic capital, are neither able to move to districts with better schools nor to finance expensive private schools for their children and are thus left behind in the under-equipped public schools with a poorer quality of school education (cf. for this purpose, see Textbox 36, a–c; see also Sect. 6.4.3). Even if a subliminal racism would be at work here, self-affirmation would often be used as 'counterevidence': "they can say 'well I don't mind going to a restaurant and eating with a restaurant with a black person... so I'm not a racist', you know, 'I go to a store and somebody sells me something, they're a black person, I don't mind that... so I'm not a racist', but there is generally a tolerance for this continued segregation", but in relation to the school system a way has been found that "tolerates that we found ways to create new ways of segregating students, so people can feel comfortable with it" (BR-19; Textbox 36, c). Thus, an institutionalized inequality of opportunities is perpetuated by the fact that the social structure of the school system organizes life chances at an early stage – to the disadvantage of the less privileged parts of the population, since access to options depends predominantly on social contexts.

Textbox 36: Schools as segregation instead of integration

a) "Baton Rouge schools, well, I guess **they're predominantly black and not as good** and [my husband] was talking about **how people are leaving to go to the suburbs for better schools** that type of thing. So, it's accurate." BR-02

b) "In South Louisiana we've always had a lot of Catholic schools because there's a fairly sizable number so there was already a parochial school system. But after integration was really finally forced on people what happened was you had to process the work. To create 'resegregation'. **We have a new type of segregation before it was neighborhood schools. You had residential segregation and neighborhood schools that were completely segregate**. With integration really, in the late 60 s, early 70 s, I mean took us almost 16, 15… 16 years to get to the point where we accepted the court orders in Louisiana. But **the 'resegregation' occurred with most people sending their kids to church schools**. And the creation of a whole new group of Baptist Academies. These were not Catholic Schools but Protestant Schools. That **allowed white children to go to schools with white children with white teachers. And so that was going within the city**. But then, in addition you had a lot of people moving to Denham Springs or to Ascension Parish which at the time were 80 to 90 percent white. So you had people **from white flight from the central city to more predominately white suburbs and also those two parishes have grown tremendously since 1960, 1970**. And this is people going there so, again, **you could have white children going to school with white children with white teachers, and that left behind large numbers of African-American students in the public schools in the inner city.**" BR-19

c) "Yeah I think for a lot of people they feel, like, OK, **they can say well I don't mind going to a restaurant and eating with a restaurant with a black person… so I'm not a racist,** you know, 'I go to a store and somebody sells me something, they're a black person, I don't mind that… so I'm not a racist', **but there is generally a tolerance for this continued segregation because people care deeply about their kids going to school where they can get a good education and that typically means where you have more children who have the same level of the same starting place**. Their goal leaving from the same level when they go to grade school and **education opportunities have always been better in white schools in the state, always remain that way**. We have a lot of charter schools which kind of really, again, kind of tolerate some minimal integration but for people who are really good, they can go to these special schools, **but it tolerates that we found ways to create new ways of segregating students, so people can feel comfortable with it**. But they feel like they're not contributing to segregation. But in fact they're tolerating it. So this is very much... **it is still very much a reality.**" BR-19

Thus, already in an early stage of the socialization of the Baton Rouge, a social and ethnic division takes place and manifests itself, which is often also understood as 'given' and 'normal' and subsequently only irritates in situations of direct confrontation and is accordingly – at least to some extent – also questioned, or even modified or even newly designed, as will be shown in the following.

The extent to which the city is also culturally divided becomes clear in the area of rap music: while the interviewed, especially young African-Americans, were aware of the success and strong musical influence of Baton Rouge rappers (cf. Brillon 2015) in the nationwide US music scene and the breakthrough of rappers from the poorest backgrounds, which they also address in their songs, so that "people's voices can be heard, especially a lot of people [...] didn't grow up in the most prosperous financial circumstances, so, they're getting a release from that" (BR-06; cf. also Textbox 37, a–b). The white people interviewed say they have either never heard of a successful Baton Rouge rap scene or are not aware of its influence. One of them reports "that's so fascinating to me because I know there is a Baton Rouge rap culture and yet I am so separated from" (BR-18) and does not see himself alone in this. He estimates that "most white people of Baton Rouge are ignorant of that" (BR-18; Textbox 37, c), thus referring once again to the ethnic and social division of the city: "And really that's [...] the kind of... not putting words in your mouth but the really negative connotation sometimes a glorification of violence and the gun culture. So, yeah again the segregated community. Just cut off from one another. It's almost like two different worlds, right?" (BR-18; Textbox 37, c). Thus, two completely separated communication bubbles emerge, each of which is further differentiated between them, but which are completely disjointed from each other while also manifesting themselves spatially.

Textbox 37: Rap music as the voice of black people from contexts of reduced life chances apart of white people of Baton Rouge

a) "I think it's cool [Rap Music; remark of authors]. **It's really great that [...] people's voices can be heard, especially a lot of people [...] didn't grow up in the most prosperous financial circumstances, so, they're getting a release from that. They're earning a little bit of money.** [...] But, I say... it's... It's nice that you wouldn't think of a city like **Baton Rouge that's not the largest city to have such an influence on music across the nation. And it's cool saying some people that your friends know or something like that grew up in doing something and being on stage performing,** so it's cool." BR-06

b) "If I go, like, **will go to a different state or something, they say ,Oh, you from Baton Rouge? You from Baton Rouge, Louisiana?** 'They're going to say, or the majority of the people will say, "Oh, you know..." it's a big rapper his name's [Name] Oh, you know him, like, he's a big rapper, they ask stuff like, how we talk and things of that sort [...] oh really, like I say they ask us about music-wise, rappers [...] 'cause **we have a lot of, you know, big rappers that came from Baton Rouge.**" BR-03

c) "Yeah, **that's so fascinating to me because I know there is a Baton Rouge rap culture and yet I am so separated from,** right, and I suspect most middle and upper middle income... well, not even upper middle income, **I would say most white people of Baton Rouge are ignorant of that.** And really that's [...] the kind of... not putting words in your **mouth but the the really negative connotation sometimes a glorification of violence and the gun culture.** So, yeah again the segregated community**. Just cut off from one another. It's almost like two different worlds, right?**" BR-18

In the following, conceptions, anxieties, and traditional narratives will be focused on the neighborhoods of the segregated city that are regarded as socially, criminally, and infrastructurally 'problematic'.

8.2.1 Conceptions, Anxieties, and Passed Down Narratives

Avoiding certain neighborhoods or areas of the city is closely related to conceptions and anxieties which, at least among the white respondents – so they report – tend to result more from following news broadcasts or passed on narratives than from direct confrontation. The black neighborhoods of North Baton Rouge, for example, are perceived by black respondents as "a whole different environment, a whole different feeling" (BR-06), which are considered to be culturally valuable despite the limited life chances and the poorly maintained buildings and infrastructure (BR-06; cf. Textbox 38, a, and on the appreciation of culture and lifestyle, Sect. 8.1.1). The white people interviewed reported that although they do not consciously avoid these neighborhoods, they perceive an increased danger in the neighborhoods, especially through crime reports in news broadcasts and newspapers, and consequently tend to avoid them, especially at night: "There's two different reasons to avoid parts of Baton Rouge. One is because we think there's a perception of crime. And other is because traffic's bad" (BR-01). Even though no areas of the city are deliberately avoided among the interviewees, this is dependent on the circumstances and time of day. For example, "no middle aged woman had ever had any problems out there" (BR-16; Textbox 38, b) and "I don't think there's any place that I wouldn't go to in the middle of the day. But there are a lot of places I wouldn't go to at night" (BR-14; Textbox 38, c).

In many cases, respondents reported how passed down narratives and crime reports overlap the actions of many of the people in Baton Rouge in such a way that they shy away from personal experiences, for example, when a busy and much more dangerous road for cyclists is preferred over a quiet and safe road in a black neighborhood for fear of being shot, "and you won't even come this way to find out that you're wrong" (BR-14; see also Textbox 38, c–e). If passed down concerns are overcome and confronted with personal encounters, the passed down preconceptions are often disturbed by diminishing the strong preconceptions or even abandoning and reinterpreting previously valid patterns of behavior. If this cycle of behavior is broken, diverse perspectives and change become possible, within which friendships can develop on a social level and potential for change with regard to physical surroundings (see Textbox 38, f–g).

Textbox 38: North Baton Rouge – classification of crime, persistence of (pre-)judgement, and the fear of personal experience

a) **The North Baton Rouge it's a whole different environment, a whole different feeling. It's a whole different setting than what is generally portrayed and what you think of when you think of Baton Rouge**. And it's nice to get the culture, the different cultures sides of Baton Rouge especially the black neighborhoods as well." BR-06

b) "And then then **in the bad neighborhood** that way, I read a report one time it said **it's usually like black on black crime**, you know [...] and I read a report that said **no middle aged woman had ever had any problems out there** and I thought, all right." BR-16

c) "We had a little event [...], a bicycle ride, from these green companies instead of getting in our cars and or their cars. Part of it was downtown and part of it was in another part of town: mid-city. So, you have to get from here to there. One of the main streets which is very dangerous is Government Street. There's a street a little further south that has no traffic. **Beautiful street, light traffic, recently repaved, beautiful.A lot of people would not ride there because they knew it was in the black part of town and they were going to get shot**. 'When's the last time you were there?' **'Well, I've never been there. But I know that's the way it is.'** I'm like: 'I ride that every couple of days and it is exactly the opposite'. And there's no traffic. **Why do you want your idea of something to overwhelm your sense of safety so that you're willing to ride on a very dangerous road? Which is really there and really dangerous**. Instead of coming down here into this community that is safe, because you think it's dangerous. [...] **And you won't even come this way to find out that you're wrong**. So, you go ride over there and I hope you don't get here. I'm going to go over here and take my time and relax. See ya. Look, **this is such a strong thing**, you know. It's just**, it's it really holds the city back. Because it doesn't allow things to happen**." BR-14

d) And then though, the white students that were in it, I mean, one of them was **a local student** who is very conservative and we had trouble with him. You know, **he grew up with this understanding that Plank Road is kind of a dangerous place** because there is a lot of crime reports that come out of the area. Some people would argue whether that's because there's more crime there, because there's more policing that happens there. But for instance, **when they went out to [...] just to walk the road** to see what it's like, he was **carrying a baseball bat**, like, 'what was I going to do to you if somebody tries to mug you?' " BR-20

e) "**But our next door neighbor [...] his father told him never to go above North Street**... [...] [my husband] was talking about a place [...] Tabby's Blues Box. [...] and [my husband] said, "Let's go!". But it was in North Baton Rouge so they went, and he just loved it. [...]" BR-16; "But this is a man who was like in his 50s and he'd never done things in North Baton. [...] He grew up here. [...] And he knew all of the history of the building that he wanted to go that we went to. He knew how many different people owned it. What they did when they had it. I mean knew everything about the area. **But his father had so pressured him not to get involved in one of these 'negro places'. He couldn't do it. [...]** and he went there... [...] and there were two other people in the bar. And one of the other people left because they were going to charge like a dollar to listen to the band. And so they were only just the two of us or one other person listening to this band. [...] So he went back he talked to the staff there. **And he went back and he got all of Tabby's recordings he had made**, because he used to do a little program with the high school. [...] **And they became the best of friends**." BR-15

f) "So that's... it's getting that... **These shields that people have for things. Like not going into a neighborhood or not getting to know those people. If you can get rid of those shields the changes.**" BR-15

g) "Yeah I was never to Plank Road right until December this past year. As an example. [...] **I've had my students work in some areas that most peoples... 'what you take your students there?!'**[...] I mean they see... mean they see the potential, **they see the potential of property that is not... has suffered from disinvestment and low economic and social problems.** They see the potential and have developed plans to show what the potential could be." BR-17

In addition to the associated danger, it is also the lack of services, recreational opportunities, and other attractions that is associated with the diminished life chances of North

Baton Rouge residents, for example, in the form of reduced walkability of sidewalks, access to education and healthy food, and lack of appeal to people outside the neighborhood without social connections (see also Sect. 6.4). These effects of reduced life chances also indirectly affect the interviewed whites in that, at least for the interviewed whites, visiting the black neighborhoods in northern Baton Rouge appears to be less appealing or is not even considered at all: "I don't really avoid an area or person per say but [...] there's no reason to go over there if you don't live there" (BR-06, Textbox 39, a), because "there are no services or anything that I would ever need to be in those neighborhoods" (BR-18; Textbox 39, b).

The lack of investment in the economic development of North Baton Rouge is seen in particular with the low level of resources available to blacks living there, in general, but also in the small number of African-American investors in particular, which could compensate for the lack of capital investment (cf. Textbox 39, c), so that possible development scenarios for the district are more likely to be in the private sector than in government action.

Textbox 39: North Baton Rouge with very few attractions or services for people outside

a) "I don't have a lot of transportation itself. So, maybe me not per say but I know there's a lot of areas even just right over that hump in Northern Baton Rouge, Scotlandville is probably not the safest area but I don't have any beef with anybody. So, **I don't really avoid an area or person** per se but... and I've also heard of there's not the nicest areas in South Baton Rouge around certain areas, certain streets that there probably won't be (snickers) in going by. But it's... again **there's no reason to go over there if you don't live there**, so." BR-06

b) "Oh sure. You know, there there are **parts of North Baton Rouge** that well I always tell you this is really almost more of the truth. **I don't have to consciously avoid those areas because there are no services or anything that I would ever need** to be in those neighborhoods. So, you know, it's it's almost not by choice. It's just unless you live in certain parts of Baton Rouge... North Baton Rouge you would have no reason to, **very few attractions for people outside. Which is the exact opposite of say what's happening on Government Street. There are reasons why people come to Government Street,** north of Florida, very few reasons, very few reasons." BR-18

c) "And... so there would be hesitancy on the part of entrepreneurs to invest and **you just don't have as many African-Americans who are big players in the entrepreneurial arena**. There aren't as many African-Americans with capital to invest in economic development projects. **There's not as much money available for African-Americans to invest in an Afro-American Neighborhood.**" BR-19

8.2.2 The Elephant in the Living Room: Racism and Persistencies in the Everyday World

In connection with the ethnic, social, and spatial segregation of Baton Rouge, sublime to overt racism towards African-American populations is often observed by the white interviewees, which is equally strongly criticized and rejected by all interviewees, whereas this was remarkably not mentioned or addressed by a single syllable by the African-Americans interviewed. From the point of view of white respondents, for example, neighborhoods with a predominantly black population are often subject to a "negative perception" that is "purely race-based" (BR-01; Textbox 40, a), and the warning about certain districts "'don't go to that part of town!' that would probably be driven by that's where black people live and people associate crime with that" (BR-01; Textbox 40, a). This can lead in part to the point that – as one interviewee reports – the mere observation, by white people, of black people traversing through predominantly white neighborhoods for a harmless visit to relatives is deemed as 'suspicious', and this subsequently triggers impulses for isolation on the part of the white resident population (BR-16; Textbox 40, b).

This shows the central importance of the distinctions and interpretations (re)produced in World 3 in the process of ordering, assigning meaning, and interpreting the phenomena of everyday life and their construction to 'reality' in World 2 and how these affect World 1 of the spatial-physical initial substrates – here the different neighborhoods. Thus, it becomes clear that in the interpretation of the world of correspondingly black neighborhoods, whites are less experienced from the direct personal interaction, since as already mentioned, there are often hardly any goods or services to offer which could be of interest to white higher-privileged people. On the other hand, however, it is also (re)produced in media reporting, especially in connection with crime – such as in the national press of the New York Times or the video platform YouTube – and in this context is updated by the reader/viewer, thus constituting and manifesting a subliminal impulse towards – physically, spatially, and socially – segregation of the white sections of the population from black neighborhoods.

Textbox 40: The long arm of history – on the unequal treatment of African-American citizens, their neighborhoods, and the schooling of their children

a) "If you go [...] up Perkins road a little bit [...], like, there's a small neighborhood, I don't think there's anything wrong with it but I'm positive if you ask some people, like, they might have a negative perception of it and that is gonna be, like, purely race-based and I think that's probably if anybody said, 'don't go to that part of town!' that would probably be driven by that's where black people live and people associate crime with that. I mean, anytime you see somebody get shot, I feel like it's a black person, eight times out of ten. And I don't know like, where you are [North Florida Boulevard; remark of authors], I would never go where your hotel is and never go up there, not on purpose. I've driven up there. During the day I would be fine, at night I'd be afraid a cop would stop." BR-01

b) "If we if we walk down the side street here and we go over one block we are in a mostly African-American neighborhood, **it's only three blocks but it's very poor African-American people**. We're just a block away. And people that live in that neighborhood walk through our neighborhood to visit with relatives who live when you go four blocks this way. You're in almost all African-American neighborhood there. If you go this way it's mostly African-American. They live over this way, and so. **And the people that walk through our neighborhood to visit with relatives live in other parts of the surrounding area**, and **there's some folks in the neighborhood who would rather have fences or whatever to keep you from walking through there**, that's not going to happen." BR-16

8.2.3 Incorporation of the City of St. George as 'Highlighting Something that Everyone Already Saw Before'

The need for protection or the safeguarding or extension of existing privileges is particularly reflected in the process of the separation of the City of St. George from East Baton Rouge Parish (see Sect, 6.5), in which the interviewees see the ailing public school system and the efforts to improve it through public funding, while at the same time excluding financially weaker sections of the population, none of the interviewees living in the City of St. George, but without exception – if addressed – are following the process with great concern and criticism (see Textbox 41 and Textbox 42 for details).

From the perspective of the interviewees, the white middle class had successively abandoned the public school system over decades and instead of insisting on improving public schools, preferred to invest in private schools for their own children "because they didn't have children in the public schools they didn't care what happened to them" (BR-14; Textbox 41, a). Now, however, the cost of private schools increasingly exceeded the financial resources of some families in the white suburbs "but they won't send them to the public schools because now the public schools are largely black" (BR-14; Textbox 41, a). Accordingly, "they want all their tax money to stay in their community so they can improve their education. So, people are interested in better education, but they don't want everybody to be educated, they just want their kids to be educated" (BR-19; Textbox 41, b). Thus, although one would want a good public school education, one fears "having black kids in school because many of them come from homes where education has not been valued or attained in the past" (BR-19; Textbox 41, b). Students from African-American families with low symbolic capital are no less capable than those from white families, but they often lack "a foundation when they went to school and they don't have the support at home, they don't have the computers and the learning tools that so many other kids these days have" (BR-19; see Textbox 41, b).

Thus, the efforts to incorporate the City of St. George are understood by the interviewees of all stripes, who do not live in the City of St. George, as a denunciation of the last remaining fundamental common undertaking to enable the maximization of life chances for future generations of all those living in Baton Rouge – regardless of their origin – with the help of public funds.

Textbox 41: Own school system as a central concern

a) "Their purported reason is... **because the public schools are bad. They and people like them over the decades have moved away from the public schools and has spent their money on private schools and then because they didn't have children in the public schools they didn't care what happened to them.** And so this is part of their feeding into the downward gyre of the condition of the schools. All right. So rather than in my opinion rather than staying in the city and working to improve the schools. They, because they had the money, moved away and spent much more money on private schools than they would have put into the public schools. So now [...] they want to start their own school system. Because **allegedly I think part of it is that they can't afford to keep sending their kids to private school anymore but they won't send them to the public schools because now the public schools are largely black.**" BR-14

b) "St. George wants to break away because they don't want that – that's a middle class area of town and part of their taxes goes to pay for Public Schools in the low income parts of the city – **they want all their tax money to stay in their community so they can improve their education. So, people are interested in better education but they don't want everybody to be educated, they just want *their* kids to be educated.** And these people are not affluent enough to be able to pay to send their kids to Catholic or Private Schools. So **they want good public education and they fear having black kids in school because many of them come from homes where education has not been valued or attained in the past.** They don't have a good foundation when they go to school. They do poorly and then that makes them disruptive because, you know, they're not getting much out of it and they don't understand what's going on necessarily, **not that they're not capable but they just didn't have a foundation when they went to school and they don't have the support at home, they don't have the computers and the learning tools** that so many other kids these days have. So, it's difficult." BR-19

The efforts to incorporate the City of St. George are not only driven by the desire for an improved public school system. Many of the interviewees also see this as a culmination of the process of isolating a white middle class from the problems of the less privileged black population: "[T]hey [the supporters of the incorporation efforts; author's note] want to set up a city that doesn't have that many poor people in it so they don't have to deal with all of the issues that come with trying to help these poor people. So, it's race that is the major driver" (BR-16; Textbox 42, a). Other interviewees are also disappointed by the white middle class focusing on their own interests instead of a further development of Baton Rouge as a whole, which is understood as a common task: "a lot of it again you can definitely tie that back to racial issues and people are worried about themselves more than they're worried about everybody" (BR-01; Textbox 42, b). Thus, in a city with a majority of African-American citizens, as a community that "would be predominantly white, be much more affluent than Baton Rouge" (BR-18; Textbox 42, c). Correspondingly, the process of incorporating the City of St. George is "a source of great concern for those of us who want to see Baton Rouge improve" (BR-18; see also Textbox 42, c-e), but at the same time also underlines "something that everyone already saw before" and "kind of like brought that sort of animosity or that sort of division a little bit more closer to the surface" (BR-11; Textbox 42, f).

Textbox 42: St George – ethnic dimension of the conflict

a) "It's about economics and… and race. They uh... **there are a bunch of people who feel that too many poor people are in Baton Rouge,** in Baton Rouge Parish and **they want to set up a city that doesn't have that many poor people in itso they don't have to deal with all of the issues** that come with **trying to help these poor people. Soit's race that is the major driver**." BR-16

b) "It's in the southeast portion of the parish going out towards the more rural parish along I-10. But where there's still housing developments going on. And so, there are people who live out there trying to break away from the rest of the parish and be their own taxing districts so they can have their own schools. And **a lot of it again you can definitely tie that back to racial issues and people are worried about themselves more than they're worried about everybody**." BR-01

c) "Yeah I mean it would be disastrous for Baton Rouge, it would really be awful. It... it began with a desire to have more say over the school system [...]. The school board and others have tried to respond to this and tried to respond to those concerns. **There is... there is an undercurrent of racial issues regarding that, you know, the City of Baton Rouge is... 'Majority Minority'. The City of St. George would not be 'Majority Minority'. It would be predominantly white**, be much more affluent than Baton Rouge. So, it's... it's... it's a **source of great concern for those of us who want to see Baton Rouge improve**. This would be this would be a real blow if it's successful. And I if I had to bet I think it will be." BR-18

d) "**But there's an interest in keeping tax dollars or restricting tax dollars to, you know, kind of white middle class. That's very much racial, there's very much a racial dimension to that conflict.**" BR-19

e) "The person that is behind that to a large extent [...] is basically an American Fascist. [...] And so **this is basically a right wing separatist maneuver initiative. [...] It is a white flight people who don't want to be in Baton Rouge**. Yeah, they're just like the Trump voters, **they have legitimate grievances**, there's no question about that. But I think they're seriously **misguided**." BR-13
f) "That's sort of over the last couple of years has become such a contentious subject. **And it's only highlighted something that everyone already saw before.** You know, kind of like **brought that sort of animosity or that sort of division a little bit more closer to the surface**." BR-11

8.3 Wishing for Baton Rouge's Future

However, the detailed examination of a social conception of Baton Rouge by its residents did not only focus on currently experienced everyday worlds, but also addressed future perspectives, which dealt with the question of what potential (Sect. 8.3.1) and also what restrictions exist with regard to further development of Baton Rouge (Sect. 8.3.2) and what wishes the interviewed Baton Rouge residents have for their city (Sect. 8.3.3), which will be focused on in the following.

8.3.1 Of Baton Rouge's Potentials or Rather the Ignoration of Them

When asked about the potentials of their city, the interviewees, who are particularly familiar with the city from a professional or voluntary point of view, often primarily cite unused potentials and subject the city to a 'pessimistic view' with regard to possible potentials: in particular, Baton Rouge's geographical location is considered to be particularly favorable compared to other metropolitan areas. It is far enough away from the coast with its dangers of severe hurricanes, yet, its direct location on the Mississippi River and the Port of Greater Baton Rouge give it access to the Gulf of Mexico and allows it to be reached from there. Even if the city has been "successful in attracting some businesses that don't ever want to have to contend with another storm like Katrina", the possibilities in this respect have not yet been exhausted from the perspective of the interviewees: "But I think we're a city still of unfulfilled potential" (BR-18; Textbox 43, a). The Innovation Parks are also "an asset that is underutilized" (BR-17; Textbox 43, b). This is accompanied by the loss of young, talented, and well-educated people, who can only develop their own potential insufficiently or hardly at all in the rather slow and restrictive structures of Baton Rouge, thus leaving the city when the first opportunity arises: "it's smart, ambitious, young people, the first opportunity they leave: Dallas,

Atlanta, Houston, Chicago… and many of them never come back" (BR-18; Textbox 43, c). However, the city also has a high degree of localization and a sense of community that brings many exiled Baton Rouge residents back to the city: "But there is a phenomenon of when people reach a certain age where they start having children, they want to come back. […] So that phenomenon exists. So that speaks well of us in a lot of ways a sense of community and and family and those kinds of things" (BR-18; Textbox 43, c). Besides social potential, physical-spatial aspects are also mentioned. There is also a "really good wildlife", the weather is good except for the conditions in summer and "a wonderful beautiful river is right here, its terrific history and I'm right next to it every day" (BR-14; Textbox 43, d).

Textbox 43: Unfulfilled potentials – natural assets, sense of community and ignored talent

a) "We've been successful in **attracting some businesses that don't ever want to have to contend with another storm like Katrina.** So, Baton Rouge **geographically has that sort of advantage.** But I think **we're a city still of unfulfilled potential.**" BR-18

b) "And **it's going to go at a slower pace. But, you know, we have some really valuable assets that are not being taken advantage of yet, but they're still there. […] And I think it's possible, I think our location, Port, Mississippi, the Gulf of Mexico,** […], so there's two entities there's LSU innovative Innovation Park […] and it's an asset that is underutilized. They acquired or were given some buildings and a huge tract of land that unfortunately the university has not invested in but has a tremendous potential in innovative, you know, incubator new initiatives. […] **The Innovation Park on Florida,** there's talk between that group and this group here of maybe somehow joining, if they join, I think it's desirable that they do. And I think that is going to be a catalyst for things to happen." BR-17

c) "We have a lot of assets. And, you know, **it's smart, ambitious, young people, the first opportunity they leave: Dallas, Atlanta, Houston, Chicago… and many of them never come back.** But there is a phenomenon of **when people reach a certain age where they start having children, they want to come back** for their parents, the grandparents and that sort of **sense of family and community.** […] Those impulses don't resonate with me but I know they're important for a lot of people, which **that speaks highly of Baton Rouge.** That people would go away to a place of you know seemingly more opportunity but once they start having children and think about raising their kids in a particular community, they want to come home. So that phenomenon exists. So that speaks well of us in a lot of ways **a sense of community and and family and those kinds of things.**" BR-18

d) "So, yeah there are many aspects of the city, despite the fact that I don't like this place, there are a lot of things about it I do like, **there's a lot of really good wildlife**. It's amazing how many birds we have. Different kinds of birds. **The weather** is usually except for the summer good, **it's flat, who could ask for more?** There's **a wonderful beautiful river** right here, its **terrific history and I'm right next to it every day**, you know, and I see all the changes that are going on and all the stuff that happens with it. **I could be out downtown and go outside and it's quiet because there's nobody there**, you know." BR-14

8.3.2 Restrictions – Mindsettings and the Resistance of the Status Quo

The interviewees do not cite material aspects such as low financial resources, or similar, as central restrictions on the unfolding of its potential and decisive further development of Baton Rouge as a city, but rather patterns of thought and action such as the slowness with which things are approached and the backwardness that would result when compared vis-à-vis with other cities. The slowness of change and the perceived backwardness is based on the overwhelming dominance of conservative and restrictive thinking styles. Particularly in the perception of differences and references to cities in other states, corresponding comparisons are made, which state that "the most frustrating thing about it is the slow pace at which it is in. I just feel like that some of those things that were happening 20 years ago other places are just now getting here" (BR-11; Textbox 44, a). This 'mental slowness' "it sort of hinders Baton Rouge, you know, as a place that attract young new talented, you know, smart people" (BR-11; Textbox 44, a), "[b]ut that's typical of Baton Rouge" (BR-18, Textbox 44, b).

Textbox 44: Baton Rouge – the discovery of slowness

a) "Mainly I would say **the most frustrating thing about it is the slow pace at which it is in**, you know. [...] I mean, that's just like I have traveled a good bit, you know, I've seen a lot parts of the country that you know, have been to bigger cities and seen those change over the last decade or more. And **I just feel like that some of those things that were happening 20 years ago other places are just now getting here**, you know. And they could have been a lot further along. I think that in some ways **it sort of hinders Baton Rouge, you know, as a place that attract young new talented, you know, smart people**. That's one of the challenges." BR-11

b) "I mean, **everything takes twice as long in Baton Rouge as it would in any other community.** So, taken, you know, two decades plus for things to really start to now work on their own. But that's typical of Baton Rouge." BR-18

In this context, the conservative patterns of thought that were noted by interviewees as being critical and which hamper, if not contradict, innovation and change are also to be found in "The worldview tends to be conservative and inward looking rather than liberal and outward looking" (BR-14; see Textbox 45, a). In particular, the low degree of observation of an 'outside' and its contemplation is painful, because "they delude themselves about the stability because they don't look outward. They don't see how much things are changing. (BR-14; see Textbox 45, b-c). This is also favored by a low population fluctuation, because "we have people who never leave and are never exposed to other cities. And so, that very much affects I think our outlook on a lot of things" (BR-18; cf. Textbox 45, d). According to one respondent, citizens should "be more impatient about change here than they are. And I think, you know, that's one of my great frustrations with Baton Rouge is that we seem to settle for the status quo and, you know, we shouldn't" (BR-18; Textbox 45, d). Only this attitude of mind makes possible "the fact that we put up with things that would just be unacceptable in other cities" (BR-18; Textbox 45, g). One of the interviewees describes the mental situation of the city as follows: "There is this sort of thick resistant over-culture that resists innovation and change but in it is the opposite: it kind of makes people who are from here [...] want to be creative and outward looking, and most of them leave" (BR-14, e; cf. detailed Textbox 45, a-g). Here, too, the reference to a city as a community hybrid between the poles of urban–rural becomes clear in the context of a multivillage-metropolis, when "the kind of parochial attitudes and lack of sophistication and lack of progress" (BR-18; Textbox 45, g) is (re)produced as the predominant mindset, which in the movement between the poles of the forces of change or design and the forces of continuity or preservation primarily refers back to.

Textbox 45: Mindsettings – the insistence of persistence and the lack of progress

a) "A lot of people who live here are from here and they haven't been anywhere else. Have no desire to go anywhere else. I don't know what percentage of the population has been here for more than two generations but I suspect it's very high. [...] And so the **world view is very restricted. It's a very politically conservative place**. And so that world view **is also very constricted in my opinion**. It's very capitalist. So economically it's also very conservative. **But as also has this libertarian neoliberal bent which is based on the idea of what's mine is mine**. And everybody else can get away. Which also is a very constricting view. So, there isn't much of a mix of different cultures here. **There are different cultures but it's not mixed. The worldview tends to be conservative and inward looking rather than liberal and outward looking**. It's a very religious area with very strong Catholic and Protestant ties. Very, very, strong and more conservative than liberal on those." BR-14

b) “It's known, it's safe, it's comfortable, you know what to expect. **There's not a lot of new ideas and people that are upsetting because they have new ideas and they want to do things that we don't do**. It's.. it's… a comfortable stayed pleasant, more or less, place as an idea. **[…] I think they delude themselves about the stability because they don't look outward.** They don't see how much things are changing. And so, what they have is what they think is what they've always had and what they will always have. And **so they don't pay attention to the things that are going on.**“ BR-14

c) “**A conservative world view that looks to the past and maintains an ideal that doesn't really exist and in order to maintain that**. In order to maintain this illusion of a past that is glorious, you don't want to move too far from it. Because then it, you know, it's not glorious… it's way bad past. So what are we doing now, and I think it also is part of that thing about how most of the people from the state in the state are from this state, so they haven't been elsewhere, and they don't see that, that view. And that if it's the ‘Quality of Life’: ‘if it's work for my family for the last three generations what's wrong with it!?! There's nothing to change because it's just fine, thank you very much!’“ BR-14

d) “And so I have the perspective about boundaries as someone who grew up here, went away, and then came back. Which makes me a somewhat harsh critic of our city in many ways. And **it's the value of of leaving and seeing how other cities are successful in areas where we continue to lag**. […] I think this speaks so much to the mindset of people in Baton Rouge is the realities that we have one of the highest percentage of native born residents of any community in the country. […] **So, we have people who never leave and are never exposed to other cities. And so, that very much affects I think our outlook on a lot of things.** And so that's why you know like I said things take so much longer. **People should be more impatient about change here than they are**. And I think, you know, **that's one of my great frustrations with Baton Rouge is that we seem to settle for the status quo. And, you know, we shouldn't.**“ BR-18

e) “Well I think it is a fascinating place… in that there are, like, the thing about the pudding. **There is this sort of thick resistant over-culture that resists innovation and change but in it is the opposite: it kind of makes people who are from here […] want to be creative and outward looking, and most of them leave.** But while they're here they add a little bit more energy to the system. They'll never had enough to come in. But it's that thick sludgy layer that makes this other more energetic thing happen.“ BR-14

f) “I like to study what's called a southern studies and Baton Rouge is interesting because it's a very conservative. I mean, like a lot of the South is a very conservative culture. But **what that means is they're very local populations, very averse to regulation.**“ BR-20

g) "Well, I felt very positive about the work that I was doing. But I felt… all within a week or two everything about Baton Rouge that came back to me, you know, **the kind of parochial attitudes and lack of sophistication and lack of progress and, you know, the fact that we put up with things that would just be unacceptable in other cities.**" BR-18

8.3.3 Desirable Prospects – Baton Rouge as 'a City that Dreams Bigger'

The wishes of some in Baton Rouge for 'their' Baton Rouge of the future are often aimed at the realization of inverse compartments of physical-spatial as well as socio-economic urban development – such as an improvement of the infrastructure (both in terms of transportation and building structure) especially in the less privileged African-American neighborhoods, a strengthening of social and societal (co)life through festivals, improved access to healthier food, the promotion of tourism, and an increasing convergence of basic mental attitudes towards the poles of change and development of shaping.

Thus, when asked what the respondents imagine for the future of Baton Rouge, the first thing that is expressed is the desire for development of North Baton Rouge in economic, infrastructural, and building terms. It is important "to develop North Baton Rouge so it's not so blighted so it's economically better" (BR-17; Textbox 46, a). In particular, the interviewees living in the district make a comparison with other prospering districts and use them as a benchmark: "I want to see that bustling with a lot of Black Businesses and things like that… I want to see it as prosperous as the LSU side of town" (BR-06; Textbox 46, b). The focus is less on the demands for public funds and more on concentrating on one's own resources and efforts: "You can buy up an area in a place like Baton Rouge and make it something special, make it something great. You don't have to move out into the suburbs and enhance that area with your dollar with your efforts" (BR-06; Textbox 46, c).

Textbox 46: Development North Baton Rouge

a) "The mayor right now is trying to develop **North Baton Rouge so it's not so blighted so it's economically better.**" BR-17

b) "I want to see this strip [North Baton Rouge; remark of authors], how it is right over the hump, **I want to see that bustling with a lot of Black Businesses** and things like that... **I want to see it as prosperous as the LSU side of town.** If not, making progress towards it so when **I see a lot of businesses, a lot of jobs, a lot of opportunities for the people and of that sort**." BR-06

c) "The main constraints are poverty, of course, this area [North Baton Rouge; remark of authors] right now isn't the best financially and economically so it's hard to pull yourself from out of that hole once you've been in it, especially for generations. So, it's just something that's the main biggest goal and effort. [...] Just because you graduate and made it out doesn't mean you have to move into the suburbs. **You can buy up an area in a place like Baton Rouge and make it something special, make it something great. You don't have to move out into the suburbs** and enhance that area with your dollar with your efforts. **You could do the same thing here and find some that doesn't look as pretty into something beautiful.**" BR-06

A further inverse compartment of the development of Baton Rouge includes the establishment and expansion of the only rudimentary existing public transportation system and an adaptation of this to the various mobility needs of the Baton Rouge citizens as an attractive alternative to motorized individual transportation or as a combination of different means of transportation. This would correspond to a series of future challenges, which, in addition to personal advantages, would also include the requirements of climate change (cf. Textbox 47, a–b; Fig. 8.2). In addition, it points to a social development in the United States as well, in which the focus on motorized private transportation is tending to decline, especially among younger people (see, for example, Sivak and Schoettle 2016), making alternative forms of mobility more attractive.

Textbox 47: A functioning and attractive mass transport system

a) "So what I would change about Baton Rouge… is probably infrastructure, like, transportation that would be helpful. I mean, it's not like we could have a subway system or anything like that but I mean **if we had better bus system or something.** [...] It takes me less than 10 min to get to work, it's like a five or six mile drive, but I have to drive. There is no safe way for me to ride a bicycle to get to work, which is stupid. That's just insane. And, there's no opportunity for me to ride my bike to a bus stop and then throw my bike on the front of a bus and then you know **have a multi-modal transportation thing**.“ BR-01

b) "Yeah, I think **transportation and mass transit and dealing with climate change**… and an understanding that this is very serious. This is an existential threat to the entire life on the planet and that people in Baton Rouge have a particular responsibility because ExxonMobil has, I believe, still their largest plant here in Baton Rouge. And so people in Louisiana have to step up and confront the reality of the oil companies and the hydrocarbon crisis.“ BR-13

Fig. 8.2 An image that may be considered symptomatic for users of the Baton Rouge bus system: OUT OF SERVICE. Own photograph

In addition to access to healthier food, the desire for an "ability to mix more as a society" (BR-20; Textbox 48, a) is also evident with regard to social life, in the context of which there is also the desire for a socially functioning downtown area that is not only built as a backdrop but is also visited and actively used by people in Baton Rouge. In this context, there is also the desire for a more comprehensive festival and event culture, which should bring tourists and businesses into the city, "but at the same time still have this sort of small town. Not a small town but like smaller cities sort of quality to it" (BR-08; Textbox 48, b), so that in this context, too, the difficult degree between metropolitan growth on the one hand, and maintaining a small town atmosphere on the other, becomes clear. At the same time, in view of the rather modest tourism volume for a capital city, to date, there is also a desire for an increase in the number of visitors and overnight guests and, associated with this, a desire for a metropolitan development of the capital with supra-local significance, but the existing infrastructure has so far been inadequate and requires fundamental adaptation (cf. Textbox 48, c–d).

Textbox 48: The Multivillage-Metropolis – maintaining a 'sort of small town' as opposed to expanding into a metropolis

a) "**Access to better food would be nice**. We have a lot of food, but it's not very healthy or very good food. [...] I think we have an incredible amount of natural resources, not just the oil and the Mississippi River, but the different industries that can be used from harvesting... in the estuary and the coastal environments that we don't... I mean, **we take the natural resources and then we fry it**, you know, and **there's so many other things that we can do with it**. And then the ability to **mix more as a society**, you know, is one thing that would be... Oh, to have a **proper downtown** to be fierce, everybody says that's where you start. [...] That's not where you're going to make the change, but to have a proper downtown culture. I should say **not just the physical environment, but the people living there** and not just coming in for fireworks on a holiday every night." BR-20

b) "But I would think it'd be nice if we had some more, maybe **even more festivals** to be more around Baton Rouge or you know **just certain attractions** that I think would bring in more tourists and therefore more businesses, **but at the same time still have this sort of small town**. Not a small town but like smaller cities sort of quality to it." BR-08

c) "So, I think the city will grow and it will develop more, and **since this is a capital, it should grow more**. Yeah, except like right now we have a very small airport. So it's the airport needs to be bigger, they need to make it bigger as more flights need to come. But for that even you need that kind of a flow. So, **if you create more tourism, more work here, so people will more will travel**." BR-04

d) "I guess **more tourism**. [...] **We don't have a whole lot of things, we have things around us**. But we don't have too many things for our city, at least.

Because **if you go to other capital cities in other states, like, their cities they have a lot of stuff to do with the tourists there.**" BR-05

The interviewees also consider it a central concern to loosen-up the rather conservative basic attitude towards more social liberalism, tolerance, and alternative thinking, which will make the city attractive for young creative people and well-educated specialists in Baton Rouge. For example, one's own self-image is a hindrance to the development of future perspectives and one wishes for "a city that could have a larger… a bigger dream for what it could be and not simply to accept the way things are and accept that things are done this way or a certain way because they've always been that way. That's a horrible way to think" (BR-18; Textbox 49, a). However, the pronounced desire for change in the face of thought patterns constructed as persistent and resistant is accompanied by a certain resignation and a fatalistic approach: "[…] I would like to see be a more politically tolerant and liberal but that's not going to happen" (BR-19; Textbox 49, b). Changes towards a more ecologically sustainable society are also extremely unlikely: "Highly unlikely but that it could become a center of ecologically responsible stewardship of the urban environment" (BR-13; Textbox 49, c), because "[t]he people in Baton Rouge they're not the most conservative people in the entire United States but they're certainly contenders for that title." (BR-13; Textbox 49, c).

Textbox 49: Paradigm shift, politically more tolerant and liberal

a) "For the future? I would say that, you know, **a city that dreams bigger**, **that does not settle simply for how things are because that's the way they've always been**. So, **it would be wonderful to think that more young people, creative type people would stay or come to Baton Rouge**. […] Our economy is doing well... but for the creative types… I don't know if they're outside of Baton Rouge, necessarily, what would be the advantage of them coming here. But if we can get more young people to stay and not leave but they'll only stay if they're economic opportunities for them, I mean, that's just a fact. And solving that riddle is not easy but **I would say a city that could have a larger...a bigger dream for what it could beand not simply to accept the way things are** and accept that things are done this way or a certain way because they've always been that way. That's a horrible way to think. And so that would be that would be I would hope my vision for that marriage." BR-18
b) "I don't think there's a Magic Wand big enough to make the city the city I would like it to be. I would like... I would like to see be a **more politically tolerant and liberal but that's not going to happen**." BR-19
c) "Highly unlikely but that **it could become a center of ecological responsible stewardship of the urban environment**. That would include mass transit, that would include bike lanes, that would include people... highly unlikely realizing

that yards and their houses are ecologically completely indefensible. They would let trees grow as they grow in their yards to produce oxygen. They would start stop using chemicals on their yards. **The people in Baton Rouge they're not the most conservative people in the entire United States but they're certainly contenders for that title.**" BR-13

8.4 Baton Rouge as Living Environment: "The Rice Pudding with the Sugar-Sweet Raisins in It" – An Interim Conclusion on the 'Multivillage-Metropolis'

In the process of ordering, assigning meaning, and interpreting the phenomena of everyday life and their construction of 'reality' in and around Baton Rouge, a fundamental importance of strong social references becomes apparent, which are particularly effective on an emotional level and tend to push cognitive aspects into the background. In addition to job offers, it is the relationships in the local social networks around partners, family, friends, and work colleagues, the generally perceived friendliness and helpfulness of the Baton Rouge people and the 'big-city small-town' atmosphere that are cited as decisive for the choice of Baton Rouge as the center of life, while the awareness on a cognitive level of a more polluted and health-endangering environment or a strong inequality of life chances is receding. A fatalistic approach is often chosen here, in that grievances in this context are constructed as 'familiar evils' and accepted as 'objectively given' and 'normal', which on the other hand also represents the decomplexation and reduction of the world that makes a routinized construction of everyday life possible. For the innovative 'Other' and the potential change associated with it is often interpreted as a threat that endangers passed down patterns of vision and interpretation and threatens to call into question routinized construction practices of everyday life as the 'uncertain new', often causing skepticism and defensive reactions, which is also framed as one of the central impairments in the further development and continuation of the city – especially with regard to ethnic issues, social and spatial segregation, and the resulting unequal distribution of life chances, the economic future, disinvestment in infrastructure and building stock, dealing with further differentiating lifestyles and the shameful education of future generations of young Baton Rouge individuals, to name but a few areas.

This gives the impression, both from a social-community and physical-spatial point of view (orientation towards small neighborhoods instead of a dominant center; wilderness and rusticity in the middle of the city; etc.), not to be moving in a metropolitan area, because neither metropolitan distance standards seem to apply nor does an innovative power radiating from urban centers or metropolises seem to be effective. Accordingly, dichotomies between the urban and the rural dissolve and the compartments of the spatial pastiche thus evolve into a multivillage-metropolis, formed by units that are strongly influenced by the community and the interior. The physical space thus created allows little impressions of unambiguity beyond the small social-spatial compartments, but rather challenges the inherited patterns of vision and interpretation of urbanized spaces from a social, but also physical-spatial perspective. In Baton Rouge, for example, stereotypical visual expectations of landscape and postmodern aesthetics of reinterpreting buildings that have fallen into disuse or are in an advanced state of decay are combined in a very small space. A sight that is often completely normalized for those who have formed their native-normal landscape there.

Looking to the future, Baton Rouge is often framed as the city that is unable to make decisive use of the opportunities and potential it offers – whether in terms of its physical-spatial conditions, such as its favorable geographical location and access to the Gulf of Mexico, or in social-community terms, particularly through the local higher education institutions, its talented graduates and its cultural diversity – and stands in its own way through its resistance to change. Thus the city, which on the one hand often falls short of the expectations and hopes of its inhabitants, but at the same time is valued by (analogous) social networks and individual talent, resembles a rice pudding, in which the rather inconspicuous and viscous mass also always has something special, surprising, pleasant about it: "So I think of Baton Rouge as a rice pudding with raisins. Most of it is kind of bland and a little lumpy and it kind of looks the same. But every once in a while, you get this really sweet burst of stuff. Because there are people in town that are actively engaged in the arts and creating things and music. And the talent in this city is amazing – and ignored. It is amazing how much the talent in this town is ignored, to be successful people have to leave here. It's crazy. But, you know, it's an interesting thing about the city: If you look at, you have to dig out places, you have to find the raisin – if you like raisins. If you don't like raisins you have the sweet treat of your choice, chocolate chips" (BR-14).

9 Conclusions - of Developments, Reflections, and Perspectives on the 'Multivillage-Metropolis'

The objectives of the present study were named in the introductory chapter and are discussed in more detail in this conclusion:

1. The operationalization and viability of the approach of 'neopragmatic landscape research' and the possibilities of further developing the approach of 'urban landscape biographies' are discussed in Sect. 9.2.
2. A reflection and evaluation of the developmental strands, events, and supra-local influences on the biography of Baton Rouge is given in Sect. 9.1.
3. The media presentation of Baton Rouge on the Internet - as an evolving component of a landscape biography - is also integrated into this reflection (also Sect. 9.1).
4. Reflections on life in Baton Rouge in its attributions of meaning and significance are also made in Sect. 9.1. Here, specifics of life in Baton Rouge are presented - certainly with an ironic undertone - that were quasi imposed by conversations about and in Baton Rouge, but also by participant observation.

In addition to these explicitly named goals, there is a section of the conclusion (Sect. 9.3), in which we recognize, certainly also due to our view from the outside, potentials in Baton Rouge, which may elude the view of everyday life in Baton Rouge.

9.1 Conclusion I: Developments and Interpretations in and of Baton Rouge

Baton Rouge has undergone a fundamental transformation in the course of its development with three central poles of development: from a pre-modern agricultural settlement, to a Mississippi River trade-related Southern State town, later with a capital function (the

O. Kühne and C. Jenal, *The Multivillage-Metropolis Baton Rouge*, RaumFragen: Stadt – Region – Landschaft, https://doi.org/10.1007/978-3-658-30716-5_9

first pole of Baton Rouge's development). Economic modernization, on the other hand, was accelerated with the arrival of the petrochemical industry at the beginning of the twentieth century (as the second pole of Baton Rouge's development). The third pole of development next to the government and the petrochemical industry is LSU, which has been located in the south of the city since the 1920s. Beyond these poles, Baton Rouge was unable to bundle any - especially - economic development impulses, the consequence of which was that the downtown area, as the center of social life, lagged behind those of comparable settlements in terms of structure and function in the course of modernization. In other words, to this day it cannot be located very far from the pole of urbanity in the spectrum of urban–rural hybridity. Efforts to revitalize the downtown area are particularly evident in public uses, and recently urban living has also been gaining in importance, although major attractiveness factors, such as nightlife districts with a string of contemporary restaurants and bars, which are also gaining in importance again in the United States, are still rather underrepresented. Although the downtown area has been able to develop into a spot of attractive gastronomy, it remains a center that is closer to the target audience of the white population with a higher level of symbolic capital. In addition, gastronomic centers have also been established on Government Street and in the area of the Interstate 10 overpass across Perkins Road, east of the LSU campus.

In the suburbs of the inner ring and in their transition area to the downtown area, modest (postmodern) tendencies of building and social re-urbanization can be observed. However, postmodern community tendencies are more evident in the shopping infrastructures beyond the downtown area than in the downtown area, which are, however, subject to rapid wear and tear (this applies less to material wear and tear than to obsolescence due to changing customer preferences). It is precisely with these shopping infrastructures that the change from the needs of the modern age, according to which form is subordinated to function (clear structuring, e.g. the parking area for cars clearly separated from the likewise clearly structured shopping area; e.g. Renaissance Park), closed, inward-looking shopping malls of the 'grid postmodernism' (starting with the Bon Marché Mall and reaching its climax at the Louisiana Mall), and a historicizing mall (Towne Center at Cedar Lodge) that partially simulates a Mediterranean city center and is integrated into its surroundings with sidewalks as an expression of the second phase of postmodern settlement development, which is more concerned with networking than demarcation. The revitalization efforts around downtown Baton Rouge can also be placed in this context.

The downtown of Baton Rouge can currently be understood as a focus of (economic, structural, social) development. It thus shows less of a state of 'spatial hysteresis' as it was before the efforts to revitalize it in the wake of the economic stagnation of the 1980s. Also as a result of its decentralized location and the material barriers surrounding it (the Mississippi River to the west, the two Interstates to the south and east), concentrations of economic, cultural, and social life have developed alongside concentrations of economic, cultural, and social life – linearly along Government Street, small-scale along the overpass of Interstate 10 over Perkins Road, and more economically oriented along the southern Airline Highway and Siegen Lane, as well as economically-pointing shopping malls. The result is a postmodern pastiche of space that - with two exceptions that will be discussed later - is characterized far less by dichotomies than by hybridities and transitions. What is also remarkable about the pastiche formation of Baton Rouge is that even re-urbanization processes are decentralized to a large extent. As a result, dense residential areas are increasingly being developed beyond the classic downtown area. A development that goes beyond that of the URFSURBs: re-urbanization is also taking place in the surroundings of the traditional city center, but mainly in decentralized locations. It's open to interpretation: Baton Rouge remains true to its development logic, following the pattern of the multivillage-metropolis, a multi-urbanizing micro-core metropolis is emerging. Thus, Baton Rouge is also a 'multivillage-metropolis' in terms of the way people live: although the metropolitan region has almost 600,000 inhabitants, it is mostly divided into neighborhoods of manageable size, and the downtown area also has a rather manageable degree of central metropolitanism (see especially Chap. 6). Here, an interference between the physical foundations and the social formation of community is also evident: The developed area of Baton Rouge is crisscrossed by numerous bodies of water, which with their flood zones divide Baton Rouge into numerous smaller community units, the multivillages. A significant influence on the impression of not being in a metropolitan area is provided by the people of Baton Rouge themselves (and indeed all those we have met, whether as interviewees, in a professional or private context, or in several other contexts): their habitus does not correspond to that which Georg Simmel (1950) characterizes as specifically 'metropolitan'. Distance norms do not seem to apply here in the big city: everywhere (including complete strangers) people are involved in sometimes excessive conversations (beyond small talk), the person opposite is perceived less as a role bearer than as a person, a purposeful, rational approach to the world is rather unusual. In short, a habitus that Simmel would rather locate in the village. Due to the large number of manageable neighborhoods with a metropolitan population (and administrative importance as well as scientific presence), the term 'multivillage-metropolis' is justified.

LSU has a significant presence in Baton Rouge. This presence reaches far beyond an institution for education and science: through its sporting presence in the form of the Tigers (especially football, but also baseball) or the design of the campus and its surroundings, it forms a condensation point of local if not regional identification, but also integrating as well as demarcating image formation, as shown by the study of the TripAdvisor rankings, but also the 10 reasons why Baton Rouge is rejected by the population of New Orleans. LSU forms a condensation core of a place of academic development in a landscape garden and sports-related diversion, which is surrounded by sublime to picturesque (Southern States) landscapes, especially in the south, east, and west. The counterpoint is the north of Baton Rouge, with a high crime rate, violence, a lack of prospects with dilapidated buildings, insufficient access to green infrastructure, and also - in connection with this - increased environmental pollution. In the triangle between Interstate 110, the Airline Highway and Florida Street/Boulevard, there is also a low density of high schools and food outlets. In conjunction with a rather poorly developed network of footpaths and local public transport, this can be described as a triangle of reduced life chances. The southern border of the triangle, marked by Florida Street/ Boulevard, stands out as particularly striking, so that it can be called a 'social equator'. In this triangle of changed life chances, the problem of the loss of ligatures, in connection with the low granting of options by society, is particularly evident: traditional ties such as church and family reduce their binding effect, while at the same time an increase in options is systematically made more difficult for the resident population, especially through the difficulty of participating in higher education (but also in traffic and access to healthy food; in the sense of Dahrendorf 1979b). So, in Baton Rouge, life chances are extremely dependent on ethnicity, symbolic capital, and suburban origin. Participation opportunities are primarily limited to their formal dimension, i.e. the entitlement in principle to achieve a qualified occupation, but not to the substantial dimension, because people in disadvantaged parts of the city are not systematically enabled to achieve higher education (in the sense of Nussbaum 2009; Rawls 2001).

Baton Rouge is a car city, but very important - not a car-friendly city. Baton Rouge today is not the result of a 'grand design' or the consequences of the 'car-friendly town' concept. The orientation towards motor vehicles was more of a side effect of an incrementalist muddling through in response to short-term challenges, the implementation of a lowest common denominator that neither strived for urban planning ideals nor for the spatial manifestation of reform-oriented political ideas (such as the sustainable city). The result of these pragmatic, case-by-case, space-shaping decisions is an essential contribution to the 'chambers' of Baton Rouge, in which a more-or-less consistent idea of the potential, challenges, and possibilities of Baton Rouge was missing. Although such strategies have begun to be developed in recent years, these physical spaces tend to persist, especially where the economic, social, or political pressure for change remains restrained. And this is Baton Rouge in general. The extent to which it is possible to develop an emergence of the living, structural, and functional chambers of Baton Rouge also depends on the degree of political fragmentation. So far, this has been rather low and was mainly due to the Mississippi River, which separated the two Parishes from Baton Rouge. But with the incorporation of the Central City and the probably upcoming St. George, the development of brackets around the already existing multivillage nature of Baton Rouge is at least made more difficult but all within the parish. A particular challenge in this respect will probably be the question of how to deal with floods caused by heavy precipitation, the intensity of which is likely to increase rather than decrease as a result of anthropogenically intensified climate change, not only in East Baton Rouge Parish, but also for the neighboring Parishes.

In the synthetic view of the community development of Baton Rouge it can be noted: Baton Rouge has not developed ultra-modern spatial structures in many areas. It has neither a dominant center nor a strong separation of functions, and the infrastructure (especially transport) has always remained in a comparatively undeveloped state and concentrated on local traffic (apart from a few supraregional roads). Modernity's goal of controlling nature (at least to a large extent) has never been achieved: the risk of flooding due to heavy precipitation events (which cannot flow down the Mississippi) is considerable. The ecological burden on the inhabitants of Baton Rouge - a consequence of the petrochemical industry as well as the early modern close spatial interlocking of industry and residences - has always remained high. The development from pre-modernity/ early modernity - if this concept is followed - led in large parts without major detours, for example via ultra-modern urban development (an exception is certainly some office buildings in the CBD) directly into post-modernity, which can be found in particular in the current shopping malls, the modest tendencies of re-urbanization including URFSURBS, but also in the historical architectural styles (preferably with borrowings from the French past) of the newer suburban settlements. In other parts, the many different Baton Rouges have retained a pre-modern village charm. But this 'simultaneity of the non-simultaneous' can also be understood as an element of postmodern settlement development.

At the same time, a spatial pattern of preference for compartments in the Baton Rouge spatial pastiche in relation to different degrees of natural cultural hybridity and urban–rural hybridity is also evident: people with a higher endowment of symbolic capital prefer (or redesign) compartments that are more inclined towards the poles: such that the pole of urbanity and stronger anthropogenic transformation, as with the construction of lofts in the downtown and the modest URFSURBanization in the adjacent inner-ring suburbs shows. Conversely, in the more rural and the pole of naturalness-oriented segments of the Baton Rouge spatial pastiche. Persons with a low endowment of symbolic capital continue to be more indifferent to urban–rural and natural-cultural hybridity. Segments, in other words, in which no symbolic capital is used to bring them closer to the respective poles of hybridity.

The city also reveals a selectivity of inclusion, combined with a partial resistance to supra-regional phenomena, specifically linked to modernization processes, with regard to the different influences upon the biography of Baton Rouge from regional, national, and global dimensions. This ranges from a still rather marginal integration into the transport system (especially with regard to rail and air transport), the limited tendency to structure the development of communities politically and administratively, the lack of national or even international governance mechanisms, the tendency of the inhabitants not to shed a rather 'rural habitus', even though Baton Rouge is involved in globalized production and competition processes through its petrochemical industry alone, as well as through its universities in global networks of research and teaching, and is present in the blues and hip-hop scene at least nationally, but also internationally to some extent.

The spatial structures, functions and processes, interpretations and evaluations that could be extracted from the evaluation of the available literature on Baton Rouge, the interviews conducted, the evaluations of internet homepages, internet videos, articles from newspapers, phenomenological walks, can be summarized in a cartographic representation (see Fig. 9.1: Lines and bubbles that mean Baton Rouge: a space pastiche.)

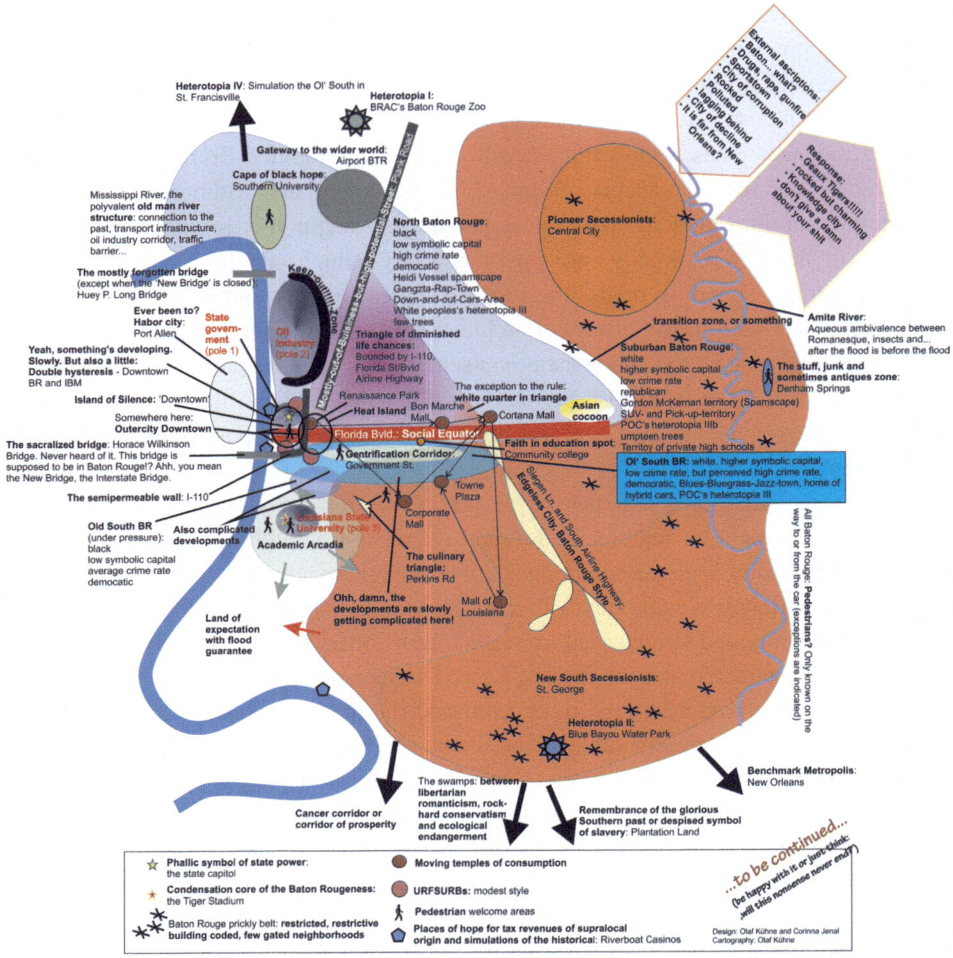

Fig. 9.1 Lines and bubbles signifying Baton Rouge: a room pastiche

9.2 Conclusion II: Reflections on the Neopragmatic Approach and the Concept of 'Urban–Rural Landscape Biographies'

The present work followed a double - mutually feedbacking - theoretical framework: 'neo-pragmatism' and the approach of 'landscape biographies'. The neopragmatic approach describes a theoretical (and related methodological and interpretative) multiperspectivity. Particularly, positivist, constructivist, and scientific theoretical perspectives that go beyond representationalism were adopted. A positivistic perspective was expressed in the mapping of material objects, such as floor heights, land use, the measurement of the distribution of air temperature and humidity and their synthesis to the equivalent temperature, but also the walkability of footpaths. But here, too, a constructivist framework is necessary: on the one hand, the definition of what is counted and measured follows professional considerations, i.e.

the patterns of expert special knowledge. On the other hand, prefabricated observation categories are based on the material world (in this case, for the sake of comparability, following Brill's categorization in 1963, especially with regard to land use and floor heights) or they are obtained inductively, such as the categorization of walkability after exploratory inspections. Other research was based on a purely constructivist perspective, such as qualitative interviews and the analysis and interpretation of lyrics on the subject of Baton Rouge. Sometimes - from a basic constructivist perspective - quantitative methods were also used (for example in the evaluation of Internet images and videos), a method that is usually assigned to a positivist approach. Here, however, a quantitative evaluation was carried out with the intention of determining thematic focuses and forms of presentation as a basis for interpreting and evaluating the media construction of Baton Rouge. We have used a more-than-representational approach with the phenomenological walks. Here, our experience of the investigated streets (sections) was the focus of our interest as a form of subjective access to Baton Rouge. An approach that fits into a synthetic investigation of a space in a meaningful way, not only because of its relationship in the history of ideas to social constructivism, but also because of its focus on the subject as a link between the social and material world. How the empirical approaches fit into the neopragmatic framework is shown in Fig. 9.2 - with reference to Fig. 2.2 and 2.3.

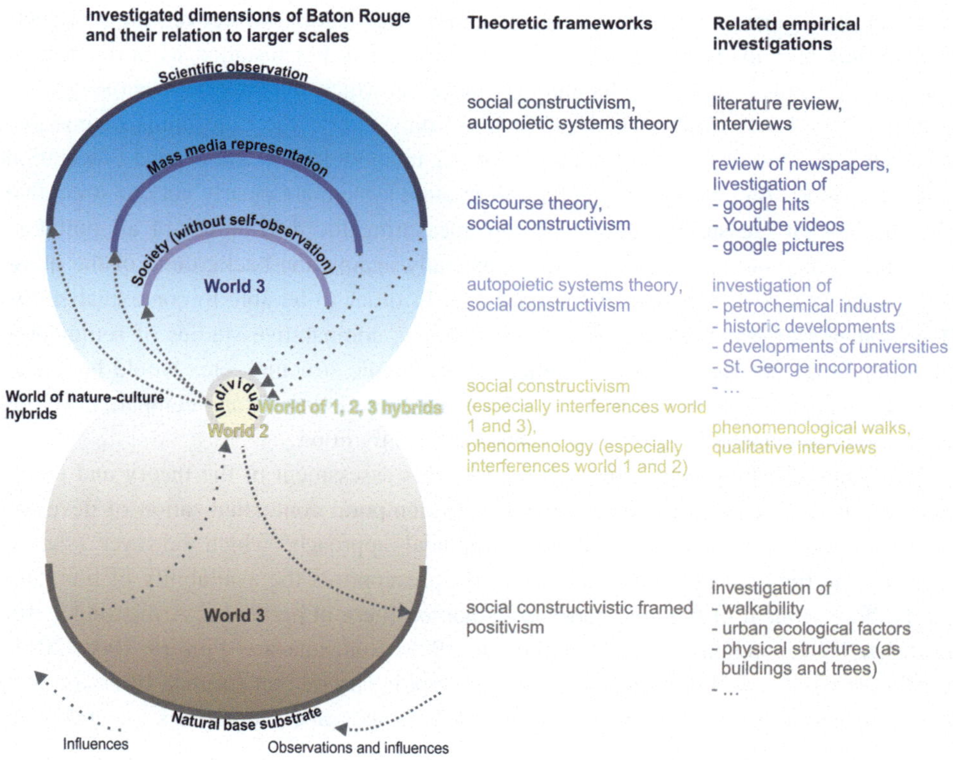

Fig. 9.2 The 'urban landscape biography' in 2.2 and 2.3, supplemented by methodological approaches to be assigned to the individual dimensions

The strengths of the approach can thus be understood as the theoretical as well as methodological diversity of perspectives of the neopragmatic approach, which enables a differentiated spatial synthesis that goes far beyond the possibilities of monotheoretical or methodological approaches and is thus able to create a suitable basis for interpretation of diverse and complex processes and interactions, especially with regard to the levels of the social, the individual, and the material. The overall social constructivist interpretation forms a framework that neither ignores nor smooths out methodological and theoretical discontinuities and incommensurabilities but accepts and reflects them. The present study on Baton Rouge also shows that a neopragmatic regional geography must not only be a processing of already generated research results, especially spatially, but also of an abstract kind, one that is following the logic of the regional synthesis, such that one's own empirical studies can be carried out, these can also be quite innovative, especially in their synopsis, as here regarding the spatial unequal distribution of life chances.

But this also results in weaknesses in the approach, of which neopragmatic-regional-geographical researchers should be aware: even if the synthetic synopsis suggests certain empirical studies with a thematic focus, these - due to their function of forming a building block of the synthetic study - fall short of focused studies in terms of scope and depth of investigation. This 'brevity' of the individual empirical building blocks in itself allows a limited horizon of explanation, which can be pragmatically justified by the fact that this is just one of many elements under investigation. Moreover, many aspects remain only marginally represented or even unexamined. For instance, as in the case of Baton Rouge, this is due to the incomplete literature, which can only be selectively compensated for by our own research. This is not only due to the time limits of possible research stays, but also to the limitations of our methodological knowledge (vegetation science or modelling with a geographic information system are clearly not our strengths) and the lack of measuring equipment (e.g. measuring the distribution of air pollutant concentrations would be very revealing, especially against the background of the question of the spatial distribution of life chances). In order to be able to contextualize the developments of Baton Rouge even more strongly, comparative studies of settlements of similar size, but also political significance, economic structure, etc., would be useful beyond quantifications to rankings, in order to be able to classify and compare development logics, especially in the landscape biographical tradition.

The landscape biography approach fits into this assessment of the theory and methodology. It adds a focus on development and a temporal contextualization of developments to the neopragmatic-regional-geographical approach, which however (should there not be time for extensive archival work) presupposes the availability of literature on the development of the area under investigation (here in particular: Armstrong 2010; Carleton 1981; Douglas 1955; Draughon Jr. 1998; Rodrigue and Phillips 2008, 2011; Ruffin 2006), if possible even a regional geographic survey that forms a basis for spatial comparisons (here in particular: Brill 1963). Landscape biographies are linked to individual (spatial) biographies. These individual biographies form a very essential aspect of spatial biographies, they also reveal interpretations of spatial development, which can be made accessible especially through thematically focused biographical interviews. Researchers (but also everyone else, such as recipients of research) should be aware of this: biography always means an interpretation of developments from retrospect. Ex post justifications dominate these biographical interpretations.

Reflecting these specifics and challenges of the approach we have chosen, we have reason to believe that we have created a professionally intersubjectively comprehensible overall picture of Baton Rouge and its development. This was a culminating result of feedback on presentations, discussions of partial aspects and individual methodological and theoretical approaches as well as the synthesis with colleagues on site. The view from the outside was also seen as helpful, which for outsiders included aspects of the development of Baton Rouge that quickly became apparent to outsiders, but which were not questioned in the everyday attention of locals, such as the statement "yes, you're right, we're a car city - now that you say so". In some cases, however, the assessment also diverged. For example, our view - possibly an outcome of the influence of European urban traditions - of the downtown area of Baton Rouge was clearly more deficiency-oriented than that of colleagues who focused more strongly on developments in the last two decades or so. We have endeavored to take both assessments into account in the analysis.

The multi-method approach chosen here, with limited partial investigations, is particularly suitable for regional geographic research when a systematic processing of a literature and database is not possible due to a lack of breadth. Depending on the available literature and databases and the time, methodological and technical (i.e. to a large extent also financial) restrictions, it therefore remains exploratory or at least partially exploratory (if essential aspects have not already been investigated). The result of this study (more generally: any such investigation) is potentially a scientifically based biography of Baton Rouge among a limited amount of biographies. It is influenced by the subject matter, but also by the social, spatial, and cultural background of the investigators. Other biographies could well look different. They could be compiled from other disciplinary perspectives, and other focuses would also be possible from the geographical-political-scientific-sociological perspective (e.g. more strongly related to economic contexts). Also, a different interpretation of the developments and spatial characteristics from a (neo-)Marxist or conservative perspective instead of the (liberal) life chances approach chosen here would be more than likely. But biographies of Baton Rouge do not have to be produced in scientific frameworks alone. Popular narratives, although theoretically and methodologically unfounded, fulfil a meaningful function in 'Baton Rougeness'.

The presented biography of Baton Rouge does not represent the 'one truth' about Baton Rouge, but rather a theoretically and methodically founded interpretation of spatial-historical developments.

9.3 Conclusion III: The View from Outside - High Potential Multivillage-Metropolis

In the following, a vision is presented that is not intended to be a utopia of Baton Rouge, because utopias tend to be all-inclusive (Dahrendorf 1968b). The vision presented here is therefore not a self-contained draft. It is intended to present possibilities of development that are geared to giving as many people as possible the greatest possible opportunities in life (life chances; Dahrendorf 1979b). Such a vision ties in with the previous spatial biographies of Baton Rouge. Accordingly, the result of the urban development of incrementalist muddling through should be accepted and not, above all, be assessed negatively. Finally, Baton Rouge has been spared the persistent side effects of the functional and car-friendly city model. Baton Rouge's small-chambered and neighborhood-centered approach, in short the quality of the multivillage-metropolis, has great potential for sustainable urban development - provided that urban policy and urban planning, together with citizens, strengthen certain (infra-)structures: for example, in relation to the 'city of short distances', which in turn presupposes the existence of continuous paths for pedestrians and cyclists. Often only a repair or a partial rededication of structures (such as protective verges to cycle paths) is necessary. This multivillage structure of Baton Rouge also has another advantage: frequently used open spaces that can be developed not only for pedestrians and cyclists, but also for green spaces, especially in and around the 'triangle of diminished life chances', community projects (for example, with a thematic reference to gardening or repairing things) could contribute early to contingencies of 'alternative-less' developments. The 'junctions' between the segments of the Baton Rouge 'villages' are also suitable for the construction of rail-bound public transport lines. Particularly regarding the downtown area, a reminiscence of 'Old Baton Rouge' from the era of early industrialization could be staged here, when the country-city flair was still present. In this way, the development of the downtown, beyond its function as CBD, could be staged in the form of a link to the great past of blues and bars, the potential for public life of the downtown could be developed and the downtown could become the pole of development not only economically, but also culturally and socially. In order to achieve this, a central light rail line could be built along the Florida Corridor to Denham Springs, on which feeder traffic (on foot, by bicycle, by car, connected to Park and Ride, but also to bus lines and, in a final stage of development, to secondary light rail lines) could be bundled from the north and south. Thus, the Florida Corridor could also partially lose its separating effect and make encounters possible. Following this idea, a second light rail line could - in parts of Plank Road - connect Southern University and Louisiana State University, crossing the downtown area on Third Street. This crossing especially could in turn revive the potential of historic Third Street as a former central business and entertainment hub. Beyond the development of spatial structures, the mutual opening of SU and LSU classes could not only facilitate the exchange of people but also perspectives.

The location of the downtown area is therefore both favorable and unfavorable for further reurbanization processes: unfavorable because it is peripheral to the settlement area, favorable because it is attractive and located directly on the Mississippi. Here, however, the possibilities of a waterfront development have so far not been fully exploited, nor has the use of numerous potential areas in the downtown area, which are currently considered to be only partially used parking spaces. Already today, parking garages are available to meet the demand, so that in addition to the development of the areas, the creation of parks can also be considered as land use, which in turn would increase the attractiveness of inner-city living, which in turn could have an inspiring effect on the downtown economy. Baton Rouge has a potential for a water/riverfront development that is beyond the scope of projects elsewhere (Brownill 2011; Hoyle 2000; Sieber 1991): the Mississippi is one of the most famous and symbolic rivers in the world, with an aura of steamship romance and adventure, so it could be built accordingly. In addition, the Mississippi River forms the western end of the downtown area (and, less important in this context, of the city as a whole), so that the aesthetic preference pattern of sunset is subject to almost constant updating. Another positive symbolic charge makes the Horace Wilkinson Bridge, which is lit up at night, favor a waterfront development. As a result of the sufficient distance to the city center, noise emissions from motor vehicle traffic are limited, as are those from shipping traffic and the loading and unloading of ships on the other side of the Mississippi.

Should the Baton Rouge route provide for increased development of the downtown area, which is also suggested by its greater safety against flooding compared to large parts of the suburban land reserves, improved accessibility - beyond automobile and possibly light rail traffic - on foot and by bicycle with or without electrical assistance appears to be advisable. Especially since the necessary construction measures for this are rather modest - traffic lights for pedestrian and bicycle traffic for the up and down ramps along Interstate 110 and partially lowered curbs. A somewhat more elaborate rehabilitation of the footpaths is required, not only along the footpaths leading to the city center, but especially in areas where people lack the capital for a dominant means of transport such as the car due to a lack of resources and are particularly dependent on functioning footpaths and cycle paths. Taking this approach further, a connection with the western Mississippi shore apart from automobile traffic would make sense for the development of Baton Rouge downtown, but also for relieving the through-traffic caused by commuter traffic. Also, in reminiscence of history, a pedestrian and cyclist ferry could be operated here, or - on a larger scale - the Horace-Wilkinson Bridge could be extended by a foot cycle path.

The strengthening of public transport, cycling, and pedestrian traffic cannot be seen solely as a measure to strengthen the urbanity of Baton Rouge, it can also be understood as a (small) contribution to the reduction of greenhouse gas emissions of Baton Rouge. However, efforts to reduce the causes of anthropogenic climate change are only one side of the coin; the other is adaptation to it. Here, measures to reduce inner-city overheating appear relatively inexpensive, such as shading by trees, i.e. maintaining them where they are found today, planting them where they are lacking today, which is also a contribution to spatial equality of opportunity (in this context, additional educational facilities and improved access to (healthy) food can also be mentioned). In order to reduce the amount of radiation, lighter surfaces, such as those of buildings and traffic areas, are also recommended. Dealing with floods caused by heavy precipitation is much more difficult due to the topography of Baton Rouge and the existing development of flood-prone areas. For further adaptation, it would be necessary to either abandon the development of flood-prone areas or to switch to flood-compatible development. The situation is more complex for the existing buildings at risk, as there are a wide range of adaptation possibilities, from individual measures (insurance) to cost-intensive poldering, which have not yet been sufficiently researched.

This aspect, in particular, makes it clear once again that the visions presented here represent possibilities and suggestions.

Post Scriptum: The Baton Rouge Style 10

From everyday encounters with the city, Baton Rouge, and the people living in Baton Rouge, from conversations with colleagues, and also from chance acquaintances, from interviews, and also from the (self-)representation of Baton Rouge on the Internet, as well as from scientific literature, the impression arose – contrary to the often expressed attribution of Baton Rouge as lacking significance – that the city and its inhabitants had some definite eccentricities and quirks, whose specific patterns of interpretation and action we have (certainly not completely) compiled in the following and ironically framed as 'Baton Rouge Style'.

10.1 Infrastructure Baton Rouge Style

- Street naming logic Baton Rouge Style: Locals know their way around. Guests are introduced to the specific logic. The rest are welcome to get lost and learn about the relativity of experiences and expectations. Examples: Florida Street and Florida Boulevard are a continuous street, North Street and North Boulevard run parallel in several blocks distance, the Scenic Highway ironically (today) runs along the Exxon plant, the streets of Baton Rouge often tend to change their names when crossing major streets, such as College Drive and Lee Drive, Acadian Throughway and Stanford Avenue when crossing Perkins Road.[1]
- Sidewalk Baton Rouge Style I: Avoid the intersections of busy roads, avoid walking in the dark, best to avoid the expectation that a sidewalk is suitable for pedestrians.

[1]Thank you, Craig, for that advice!

O. Kühne and C. Jenal, *The Multivillage-Metropolis Baton Rouge*, RaumFragen: Stadt – Region – Landschaft, https://doi.org/10.1007/978-3-658-30716-5_10

- Sidewalk Baton Rouge Style II: Don't put the crutches away from your last ankle injury – you might still need them.
- Crossing train traffic Baton Rouge Style I: Wait until the train has been shunted forward, stopped, shunted back. Even if you are in a hurry, audio books help. Don't take ones that could be too short.
- Crossing train traffic Baton Rouge Style II: You always wanted to have time to finally read the last 500 pages of Dostoyevsky's Crime and Punishment? This would be the perfect opportunity!
- Crossing train traffic Baton Rouge Style III: Gamblers Version – will the track system withstand a speed over 5 mph?
- Bus riding: Do you mean bus waiting?
- Bus stop Baton Rouge Style I: Masonic Version - located in secret places, so only highly skilled know where to find them
- Bus schedule Baton Rouge Style II: Masonic Version - only the highly skilled know the arrival times.

St_ _ts Ba_ton *R* e Styl_e
\ree/ \oug/

I
F
THE
CA*P*
I
TOL
~~W~~AS
A
ST REETS TR^E ETSTrEET
STR*E*ETSTREETSTREE_T
~~S~~TREE TsTREETSTREE_T

10.2 All Around Driving: Baton Rouge Style

- Parking Baton Rouge Style: Iron rules - 1) Never, never, never move more than 100 m away from your car unless you are into anesthetic-free amputation pain. 2) Make sure that there is a distance of at least one car length (full size car) between your parked car and other parked cars at the front, back, left and right (ideally also above and below).

- Incoming onto the Interstate, Baton Rouge Style: Drive briskly across the feeder road, see if there is any traffic in the lane you want to enter. If no one's coming… slow down, hard if someone is coming… maybe. Do not drive up until you have stopped at least one car. If the maneuver has led to an accident: take the opportunity to chat with your new acquaintances. However, strictly refuse to communicate with other drivers in terms of signaling direction changes; indicator bulbs could wear out![2]
- Driving Baton Rouge Style: Beginners - always drive slowly so that your smartphone does not slip out of your hand. Advanced - drive faster or take a second smartphone. Professionals - 'Pedal to the Metal', at least two smartphones, if possible, making calls simultaneously.
- Fixing Baton Rouge Style: Oh, this driving style has unintended side effects? Doesn't matter, that falls under patina (Baton Rouge style). Is something about to fall off the car? Take Duct Tape, or whatever else you can get your hands on and find suitable. Sometimes even a plastic bag helps. This procedure can also be applied to other objects of daily life, including pavements and power lines (see Figs. 10.1 and 10.2).
- Refueling Baton Rouge Style: Sorry, we only take cash. Try somewhere else. But even there you will have little luck. The Internet is lousy today.

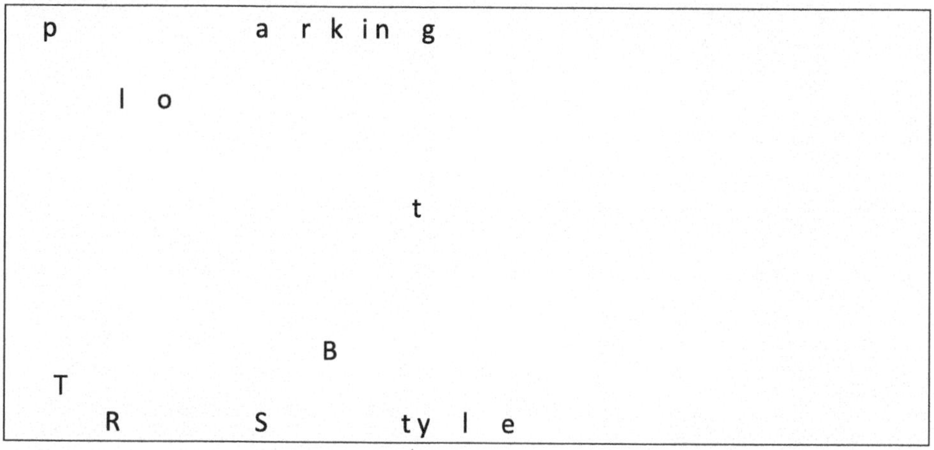

[2]Again, thanks Craig, for the tip on turn signal wear.

Fig. 10.1 Fixing Baton Rouge Style. (own photos)

Fig. 10.2 Resting Baton Rouge Style. (own photos)

10.3 Room Development and Planning Baton Rouge Style

- Downtown Shopping Baton Rouge: Who wants to pick up used office furniture? You won't find much else!
- Waterfront Baton Rouge Style: What do you think? Has anyone here seen water except when it rains?
- Urban Planning Baton Rouge Style: A measure of the smallest possible unit of political-administrative intervention in urban development.
- Gated Community Baton Rouge Style: Oh no, really with high wall and solid fence? And access control? Isn't a knee-high wall enough? Everyone knows what it means. Don't get too serious. Always stay relaxed.
- Flood Baton Rouge Style: there was no water.

Gated Communities Los Angeles Style

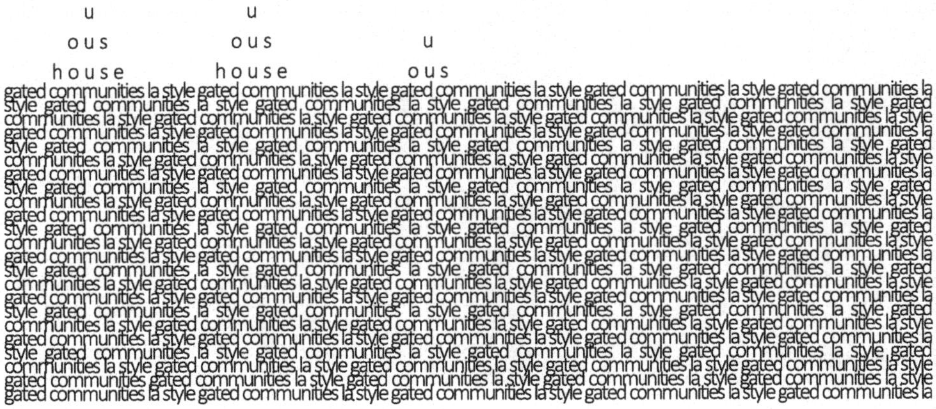

Gated Communities Baton Rouge Style

10.4 Social and Economic Life: Baton Rouge Style

- Gross Baton Rouge Product: An estimate of total value of all the bon temps and social relations in a given period owned by those of Baton Rouge and their guests
- Local Identity Baton Rouge Style: Fuck you context
- Advertising Baton Rouge style: Get it done. Or: See Fig. 10.3.
- Fitness Baton Rouge Style: Beginners - Watching an LSU Tigers football game along with Beer und BBQ.
- Fitness Baton Rouge Style: Advanced - Lifting heavy beer cans and tons of food watching an LSU Tigers football game.
- Tourism Baton Rouge: Capitol from the south, view from the Capitol. Have you looked over Capitol Lake yet? Oh, the way to the most impressive view is not accessible? Relax, the Capitol is also great from other perspectives. And: We still have an exciting cemetery!

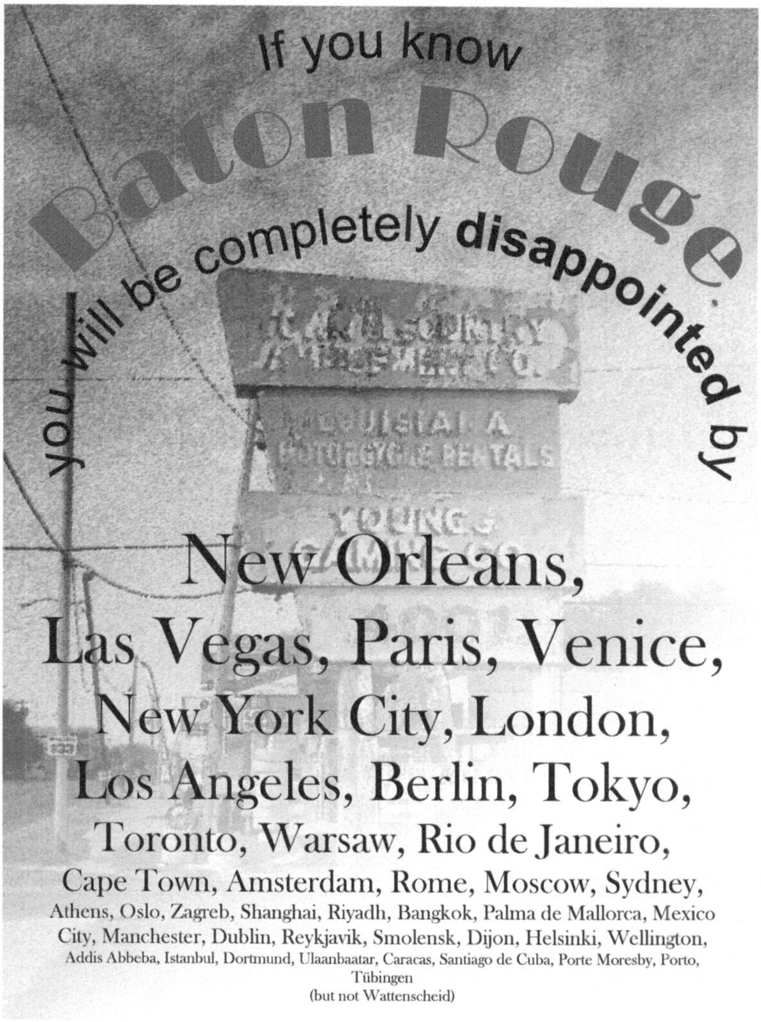

Fig. 10.3 Our contribution to Baton Rouge profiling (own design)

- Environmental Protection Baton Rouge Style: What the swamp is this?
- Gardening Baton Rouge Style I: A can of weed killer will do
- Gardening Baton Rouge Style II: Lawn trimming 24/7
- Gardening Baton Rouge Style III: Welcome to the jungle
- Garden Design Baton Rouge Style: Keep the maintenance effort low, unless gardening is your hobby, but then you might be considered strange, which is also not an issue.
- Rap Music Baton Rouge Style: Kickin' in Jigga City struttin' fo' yo' respect
- Baton Rouge Style Baton Rouge Style: Who needs stringent planning? Spontaneous and occasional is great. And if it does not correspond to the common ideas of beauty

and perfection? Then change them! Besides, it would be good to talk about it over barbecue and beer! A bar with live music would also be a possibility, of course.

- Planning System Baton Rouge style: Who cares?! We've always done it this way – it works!

Nightlife Downtown Baton Rouge

```
         ZZZ
         ZZZ
         ZZZ                    ZZ
 ZZZZZ   ZZZ              ZZZ   ZZ
 ZZZZZ   ZZZZ   ZZ      ZZZZZ   ZZ   ZZ
 ZZZZZ   ZZZZ   ZZ   ZZ ZZZZZ  ZZZ   ZZ
ZZZZZZZZZZZZZZZZZZZZZZZZZZZZZZZZZZZZZZZZZZZZZZZZ
```

…to be continued

Annex 11

11.1 Google Results

Table 11.1 Overview and systematization of the first three pages of a google keyword search for the term 'Baton Rouge' with German IP address; retrieved on 09/30/2019; own survey and presentation

No.	URL	Google result page	categorie	Abr.
1	https://de.wikipedia.org/wiki/Baton_Rouge	1	Wikipedia	go-01
2	https://en.wikipedia.org/wiki/Baton_Rouge,_Louisiana	1	Wikipedia	go-02
3	https://www.visitbatonrouge.com	1	Tourism	go-03
4	https://www.batonrougeguitars.com/home.html	1	Shopping	go-04
5	https://www.visittheusa.de/destination/baton-rouge	1	Tourism	go-05
6	https://www.tripadvisor.de/Attractions-g40024-Activities-Baton_Rouge_Louisiana.html	1	Tourism	go-06
7	https://www.baton-rouge.de/	1	Gastronomy	go-07
8	https://www.brla.gov	1	Administration	go-08
9	https://www.batonrouge.ca/en	2	Gastronomy	go-09
10	https://www.batonrouge.ca/	2	Gastronomy	go-10
11	https://www.lsu.edu/	2	Education	go-11
12	https://www.businessreport.com/	2	News	go-12

(continued)

O. Kühne and C. Jenal, *The Multivillage-Metropolis Baton Rouge*, RaumFragen: Stadt – Region – Landschaft, https://doi.org/10.1007/978-3-658-30716-5_11

Table 11.1 (continued)

No.	URL	Google result page	categorie	Abr.
13	http://www.watermarkbr.com	2	Hotel	go-13
14	http://www.brzoo.org/	2	Leisure	go-14
15	https://brac.org/	2	Campaign	go-15
16	https://www.louisianatravel.com/cities/baton-rouge	2	Tourism	go-16
17	http://www.magazinusa.com/us/states/show.aspx?state=la&doc=18&dsc=baton_rouge_hauptstadt	2	Tourism	go-17
18	https://www.braf.org/	2	Foundation	go-18
19	https://www.tourlane.de/l/usa	2	Travel Agency	go-19
20	https://www.amazon.de/	2	Shopping	go-20
21	https://www.batonrougegallery.org/	3	Culture	go-21
22	https://belleofbatonrouge.com/	3	Hotel	go-22
23	https://www.theadvocate.com/	3	News	go-23
24	http://www.therumhouse.com/	3	Gastronomy	go-24
25	https://www.ebrpl.com/	3	Education	go-25
26	https://www.agoda.com/de-de/baton-rouge-down-town/maps/baton-rouge-la-us.html?cid=-218	3	Travel Agency	go-26
27	https://www.episcopalbr.org/	3	Education	go-27
28	https://www.brso.org/	3	Culture	go-28
29	http://www.lsusports.net/	3	Sports	go-29
30	https://www.wbrz.com/news/two-wanted-after-armed-robbery-at-baton-rouge-gas-station/	3	News	go-30

11.2 Internet Videos

Table 11.2 Overview Systematization of songs for keyword search on the term 'Baton Rouge' on the video platform YouTube, accessed in March 2019; own survey and presentation

No.	Title	URL
1	Marc Broussard - Baton Rouge (Acoustic Video)	https://www.youtube.com/watch?v=ocw1TzLNytU
2	Baton Rouge Vacation Travel Guide I Expedia (4K)	https://www.youtube.com/watch?v=Ca8slCfepjg
3	Baton Rouge	https://www.youtube.com/watch?v=0KJ_sa6bzS4&t=18s

(continued)

Table 11.2 (continued)

No.	Title	URL
4	BATON ROUGE EFFECT	https://www.youtube.com/watch?v=idl8FfIa_Sg
5	Callin Baton Rouge	https://www.youtube.com/watch?v=QQK8uj01JhI
6	Travel Guide: Baton Rouge, Louisiana	https://www.youtube.com/watch?v=BFMurXcRNno
7	Top 10 reasons NOT to move to Louisiana. The weather is on the list and so Baton Rouge	https://www.youtube.com/watch?v=R469wCmoPok
8	Baton Rouge fight	https://www.youtube.com/watch?v=Sc92VmZVL_Q
9	Veltboy314 - Baton Rouge, Louisiana Whipz	https://www.youtube.com/watch?v=JqAQnpIbYfE&t=137s
10	PALATINE - BATON ROUGE	https://www.youtube.com/watch?v=fekQzQsxHJs
11	Baton Rouge Fight (He Leaking)	https://www.youtube.com/watch?v=cQ9qU1qwtU0
12	The King Of Louisiana	https://www.youtube.com/watch?v=lYc4I75IgbE
13	Baton rouge fight	https://www.youtube.com/watch?v=fYxck1ze4fs
14	Ghost Town Baton Rouge Louisiana	https://www.youtube.com/watch?v=vvPMv0TKNII&t=270s
15	Life in a US city's most dangerous postal code - BBC News	https://www.youtube.com/watch?v=0gFHZR-wOJk
16	DOWNTOWN BATON ROUGE LOUISIANA TOUR	https://www.youtube.com/watch?v=8LQodpDjX60
17	Callin' Baton Rouge	https://www.youtube.com/watch?v=qZEPXFosmMM
18	Eyewitness Captures Terrifying Video From Middle of Baton Rouge Police Ambush	https://www.youtube.com/watch?v=SDdjBgZXQts
19	Spanishtown 2019 *Baton Rouge* Louisiana	https://www.youtube.com/watch?v=DLz-Kh-xgzM
20	Sherwood Marty: explains difference between New Orleans & Baton Rouge. "New Orleans is more gangsta"	https://www.youtube.com/watch?v=1zvaQjr1UxU
21	Baton Rouge Hood Legends PT 3 Exclusive with(Vontray)	https://www.youtube.com/watch?v=UfhcBArsC7g
22	NBA Young Boy Explains The Hate Amongst Baton Rouge Rappers [unreleased]	https://www.youtube.com/watch?v=4bJgbNDLTAw
23	Callin baton rouge	https://www.youtube.com/watch?v=ps8PubLV8Fw

(continued)

Table 11.2 (continued)

No.	Title	URL
24	Issabelle Boulay - Entre Matane et Baton rouge (Lyrics)	https://www.youtube.com/watch?v=Y4O2O62iDtk
25	"Callin' Baton Rouge" as Gaeilge le Maitiú Ó Domhnaill	https://www.youtube.com/watch?v=0de-qUxHLBk
26	70th Street Carlos: "Boosie is the only Baton Rouge legend no other rapper in B.R is a legend"	https://www.youtube.com/watch?v=QUvi65RPQHQ
27	SFTP On Baton Rouge: It Was Cliques At First, We Ain't Had No Gangs	https://www.youtube.com/watch?v=rXQh8_z37xk
28	5 Things To Do in Baton Rouge, Louisiana	https://www.youtube.com/watch?v=ooHRdPhuBbw
29	Southside Baton Rouge	https://www.youtube.com/watch?v=EKVg_Kbp5UQ
30	22 Savage breaks down Baton Rouge slang "Dick & Round"	https://www.youtube.com/watch?v=HxvVOAH7UBo
31	Baton Rouge Slangs/Lingo Louisiana	https://www.youtube.com/watch?v=p2Zl15W1TR0
32	Forgotten North Baton Rouge Vol 2	https://www.youtube.com/watch?v=5hQov1_Iczk
33	Rapper Blvd Quick Shot and Killed in Baton Rouge	https://www.youtube.com/watch?v=oyTEbBoLfRo
34	Top 5 Baton Rouge Rappers Charged With Murder	https://www.youtube.com/watch?v=ploSnarWV44
35	Thuggin It & Lovin It (DVD About The Streets Of Baton Rouge)	https://www.youtube.com/watch?v=e52-Vx5PMCQ
36	Kevin Gates NOT allowed in his old neighborhood by Baton Rouge Rapper Bussetta	https://www.youtube.com/watch?v=w6XYZUPqnSk
37	Scotty Cain Explains The Hate Amongst Baton Rouge Rappers	https://www.youtube.com/watch?v=Xf2oZeBvREk
38	Baton Rouge Rapper Maine Musik FAINTS in court after sentenced 9 yrs in prison!	https://www.youtube.com/watch?v=rotCrbvOtJA
39	New Grass Revival Callin' Baton Rouge	https://www.youtube.com/watch?v=00AFX_2aDJ4
40	Old Baton Rouge Louisiana	https://www.youtube.com/watch?v=pzsMKVqjHHg
41	Baton Rouge, La-FBI Unlawful Detention 4-11-17	https://www.youtube.com/watch?v=G19IeIY_ZzY
42	The Oak Ridge Boys - Callin' Baton Rouge	https://www.youtube.com/watch?v=7mUJjFiu5ew
43	Woman RAPED while friend was on Instagram Live in Baton Rouge! I @JamesJeffersonJ	https://www.youtube.com/watch?v=DaKH1WFsj-8

(continued)

Table 11.2 (continued)

No.	Title	URL
44	Baton Rouge video goes viral on instagram+ an update on the "case"	https://www.youtube.com/watch?v=SESWAojSu7M
45	NEWS FOOTAGE : Baton Rouge Rapper "Blvd Quick" Killed (FREDO BANG & Da Real Gee Money HOMIE)	https://www.youtube.com/watch?v=xFj_dhX1G_Y
46	Old Baton Rouge Louisiana historical slide show 2	https://www.youtube.com/watch?v=Gkoxl5rlGEo
47	Police Shoot Alton Sterling in Baton Rouge I The New York Times	https://www.youtube.com/watch?v=1pr3QPNVJ9U
48	Bankstown 75th Ave Niggas (Baton Rouge Louisiana)	https://www.youtube.com/watch?v=6jsX0OYhO0A
49	Road Trip #108 - I-110 North - Baton Rouge, Louisiana	https://www.youtube.com/watch?v=9vWWB4kyBlQ
50	Crossing the Mississippi River, Baton Rouge, Louisiana	https://www.youtube.com/watch?v=idRaMhNGI1E
51	Woman in iconic Baton Rouge standoff photo breaks silence	https://www.youtube.com/watch?v=LFC6l0DjDF0
52	The Voice 2017 Natalie Stovall - The Playoffs: "Callin' Baton Rouge"	https://www.youtube.com/watch?v=sJnSkJn5v08
53	PALATINE - BATON ROUGE (Live @ Studio Pigalle)	https://www.youtube.com/watch?v=uZj5TD5yO_s
54	Fire in Baton Rouge	https://www.youtube.com/watch?v=QyIle5T5beM
55	Baton Rouge	https://www.youtube.com/watch?v=0KJ_sa6bzS4

11.3 Baton Rouge in a Song

Table 11.3 Overview systematization of songs for keyword search of the term 'Baton Rouge' in Amazon Music and YouTube; retrieved July 2019; own survey and presentation

No.	Interpret – Song	Year of release	Abr.
1	Guy Clark – Baton Rouge	1992	sg-01
2	Ron Rogers – Baton Rouge	2000	sg-02
3	Amos Garrett – Baton Rouge	1984	sg-03
4	Chuck Prophet – Baton Rouge	1993	sg-04
5	Cliff Eberhardt – Baton Rouge	1993	sg-05

(continued)

Table 11.3 (continued)

No.	Interpret – Song	Year of release	Abr.
6	Lil Boosie – Baton Rouge	2007	sg-06
7	Magnolia Summer – Baton Rouge	2003	sg-07
8	Garth Brooks – Callin‘ Baton Rouge	1993	sg-08
9	Kenny Neal – Bio on the Bayou	1988	sg-09
10	Justin Rendell – “Freedom” Baton Rouge	2015	sg-10
11	Larry Garner – Go to Baton Rouge	1995	sg-11
12	Marc Broussard – Baton Rouge	2001	sg-12
13	Justin Rendell – Baton Rouge's "Wayne Dane"	2010	sg-13
14	Goon (Louisiana Rappers) – Welcome to Jigga City	2008	sg-14
15	Lucinda Williams – Bus to Baton Rouge	2001	sg-15
16	Shalacy – Baton Rouge	2002	sg-16
17	BME Beezy – Baton Rouge	2019	sg-17
18	Palatine – Baton Rouge	2016	sg-18
19	Ricco’lavelli – Baton Rouge	2019	sg-19
20	J.J. Diamond – Baton Rouge	2005	sg-20
21	Antoine Cash – Baton Rouge	2018	sg-21
22	Young Gage – Baton Rouge	2015	sg-22
23	IW – Baton Rouge	2016	sg-23
24	Christien Bold Baton Rouge	2016	sg-24
25	Delwin The Krazyman – Baton Rouge	2019	sg-25
26	Insane & O.N.E [feat. Young Bleed] – Baton Rouge	2018	sg-26
27	Esco McCollum Jr – I love Baton Rouge	2019	sg-27
28	Gordon Bradberry My Baby and Baton Rouge	2004	sg-28
29	Zachary Flynn – Baton Rouge	2014	sg-29
30	Tim Nielsen – Baton Rouge	2000	sg-30
31	Henry Turner Jr. & Flavor – Baton Rouge Theme Song	2016	sg-31
32	Djayy Badazz – Welcome 2 Baton Rouge	2017	sg-32
33	Vigilante Salazar – Baton Rouge Hard	2019	sg-33
34	Yungins Wit a Dream – South Baton Rouge	2019	sg-34
35	Tim James – Leaving Baton Rouge	2018	sg-35
36	Wo Purple Wo – Nightmares in Baton Rouge	2018	sg-36
37	Lit Phil & Selfmade Shark – Baton Rouge	2017	sg-37
38	Castle Boy Raw – Welcome to My City (Baton Rouge)	2017	sg-38
39	Money Bagz – So Baton Rouge	2013	sg-39
40	Down Bad – I’m from Baton Rouge	2010	sg-40

(continued)

Table 11.3 (continued)

No.	Interpret – Song	Year of release	Abr.
41	Scarlet Stitch – Baton Rouge	2019	sg-41
42	Lou Reed – Baton Rouge	2000	sg-42
43	J.J. Cale – Louisiana Women	1972	sg-43
44	Kris Kristofferson – Me and Bobby McGee	1970	sg-44
45	Lil Trill – Welcome To Baton Rouge	2012	sg-45

11.4 Articles, New York Times (NYT)

Table 11.4 Overview and systematization of the articles in the New York Times in the keyword search for the term 'Baton Rouge'; retrieved in May 2019; own survey and presentation

No.	Title Article	Abr.
1	F.D.U. Five Awaits New Home; SPORTS	NYT 1978-01
2	Sites Included In F.A.A. Plan On Radar Use	NYT 1978-02
3	Around the Nation; Large Crude-Oil Tank Burning in Louisiana More Bodies Anticipated In Illinois Murder Case Sentencing Study Shows Wide Variation in Capital Striking Pressmen Ratify Contract at St. Louis Paper Two Go on Shooting Spree In Utah; One Suspect Held	NYT 1978-03
4	Five-Cent Phone Calls Ending in Louisiana	NYT 1978-04
5	Indiana, Led by Woodson, Routs Davidson, 101 to 64; Hoosiers Pull Away Pratt 78, Adelphi 70 L.S.U. 99, Montana State 89 Detroit 92, Marshall 61	NYT 1978-05
6	Evidence Questioned in White Factory Worker's Job-Discrimination Suit; Industry Programs Not Addressed 'There Was Discrimination' Order to Remand Sought Dual-Badge System Disputed Anger at National Union One-Day Hearing 'Not Desegregation Any More'	NYT 1978-06
7	Around the Nation; Four Die in a Fiery Crash On Highway in Louisiana	NYT 1978-07
8	Bias Case That May Rival Bakke's Accepted by High Court for Review; A White Factory Worker Charges Program Designed to Give Skilled Work to Blacks Is Illegal Bribery Case to Be Heard Bias Case That May Rival Bakke's Accepted by High Court for Review Reconsideration Is Urged	NYT 1978-08
9	An Expert Doubts Roman Coins Found in U.S. Are Sea-Link Clue; Vikings in Newfoundland	NYT 1978-09
10	Irish Five Romps By 105-6 Kentucky 121, West Texas 67 No. Carolina 93, Detroit 76 So. Carolina 85, Harvard 71 L.S.U. 109, Tulane 85 N.Y. Tech 62, Pratt 58 Canisius 80, Cornell 75	NYT 1978-10

(continued)

Table 11.4 (continued)

No.	Title Article	Abr.
11	St. John's Five Triumphs; Wagner Stuns Alabama; Seton Hall 78, Georgia Tech 73 L.S.U. 116, St. Francis (Pa.) 81 Maryland 107, Bucknell 97 Nebraska 64, Birmingham 55 N.C. State 81, Texas A&M 65	NYT 1978-11
12	U.S. Trying to Breed an Ideal Bee; More Resistant to Disease 'Little Defense as Possible'	NYT 1978-12
13	GASOLINE ADDITIVE CALLED CANCER RISK; National Institute Finds That EDB Causes Disease in Test Mice	NYT 1978-13
14	Louisiana's Stringent New Abortion Law Termed 'Standard-Bearer'; 'Tremendous Excitement Aroused'	NYT 1978-14
15	COMPANY NEWS	NYT 1978-15
16	Educators Welcome Bakke Ruling As Signal to Retain Current Policy; Unconcerned About Directive Expressed Relief Viewed as a Setback	NYT 1978-16
17	The Old Scout Who Found Guidry; Sports of The Times 'Shantz Couldn't Throw as Hard' 'After 41 Years in the Organization'	NYT 1978-17
18	Reverend Martin's; Son	NYT 1978-18
19	Ernest Gaines: A Conversation; Gaines	NYT 1978-19
20	Southern Blacks, Women Found in Low City Jobs; Reportedly 50 Years Behind Exempt Until 1972 Six Cities Provided Information	NYT 1978-20
21	Southern Comforts; Airships	NYT 1978-21
22	Pickets at TV Studio Protest 'Holocaust'	NYT 1978-22
23	Louisiana's Cajun Country; Prospers Amid Traditions A Family Link Evangeline Everywhere Administrator's Tour Another Sort of Feast If You Go ...	NYT 1978-23
24	Gov. Edwards Explains Passman Link	NYT 1978-24
25	5 Children Die in Blaze at House That Firemen Thought Was Empty	NYT 1978-25
26	Bridge:; 2 Top Championships Begin In League's Spring Nationals	NYT 1978-26
27	Marine Corps General Named	NYT 1978-27
28	Pop Concert: Thorogood's 'Destroyers'	NYT 1978-28
29	WAYS SOUGHT TO FIND GRAVITATIONAL WAVE; Magnetic Levitation of Big Cooled Cylinders Developed to Try and Detect Form of Radiation Almost Absolute Zero Experiments Began in 1960's Beams Are Reunited	NYT 1978-29
30	Arkansas, Marquette Take Over at Top; Arkansas, Marquette No. 1 O'Koren Is Sidelined Williams on Record Pace Appeal to the Fans	NYT 1978-30
31	Kentucky Is Upset, 95–94, By L.S.U. in Overtime; Columbia 62, Dartmouth 58	NYT 1978-31
32	L.S.U. Women Upset, 86–68	NYT 1978-32
33	Alabama Upsets Kentucky; Irish Win; King Scores 26 and Tide Uses Press	NYT 1978-33

(continued)

Table 11.4 (continued)

No.	Title Article	Abr.
34	The Ultimate in Track and Field Is Knowing You Did Your Best; You Gotta Have Heart A Feat for All Sports	NYT 1978-34
35	Unflagging Despite Setbacks, He Championed Liberalism; Unflagging Despite Disappointments,	NYT 1978-35
36	New Orleans Is Unawed by Game	NYT 1978-36
37	Superdome a Bottomless Pit for Taxpayer Funds; Bigger Than Astrodome Superdome Is a Super Drain on Public Funds Project Assailed	NYT 1978-37
38	Standards Are a Calling For Bold SUNY Trustee	NYT 1998-01
39	TRANSACTIONS	NYT 1998-02
40	Feeling at Home South of the Mason-Dixon	NYT 1998-03
41	A Rapper Goes to the Hoop	NYT 1998-04
42	Former Louisiana Governor Is Indicted in Extortion Case	NYT 1998-05
43	U.S. Suing Louisiana On Prison Ills	NYT 1998-06
44	LOUISIANA	NYT 1998-07
45	LOUISIANA	NYT 1998-08
46	JACKPOT ENTERPRISES IN STOCK DEAL FOR CRC HOLDINGS	NYT 1998-09
47	UNITED COMPANIES FINANCIAL TO CUT ITS BUSINESS BY 30%	NYT 1998-10
48	Market Puts Price Tags on the Priceless	NYT 1998-11
49	Glenn William Poorman, 90, An Executive at Standard Oil	NYT 1998-12
50	Credit Offerings Planned During the Week	NYT 1998-13
51	Spending Bill, Laden With Pork, Is Signed Into Law	NYT 1998-14
52	TRANSACTIONS	NYT 1998-15
53	Sports of The Times; Applying Stern Measures for Eddie D. and the Steinbrenner Yardstick	NYT 1998-16
54	Owner of N.F.L. Team Ties Ex-Governor to Extortion	NYT 1998-17
55	Gulf Coast Feeling Lucky Hurricane Wasn't Worse	NYT 1998-18
56	Ghosts and Ruins Along the Mississippi	NYT 1998-19
57	World of the Bayou And the Plantation	NYT 1998-20
58	Johnny Adams, 67, Vocalist; One of the Legends in Blues	NYT 1998-21
59	THE NEW YORK TIMES PRESEASON TOP 20	NYT 1998-22
60	LAMAR ADVERTISING TO BUY OUTDOOR COMMUNICATIONS INC.	NYT 1998-23
61	Reunion Hails Legacy Of Huey and Earl Long; Relatives Remember Louisiana Legends	NYT 1998-24
62	UNITED COMPANIES LOOKS FOR STRATEGIC PARTNER	NYT 1998-25

(continued)

Table 11.4 (continued)

No.	Title Article	Abr.
63	HARD TIME: A special report.; Profits at a Juvenile Prison Come With a Chilling Cost	NYT 1998-26
64	A FAMILY PORTRAIT: A special report.; A Twisted Tale of Deceit, Fraud and Violence	NYT 1998-27
65	Both Home and Prison, Leprosy Site May Shut	NYT 1998-28
66	Seems the Oh-So-Serious Phase Is Over	NYT 1998-29
67	SUNDAY: May 24, 1998: Resting Places; If You Bury It, He Will Come	NYT 1998-30
68	Blanche Revere Long, 93, Louisiana Legend	NYT 1998-31
69	Clyde Connell, 97, Sculptor Inspired by Louisiana Bayous	NYT 1998-32
70	POLLUTION POLICY IS UNFAIR BURDEN, STATES TELL E.P.A.	NYT 1998-33
71	Weary Hands At the Throttle	NYT 1998-34
72	Reason for Dramatic Drop In Crime Puzzles the Experts	NYT 1998-35
73	Strong Stuff	NYT 1998-36
74	The Heat Is On; Hot sauces are burning their way across America	NYT 1998-37
75	Atlanta at Crossroads Of Gas Price Paradise	NYT 1998-38
76	Bayou Country	NYT 1998-39
77	New Orleans Mayor Thrives on Lower Crime and Lifted Spirit	NYT 1998-40
78	TRANSACTIONS	NYT 1998-41
79	Third Base or Quarterback? Marlins' Josh Booty Could Be a Bit of Both	NYT 1998-42
80	Cutting Loose in Cajun Country	NYT 1998-43
81	Digital Data Gives Billboard Owners More Reason to 'Love a Good Traffic Jam'	NYT 2018-01
82	Andrew Frierson, Pioneering Black Opera Singer, Dies at 94	NYT 2018-02
83	More Floods and More Droughts: Climate Change Delivers Both	NYT 2018-03
84	What We Know About Diet and Weight Loss	NYT 2018-04
85	For Shame: A Brief History of the Perp Walk	NYT 2018-05
86	These 3 Teams Will Help Shape the College Football Playoff	NYT 2018-06
87	Andrew Fitzgerald, Who Saved Men From a Storm-Sundered Ship, Dies at 87	NYT 2018-07
88	Kansas, Seeking Lift in Football, Hires Les Miles as Coach	NYT 2018-08
89	Alabama Becomes the Golden State Warriors	NYT 2018-09
90	Alabama Stuffs L.S.U., Putting an Exclamation Point on Its Top Ranking	NYT 2018-10
91	Kentucky and Northwestern Become Surprise Contenders	NYT 2018-11
92	American ISIS Suspect Is Freed After Being Held More Than a Year	NYT 2018-12
93	After the Hurricane Came the Con	NYT 2018-13

(continued)

Table 11.4 (continued)

No.	Title Article	Abr.
94	Capturing the Struggle for Racial Equality, Past and Present	NYT 2018-14
95	Jim Taylor, Hall of Fame Fullback for the Green Bay Packers, Dies at 83	NYT 2018-15
96	Louisiana State's $2.5 Million Man Isn't Even the Head Coach	NYT 2018-16
97	Cleveland Officer Who Killed Tamir Rice Is Hired by an Ohio Police Department	NYT 2018-17
98	Lil Wayne's Long-Delayed 'Tha Carter V' Is a Streaming Blockbuster	NYT 2018-18
99	South Carolina Shooting Kills One Officer and Wounds 6 Others	NYT 2018-19
100	Beer: Colleges' New Way to Fill Seats, Not Couches	NYT 2018-20
101	Classical Fall Preview: Debuts, Premieres, a New Philharmonic Maestro	NYT 2018-21
102	5 New Designers to Watch This Season	NYT 2018-22
103	Revealing a Secret Art Life: A Painter's Sculptures	NYT 2018-23
104	A Cajun Seasoned Boil for a Big Party	NYT 2018-24
105	Lazy Lester, Luminary of Swamp Blues, Is Dead at 85	NYT 2018-25
106	Our Blind Spots Often Hide the Truth About America	NYT 2018-26
107	What's on TV Thursday: 'Snowfall' and 'The Great Food Truck Race'	NYT 2018-27
108	Meet New Orleans's All-Female Biker Club	NYT 2018-28
109	'I'm Not the Only Young Conservative'	NYT 2018-29
110	Americans Are Having Fewer Babies. They Told Us Why.	NYT 2018-30
111	Henry Butler, Quintessential New Orleans Pianist, Is Dead at 69	NYT 2018-31
112	One-Season Wednesdays: 'Aliens in America'	NYT 2018-32
113	The Heart Just Grew Fonder When He Was in the Middle East	NYT 2018-33
114	Clarence Fountain, 88, Dies; Led the Blind Boys of Alabama	NYT 2018-34
115	How Robert Kennedy Inspired 'Abraham, Martin and John'	NYT 2018-35
116	Billy Cannon, Football Star With a Troubled Life, Dies at 80	NYT 2018-36
117	Louisiana Deputies Are Investigated in Death of Black Man During Arrest	NYT 2018-37
118	Nashville Suspect Once Called Himself a 'Sovereign Citizen,' Police Say. What Is That?	NYT 2018-38
119	Tracy K. Smith, America's Poet Laureate, Is a Woman With a Mission	NYT 2018-39
120	The Paris Review Names a New Editor: Emily Nemens of The Southern Review	NYT 2018-40
121	Taking Aim at Gun Violence, With Personal Deterrence	NYT 2018-41
122	How to Push a Story	NYT 2018-42
123	Sacramento, Russia, Easter: Your Weekend Briefing	NYT 2018-43
124	Police Release New Body Camera Footage of Alton Sterling Shooting	NYT 2018-44

(continued)

Table 11.4 (continued)

No.	Title Article	Abr.
125	Baton Rouge Officer Is Fired in Alton Sterling Case as Police Release New Videos	NYT 2018-45
126	Transgender Woman Shot Dead in Motel Is 7th Killed in U.S. This Year, Rights Advocates Say	NYT 2018-46
127	20 Shots in Sacramento: Stephon Clark Killing Reignites a Furor	NYT 2018-47
128	Citizenship, Louisiana, North Korea: Your Tuesday Evening Briefing	NYT 2018-48
129	No Charges for Officers in Death of Alton Sterling	NYT 2018-49
130	Baton Rouge Officers Will Not Be Charged in Alton Sterling's Killing	NYT 2018-50
131	Stormy Daniels, Porn Star Suing Trump, Is Known for Her Ambition: 'She's the Boss'	NYT 2018-51
132	March for Our Lives Highlights: Students Protesting Guns Say 'Enough Is Enough'	NYT 2018-52
133	Tom Benson, Tycoon Who Danced With His Saints, Dies at 90	NYT 2018-53
134	Louisiana, 'a Failed State'	NYT 2018-54
135	Insects Feast on Louisiana Wetlands, Inviting the Gulf In	NYT 2018-55
136	Left to Louisiana's Tides, a Village Fights for Time	NYT 2018-56
137	At the Mouth of the Mississippi, a Weird and Fragile Beauty	NYT 2018-57
138	Where Did Rents Rise the Most in 2017?	NYT 2018-58
139	A Sports Rodin Works in Bronze, and in Volume	NYT 2018-59

11.5 Interview Guide

Personal References

- Do you live here in Baton Rouge and where do you come from? Since when do you live here?
- What is your favorite place/spot in Baton Rouge?

Segregation and Division of the City

- Is there a place in Baton Rouge that you consciously avoid and if so, why?
- Is there a place in Baton Rouge where you have never been, and you never thought of going to?

Challenges and Conflicts

- Is there anything that bothers you about Baton Rouge?
- In your opinion, what are the biggest challenges for Baton Rouge, at the moment?
- What are the major conflicts in your opinion in Baton Rouge?
- Would you like the petrochemical industry to withdraw from Baton Rouge

Stereotypes

- What do you think is associated with Baton Rouge beyond Baton Rouge?

Future Prospects and Wishes

- What would be your vision of a future Baton Rouge?
- Is there anything we haven't mentioned about Baton Rouge yet, but would like to mention it to you?

Correction to: Baton Rouge as a Living Environment 12

Correction to:
Chapter 8 in: O. Kühne and C. Jenal, *The Multivillage-Metropolis Baton Rouge*, RaumFragen: Stadt – Region – Landschaft, https://doi.org/10.1007/978-3-658-30716-5_8

Due to an error during production, the wrong content was accidentally included as video 1. This has now been corrected.

The updated version of this chapter can be found at https://doi.org/10.1007/978-3-658-30716-5_8

O. Kühne and C. Jenal, *The Multivillage-Metropolis Baton Rouge*, RaumFragen: Stadt – Region – Landschaft, https://doi.org/10.1007/978-3-658-30716-5_12

Bibliography

Ackermann, A. (2004). Das Eigene und das Fremde: Hybridität, Vielfalt und Kulturtransfers. In F. Jaeger, & J. Rüsen (Eds.), *Handbuch der Kulturwissenschaften. Band 3: Themen und Tendenzen* (vol. 3, pp. 139–154). Stuttgart: Metzler.

Adger, W. N. (2006). Vulnerability. *Global Environmental Change, 16*(3), 268–281. https://doi.org/10.1016/j.gloenvcha.2006.02.006.

Aglietta, M. (1976). *Régulation et crises du capitalisme. L'expérience des Etats-Unis* (Perspectives de l'economique). Paris: Calmann-Lévy.

Aitken, S. C., & Zonn, L. (1994). *Re*-presenting the place pastiche. In S. C. Aitken, & L. Zonn (Eds.), *Place, power, situation, and spectacle. A geography of film* (pp. 3–25). Lanham: Rowman & Littlefield.

Akbari, H., Pomerantz, M., & Taha, H. (2001). Cool surfaces and shade trees to reduce energy use and improve air quality in urban areas. *Solar Energy, 70*(3), 295–310.

Albert, G. (2005). *Hermeneutischer Positivismus und dialektischer Essentialismus Vilfredo Paretos*. Wiesbaden: VS Verlag.

Allen, B. (2000). The popular geography of illness in the industrial corridor. In C. E. Colten (Ed.), *Transforming New Orleans and its environs. Centuries of change* (History of the Urban Environment, pp. 178–201). Pittsburgh: University of Pittsburgh Press.

Allen, B. L. (2003). *Uneasy alchemy. Citizens and experts in Louisiana's chemical corridor disputes* (Urban and industrial environments). Cambridge: MIT Press.

Alviola, P. A., Nayga, R. M., Thomsen, M. R., & Wang, Z. (2013). Determinants of food deserts. *American Journal of Agricultural Economics, 95*(5), 1259–1265. https://doi.org/10.1093/ajae/aat029.

Amin, A., & Thrift, N. (2002). *Cities. Reimaging the urban*. Cambridge: Polity.

Anderson, G. L. (1976). *Land-grant universities and their continuing challenge*. East Lansing: Michigan State University Press.

Anderson, I. T. (2011). Are local chemical releases causing an increase in cancer rates in East Baton Rouge Parish? LSU Master's Theses. 418. https://digitalcommons.lsu.edu/gradschool_theses/418/. Accessed 1 July 2019.

Antipova, A. (2010). Land use, individual attributes, and travel behavior in Baton Rouge, Louisiana, LSU Doctoral Dissertations. 3450. https://digitalcommons.lsu.edu/gradschool_dissertations/3450/. Accessed 13 June 2019.

Appadurai, A. (1988). Introduction: Place and voice in anthropological theory. *Cultural Anthropology, 3*(1), 16–20. https://doi.org/10.1525/can.1988.3.1.02a00020.

O. Kühne and C. Jenal, *The Multivillage-Metropolis Baton Rouge,* RaumFragen: Stadt – Region – Landschaft, https://doi.org/10.1007/978-3-658-30716-5

Appadurai, A. (Ed.). (2015 [1986]). *The social life of things. Commodities in cultural perspective* (12. Aufl.). Cambridge: Cambridge University Press.

Armstrong, A. M. (2010). *Historic neighborhoods of Baton Rouge* (American Chronicles: A History Press Series). Charleston: The History Press.

Arthur, S. C. (1935). *The story of the West Florida rebellion.* St. Francisville: The St. Francisville Democrat.

Bankston, C. L., & Caldas, S. J. (2010). *A troubled dream. The promise and failure of school desegregation in Louisiana.* Nashville: Vanderbilt University Press.

Barrett, J. C., & Ko, I. (2009). A phenomenology of landscape: A crisis in British landscape archeology? *Journal of Social Archaeology, 9*(3), 275–294. https://doi.org/10.1177/1469605309338422.

Bartholomew, H. (1945–1948). *The 25 year-parish plan for metropolitan Baton Rouge. Louisiana.* Baton Rouge: Eigenverlag.

Basten, L. (2005). *Postmoderner Urbanismus. Gestaltung in der städtischen Peripherie* (Schriften des Arbeitskreises Stadtzukünfte der Deutschen Gesellschaft für Geographie, vol. 1). Münster: LIT.

Baton Rouge Area Chamber. (2019). 2015 EBR population density. https://brac.org/wp-content/uploads/population-density-EBR-Converted.pdf. Accessed 29 Nov 2019.

Bauman, Z. (1990). Modernity and ambivalence. *Theory, Culture & Society, 7*(2–3), 143–169.

Bauman, Z. (1995). *Life in fragments. Essays in postmodern morality.* Oxford: Blackwell.

Bauman, Z. (2008). *Flüchtige Zeiten. Leben in der Ungewissheit.* Hamburg: Hamburger Edition.

Beauregard, R. A. (2006). *When America became suburban.* Minneapolis: University of Minnesota Press.

Bedient, P. B., Horsak, R. D., Schlenk, D., Hovinga, R. M., & Pierson, J. D. (2005). Environmental impact of fipronil to the Louisiana crawfish industry. *Environmental Forensics, 6*(3), 289–299. https://doi.org/10.1080/15275920500194530.

Berg, M. (2017). Afroamerika – Der lange Weg zur Emanzipation. In W. Gamerith, & U. Gerhard (Eds.), *Kulturgeographie der USA. Eine Nation begreifen* (pp. 73–80). Berlin: Springer Spektrum.

Berger, P. L., & Luckmann, T. (1966). *The social construction of reality. A treatise in the sociology of knowledge.* New York: Anchor Books.

Berking, H., & Löw, M. (2008). *Die Eigenlogik der Städte – Neue Wege für die Stadtforschung* (Interdisziplinäre Stadtforschung). Frankfurt a. M.: Campus.

Berr, K., & Kühne, O. (2019). Werte und Werthaltungen in Landschaftskonflikten. In K. Berr, & C. Jenal (Eds.), *Landschaftskonflikte* (pp. 65–88). Wiesbaden: Springer VS.

Berr, K., & Kühne, O. (2020). *„Und das ungeheure Bild der Landschaft…". The Genesis of Landscape Understanding in the German-speaking Regions* (RaumFragen: Stadt – Region – Landschaft). Wiesbaden: Springer VS.

Berr, K., & Schenk, W. (2019). Begriffsgeschichte. In O. Kühne, F. Weber, K. Berr, & C. Jenal (Eds.), *Handbuch Landschaft* (pp. 23–38). Wiesbaden: Springer VS.

Berr, K., Jenal, C., Kühne, O., & Weber, F. (2019). Inter- und transdisziplinäre Landschaftsforschung. In O. Kühne, F. Weber, K. Berr, & C. Jenal (Eds.), *Handbuch Landschaft* (pp. 165–180). Wiesbaden: Springer VS.

Bertram, G. W. (2014). *Kunst als menschliche Praxis. Eine Ästhetik.* Berlin: Suhrkamp.

Beuka, R. (2004). *SuburbiaNation. Reading suburban landscape in twentieth-century American fiction and film.* New York: Palgrave Macmillan.

Beus, J. (2015). Klickwahrscheinlichkeiten in den Google SERPs. https://www.sistrix.de/news/klickwahrscheinlichkeiten-in-den-google-serps/. Accessed 20 May 2019.

Bhabha, H. K. (2000 [engl. Original 1994]). *Die Verortung der Kultur.* Tübingen: Stauffenburg.

Biographical Directory of the United States Congress. (o. J.). EDWARDS, Edwin Washington, (1927–). https://bioguide.congress.gov/scripts/biodisplay.pl?index=E000067. Accessed 17 June 2019.

Bishop, B. (2009). *The big sort. Why the clustering of like-minded American is tearing us apart.* Boston: Mariner Books (With a New Afterword).

Bittner, C. (2014). Reproduktion sozialräumlicher Differenzierungen in OpenStreetMap: das Beispiel Jerusalems. *Kartographische Nachrichten, 64*(3), 136–144.

Black, H. (2008). Unnatural disaster. Human factors in the Mississippi floods. *Environmental Health Perspectives, 116*(9), A390–A393. https://doi.org/10.1289/ehp.116-a390.

Bloch, E. (1962[1935]). *Erbschaft dieser Zeit* (Gesamtausgabe, Band 4, Erweiterte Ausgabe). Frankfurt a. M.: Suhrkamp.

Blodgett, A. D. (2006). An analysis of pollution and community advocacy in 'Cancer Alley': Setting an example for the environmental justice movement in St James Parish, Louisiana. *Local Environment, 11*(6), 647–661. https://doi.org/10.1080/13549830600853700.

Blotevogel, H. H. (1996). Aufgaben und Probleme der Regionalen Geographie heute. Überlegungen zur Theorie der Landes- und Länderkunde anläßlich des Gründungskonzepts des Instituts für Länderkunde, Leipzig. *Berichte zur deutschen Landeskunde, 70*(1), 11–40.

Bos, M., Carter, S. P., & Skiba, P. M. (2012). The pawn industry and its customers: The United States and Europe. Vanderbilt law and economics research paper: 12–26. https://papers.ssrn.com/sol3/papers.cfm?abstract_id=2149575.

Bottles, S. L. (1987). *Los Angeles and the automobile: The making of the modern city.* Berkeley: University of California Press.

Bourdieu, P. (1979 [frz. Original 1972]). *Entwurf einer Theorie der Praxis auf der ethnologischen Grundlage der kabylischen Gesellschaft.* Frankfurt a. M.: Suhrkamp.

Bourdieu, P. (1985a). *Sozialer Raum und „Klassen". Leçon sur la leçon; 2 Vorlesungen.* Frankfurt a. M.: Suhrkamp.

Bourdieu, P. (1985b). The market of symbolic goods. *Poetics, 14*(1–2), 13–44.

Bourdieu, P. (1989). Social space and symbolic power. *Sociological theory, 7*(1), 14–25.

Bourdieu, P. (1991). Physischer, sozialer und angeeigneter physischer Raum. In M. Wentz (Ed.), *Stadt-Räume* (pp. 25–34). Frankfurt a. M.: Campus.

Bourdieu, P. (1998). *Der Einzige und sein Eigenheim.* Hamburg: VSA-Verlag.

Bourdieu, P. (2016). *La distinction: Critique sociale du jugement* (Le Sens commun). Paris: Editions de Minuit.

Bourne, L. S. (1996). Reinventing the suburbs: Old myths and new realities. *Progress in Planning, 46*(3), 163–184. https://doi.org/10.1016/0305-9006(96)88868-4.

Bratzel, S. (1995). *Extreme der Mobilität. Entwicklung und Folgen der Verkehrspolitik in Los Angeles.* Basel: Birkhäuser.

Breckenfeld, G. (1977). Refilling the metropolitan doughnut. In D. C. Perry, & A. J. Watkins (Eds.), *The rise of the sunbelt cities* (Urban affairs annual reviews, vol. 14, pp. 231–258). Beverly Hills: SAGE.

Brian, J. L. (2003). Relative costs of infill vs. suburban residential developments: A case study of the greater Baton Rouge area, LSU Master's Theses. 1243. https://digitalcommons.lsu.edu/gradschool_theses/1243/. Accessed 13 June 2019.

Brill, D. (1963). *Baton Rouge, LA. Aufstieg, Funktionen und Gestalt einer jungen Großstadt des neuen Industriegebietes am unteren Mississippi* (Schriften des Geographischen Instituts der Universität Kiel, vol. 21, 2). Kiel: Selbstverlag des Geographischen Instituts der Universität Kiel.

Brillon, G. (2015). While Baton Rouge rap scene may be small, there is plenty of talent to be found. It's Bigger Than Hip-Hop. *The Advocate.* https://www.theadvocate.com/baton_rouge/entertainment_life/music/article_daf382ac-b92d-5001-8056-560295783a6d.html. Accessed 2 Dec 2019.

Brinkley, A. (1981). Huey Long, the share our wealth movement, and the limits of depression dissidence. *Louisiana History: The Journal of the Louisiana Historical Association, 22*(2), 117–134.

Brownill, S. (2011). London docklands revisited: The dynamics of waterfront development. In G. Desfor, J. Laidley, Q. Stevens, & D. Schubert (Eds.), *Transforming urban waterfronts. Fixity and flow* (Routledge Advances in Geography, vol. 3, pp. 121–142). New York: Routledge.

Bründl, W., Mayer, H., & Baumgartner, A. (1986). *Untersuchung des Einflusses von Bebauung und Bewuchs auf das Klima und die lufthygienischen Verhältnisse in bayerischen Großstädten* (Materiale Textkulturen, vol. 43). München: Bayerisches Staatsministerium für Landesentwicklung und Umweltfragen.

Bullard, R. D. (1997). Environmental justice for all. In R. D. Bullard (Ed.), *Unequal protection. Environmental justice and communities of color* (pp. 3–22). San Francisco: Sierra Club Books.

Bullard, R. D. (Ed.). (2005). *The quest for environmental justice. Human rights and the politics of pollution*. San Francisco: Sierra Club Books.

Bulmer, M. (1984). *The Chicago school of sociology. Institutionalization, diversity, and the rise of sociological research*. Chicago: University of Chicago Press.

Burby, R. J. (2000). Baton Rouge: The making (and breaking) of a petrochemical paradise. In C. E. Colten (Ed.), *Transforming New Orleans and its environs. Centuries of change* (History of the Urban Environment, pp. 160–177). Pittsburgh: University of Pittsburgh Press.

Burckhardt, L. (2004a). *Wer plant die Planung? Architektur, Politik und Mensch*. Berlin: Martin Schmitz Verlag.

Burckhardt, L. (2004b). Wert und Sinn städttebaulicher Utopien (1968). In L. Burckhardt (Ed.), *Wer plant die Planung? Architektur, Politik und Mensch* (pp. 146–162). Berlin: Martin Schmitz Verlag.

Burd-Sharps, S., Lewis, K., & Martins, E. B. (2009). A portrait of Louisiana. Louisiana human development report 2009, American human development project. https://mk0moaorgidkh7gs-b4pe.kinstacdn.com/wp-content/uploads/A_Portrait_of_Louisiana.pdf. Accessed 1 July 2019.

Campanella, R. (2002). *Time and place in New Orleans. Past geographies in the present day*. Gretna: Pelican Publishing Company.

Campanella, R. (2008). *Bienville's dilemma. A Historical geography of New Orleans*. Lafayette: Center for Louisiana Studies, University of Louisiana at Lafayette.

Carleton, M. T. (1981). *River capital. An illustrated history of Baton Rouge*. Woodland Hills: Windsor Publications Inc.

Carpenter, J. M. (2008). Consumer shopping value, satisfaction and loyalty in discount retailing. *Journal of Retailing and Consumer Services, 15*(5), 358–363. https://doi.org/10.1016/j.jretconser.2007.08.003.

Carpenter, J. M., & Moore, M. (2009). Utilitarian and hedonic shopping value in the US discount sector. *Journal of Retailing and Consumer Services, 16*(1), 68–74. https://doi.org/10.1016/j.jretconser.2008.10.002.

Carter, S. P. (2015). Payday loan and pawnshop usage: The impact of allowing payday loan rollovers. *The Journal of Consumer Affairs, 49*(2), 436–456. https://doi.org/10.1111/joca.12072.

Caughey, J., & Caughey, L. (Eds.). (1977). *Los Angeles. Biography of a city*. Berkeley: University of California Press.

Centers for Disease Control and Prevention. (2019). Leading cancer cases and deaths, male and female, 2011–2015. https://gis.cdc.gov/Cancer/USCS/DataViz.html. Accessed 12 Apr 2019.

Cerami, C. A. (2004). *Jefferson's great gamble. The remarkable story of Jefferson, Napoleon and the Men behind the Louisiana purchase*. Naperville: Sourcebooks Inc.

Champion, T. (2001). Urbanization, suburbanization, counterurbanization and reurbanization. In R. Paddison (Ed.), *Handbook of urban studies* (pp. 143–161). London: SAGE.

Chaney, D. (1997). Authenticity and suburbia. In S. Westwood, & J. Williams (Eds.), *Imagining cities. Scripts, signs, memories* (pp. 137–147). London: Routledge.

Chilla, T., Evrard, E., & Schulz, C. (2012). On the territoriality of cross-border cooperation: "Institutional Mapping" in a multi-level context. *European Planning Studies, 20*(6), 961–980. https://doi.org/10.1080/09654313.2012.673563.

Chilla, T., Kühne, O., Weber, F., & Weber, F. (2015). „Neopragmatische" Argumente zur Vereinbarkeit von konzeptioneller Diskussion und Praxis der Regionalentwicklung. In O. Kühne, & F. Weber (Eds.), *Bausteine der Regionalentwicklung* (pp. 13–24). Wiesbaden: Springer VS.

Chilla, T., Kühne, O., & Neufeld, M. (2016). *Regionalentwicklung* (UTB, vol. 4566). Stuttgart: Ulmer.

Christy, R. D., & Williamson, L. (Eds.). (2017). *A century of service. Land-Grant colleges and universities, 1890–1990.* London: Routledge.

City of Baton Rouge. (2019). Future Baton Rouge. https://www.brla.gov/662/FUTUREBR. Accessed 1 July 2019.

City of Baton Rouge and Parish of East Baton Rouge. (2015). Consolidated plan and strategy 2015–2019 and year 2015 action plan. https://www.brla.gov/DocumentCenter/View/2518/2015-2019-Consolidated-Plan-and-Strategy-with-Year-One-2015-Action-Plan-PDF. Accessed 1 July 2019.

Clark, T. D. (1946). The furnishing and supply system in southern agriculture since 1865. *The Journal of Southern History, 12*(1), 24–44. https://doi.org/10.2307/2197729.

Clarke, G. (Ed.). (1993). *The American landscape. Literary sources & documents*. Mountfield: Helm Information.

Colten, C. E. (2000). Too much of a good thing: Industrial pollution in the lower Mississippi river. In C. E. Colten (Ed.), *Transforming New Orleans and its environs. Centuries of change* (History of the Urban Environment, pp. 141–159). Pittsburgh: University of Pittsburgh Press.

Colten, C. E. (2006a). *An unnatural metropolis. Wresting New Orleans from nature*. Baton Rouge: LSU Press.

Colten, C. E. (2006b). The rusting of the chemical corridor. *Technology and Culture, 47*(1), 95–101.

Colten, C. E. (2012). An incomplete solution: Oil and water in Louisiana. *The Journal of American History, 99*(1), 91–99. https://doi.org/10.1093/jahist/jas023.

Colten, C. E. (2014a). Industrie und Umwelt in den Südstaaten der USA. Eine historisch-geographische Betrachtung. *Geographische Rundschau, 66*(3), 53–37.

Colten, C. E. (2014b). *Southern waters. The limits to abundance*. Baton Rouge: Louisiana State University Press.

Combs, H. J. (2003). The antiques trade. The diffusion of America's cultural patrimony. *Material Culture, 35*(1), 42–54.

Conzen, M. P. (Ed.). (2010). *The making of the American landscape* (2nd ed.). New York: Routledge.

Conzen, M. P. (2017). Reflection: Cultural forces in the American landscapes. In W. Gamerith, & U. Gerhard (Eds.), *Kulturgeographie der USA. Eine Nation begreifen* (pp. 15–20). Berlin: Springer Spektrum.

Cosgrove, D. E. (1984). *Social formation and symbolic landscape*. London: University of Wisconsin Press.

Couvillion, B. R., Barras, J. A., Steyer, G. D., Sleavin, W., Fischer, M., & Beck, H., et al. (2011). Land area change in coastal Louisianafrom 1932 to 2010. https://pubs.usgs.gov/sim/3164/downloads/SIM3164_Pamphlet.pdf. Accessed 24 Jan 2020.

Couvillion, B. R., Beck, H., Schoolmaster, D., & Fischer, M. (2017). Land area change in coastal Louisiana (1932 to 2016). https://pubs.usgs.gov/sim/3381/sim3381_pamphlet.pdf. Accessed 24 Jan 2020.

Cowan, B. (2013). *Louisiana state university* (The Campus History Series). Charleston: Arcadia Publishing.

Crilley, D. (1993). Architecture as advertising: Constructing the image of redevelopment. In G. Kearns, & C. Philo (Eds.), *Selling places. The city as cultural capital, past and present* (Policy, planning, and critical theory, pp. 231–252). Oxford: Pergamon Press.

Cross, C. F., II. (1999). *Justin Smith Morrill. Father of the Land-Grant colleges.* East Lansing: Michigan State University Press.

Cuff, D. (2000). *The provisional city. Los Angeles stories of architecture and urbanism.* Cambridge: MIT Press.

Culver, L. (2010). *The frontier of leisure. Southern California and the shaping of modern America.* Oxford: Oxford University Press.

Culver, G. (2017). Mobilität und die amerikanische Stadt. In W. Gamerith, & U. Gerhard (Eds.), *Kulturgeographie der USA. Eine Nation begreifen* (pp. 177–186). Berlin: Springer Spektrum.

Cummins, L. T. (2014). Part one. In B. H. Wall, & J. C. Rodrigue (Eds.), *Louisiana. A history* (6th ed., pp. 7–104). Chichester: Wiley-Blackwell.

Cummins, S., & Macintyre, S. (2002). "Food deserts" – Evidence and assumption in health policy making. *BMJ British Medical Journal, 325*(7361), 436–438. https://doi.org/10.1136/bmj.325.7361.436.

Curran, J. (Ed.). (1997). *Mass media and society* (2. edition, 3. impression). London: Arnold.

Cutter, S. L., Boruff, B. J., & Shirley, W. L. (2003). Social vulnerability to environmental hazards. *SSQ – Social Science Quaterly, 84*(2), 242–261. https://doi.org/10.1111/1540-6237.8402002.

Dahrendorf, R. (1961). *Gesellschaft und Freiheit. Zur soziologischen Analyse der Gegenwart.* München: Piper.

Dahrendorf, R. (1963). *Die angewandte Aufklärung. Gesellschaft und Soziologie in Amerika.* München: Piper.

Dahrendorf, R. (1965). *Industrie- und Betriebssoziologie.* Berlin: de Gruyter.

Dahrendorf, R. (1968a). *Bildung ist Bürgerrecht. Plädoyer für eine aktive Bildungspolitik* (Die Zeit Bücher, Neuauflage). Hamburg: Christian Wegner.

Dahrendorf, R. (1968b). *Pfade aus Utopia. Arbeiten zur Theorie und Methode der Soziologie.* München: Piper.

Dahrendorf, R. (1972). *Konflikt und Freiheit. Auf dem Weg zur Dienstklassengesellschaft.* München: Piper.

Dahrendorf, R. (1979a). Frieden durch Politik. In Volksbund Deutsche Kriegsgräberfürsorge Landesverband Baden-Württemberg (Ed.), *Rückblick für die Zukunft* (pp. 11–22). Konstanz: Volksbund Deutsche Kriegsgräberfürsorge Landesverband Baden-Württemberg.

Dahrendorf, R. (1979b). *Lebenschancen. Anläufe zur sozialen und politischen Theorie* (Suhrkamp-Taschenbuch, vol. 559). Frankfurt a. M.: Suhrkamp.

Dahrendorf, R. (1983). *Die Chancen der Krise. Über die Zukunft des Liberalismus.* Stuttgart: Deutsche Verlags-Anstalt.

Dahrendorf, R. (1987). *Fragmente eines neuen Liberalismus.* Stuttgart: Deutsche Verlags-Anstalt.

Dahrendorf, R. (1991). Die gefährdete Civil Society. In K. Michalski (Ed.), *Europa und die Civil Society. Castelgandolfo-Gespräche 1989* (pp. 247–263). Stuttgart: Klett-Cotta.

Dahrendorf, R. (1994). *Der moderne soziale Konflikt. Essay zur Politik der Freiheit.* München: Deutscher Taschenbuch Verlag.

Dahrendorf, R. (2007). *Auf der Suche nach einer neuen Ordnung. Vorlesungen zur Politik der Freiheit im 21. Jahrhundert* (Krupp-Vorlesungen zu Politik und Geschichte am Kulturwissenschaftlichen Institut im Wissenschaftszentrum Nordrhein-Westfalen, vol. 3, 4. Aufl.). München: Beck.

Davis, M. (2004). *Ökologie der Angst. Das Leben mit der Katastrophe* (Serie Piper, vol. 3819). München: Piper.

Davison, P. (2012). The language of internet memes. *The social media reader,* 120–134. New York University Press: New York.

Dawkins, R. (1978). *The selfish gene*. London: Paladin.

de Landa, M. (2006). *A new philosophy of society. Assemblage theory and social complexity*. London: continuum.

Dear, M. J. (Ed.). (2004). *From Chicago to L.A. making sense of urban theory*. Thousand Oaks: SAGE.

Dear, M. (2005). Die Los Angeles School of Urbanism. *Geographische Rundschau, 57*(1), 30–36.

Dear, M., & Flusty, S. (1998). Postmodern urbanism. *Annals of the Association of American Geographers, 88*(1), 50–72.

Degen, M. M. (2008). *Sensing cities. Regenerating public life in Barcelona and Manchester*. London: Routledge.

Desmond, J. M., Raymond, G., Prescott, M., Simon, C., Kerry, B., Simon, D., et al. (2005). *Multifamily housing opportunites in downtown Baton Rouge*. Baton Rouge: LSU School of Architecture.

Diaz, J. H. (2011). The legacy of the Gulf oil spill: Analyzing acute public health effects and predicting chronic ones in Louisiana. *American Journal of Disaster Medicine, 6*(1), 5–22.

Dill, K. E. (2009). *How fantasy becomes reality: Seeing through media influence*. New York: Oxford University Press.

Dimaggio, P., & Useem, M. (1978). Social class and arts consumption. The origins and consequences of class differences in exposure to the arts in America. *Theory and Society, 5*(2), 141–161. https://doi.org/10.1007/BF01702159.

Dörfler, T., & Rothfuß, E. (2018). Lebenswelt, Leiblichkeit und Resonanz: Eine raumphänomenologisch-rekonstruktive Perspektive auf Geographien der Alltäglichkeit. *Geographica Helvetica, 73*(1), 95–107. https://doi.org/10.5194/gh-73-95-2018.

Dottle, R. (FiveThirtyEight, Ed.). (2019). Where democrats and republicans live in your city. Republicans and democrats tend not to live side-by-side, even when they live in the same city. https://projects.fivethirtyeight.com/republicans-democrats-cities/. Accessed 13 Dec 2019.

Douglas, M. L. (1955). *Some aspects of the social history of Baton Rouge from 1830 to 1850*. Baton Rouge: Eigenverlag.

Draughon Jr., R. (1998). *Down by the river. A history of the Baton Rouge riverfront* (Preserving Louisiana's Heritage). New Orleans: US Army Corps of Engineers, New Orleans District.

Driskell, L. (2010). Mapping the digital divide in neighborhoods: Wi-Fi access in Baton Rouge, Louisiana, LSU Master's Theses. 3631. https://digitalcommons.lsu.edu/gradschool_theses/3631/. Accessed 1 July 2019.

Driskell, L., & Wang, F. (2009). Mapping digital divide in neighborhoods: Wi-Fi access in Baton Rouge, Louisiana. *Annals of GIS, 15*(1), 35–46.

Duany, A., Plater-Zyberk, E., & Speck, J. (2000). *Suburban nation. The rise of sprawl and the decline of the American dream*. New York: North Point Press.

Duany Plater-Zyberk & Company. (1999). Plan Baton Rouge. Updated by plan Baton Rouge 2002. https://static1.squarespace.com/static/536d55f1e4b07afeea8cef61/t/589202ea20099ec08ca823fa/1485964024093/Plan_BR.pdf. Accessed 29 Nov 2019.

Duncan, J. S. (1980). The superorganic in American cultural geography. *Annals of the Association of American Geographers, 70*(2), 181–198.

Duncan, J. S. (1990). *The city as text: The politics of landscape interpretation in the Kandyan Kingdom*. Cambridge: Cambridge University Press.

Duncan, J. S., & Duncan, N. (2004). *Landscapes of privilege. The politics of the aesthetics in an American suburb*. New York: Routledge.

Duncan, N., & Duncan, J. (2009). Doing landscape interpretation. In D. DeLyser, S. Herbert, S. Aitken, M. A. Crang, & L. McDowell (Eds.), *The SAGE handbook of qualitative geography.*

London: SAGE. https://www.sage-ereference.com/hdbk_qualgeography/Article_n13.html. Accessed 7 Jan 2019.

Düttmann, S. (2000). *Ästhetische Lernprozesse. Annäherungen an atmosphärische Wahrnehmungen von LernRäumen*. Marburg: Tectum Verlag.

Eckardt, F. (2014). *Stadtforschung. Gegenstand und Methoden*. Wiesbaden: Springer VS.

Edensor, T. (2005). *Industrial ruins. Spaces, aesthetics and materiality*. Oxford: Berg.

Edler, D., Kühne, O., Jenal, C., Vetter, M., & Dickmann, F. (2018). Potenziale der Raumvisualisierung in Virtual Reality (VR) für die sozialkonstruktivistische Landschaftsforschung. *Kartographische Nachrichten, 68*(5), 245–254.

Eisel, U. (2009). *Landschaft und Gesellschaft. Räumliches Denken im Visier* (Raumproduktionen: Theorie und gesellschaftliche Praxis, Bd. 5). Münster: Westfälisches Dampfboot.

Ellis, C. (2002). The new urbanism: Critiques and rebuttals. *Journal of Urban Design, 7*(3), 261–291.

EMPORIS. (o. J.). Baton Rouge. Buildings of the city. https://web.archive.org/web/20101225001940/http://www.emporis.com/en/wm/ci/bu/?id=102363. Accessed 14 June 2019.

Entrikin, J. N. (1996). Place and region 2. *Progress in Human Geography, 20*(2), 215–221. https://doi.org/10.1177/030913259602000206.

Färber, A. (2014). Potenziale freisetzen: Akteur-Netzwerk-Theorie und Assemblageforschung in der interdisziplinären kritischen Stadtforschung. *sub\urban zeitschrift für kritische stadtforschung, 2*(1), 95–103.

Fertig, G. (2005). Zwischen Xenophobie und Freundschaftspreis: Landmarkt und familiäre Beziehungen in Westfalen, 1830–1866. *Jahrbuch für Wirtschaftsgeschichte/Economic History Yearbook, 46*(1), 53–76. https://doi.org/10.1524/jbwg.2005.46.1.53.

Fine, A. (2000). Der Blickpunkt von niemand besonderen. In M. Sandbothe (Ed.), *Die Renaissance des Pragmatismus. Aktuelle Verflechtungen zwischen analytischer und kontinentaler Philosophie* (pp. 59–77). Weilerswist: Velbrück.

Finlay, L. (2012). Debating phenomenological methods. In N. Friesen, C. Henriksson, & T. Saevi (Eds.), *Hermeneutic phenomenology in education. Method and practice* (Practice of Research Method, vol. 4, pp. 17–37). Rotterdam: SensePublishers.

Flach, E. (1957). Grundbegriffe und Grundtatsachen der Bioklimatologie. In F. Baur (Ed.), *Linkes Meterologisches Taschenbuch. Neue Ausgabe III* (pp. 178–271). Leipzig: Akademische Verlagsgesellschaft Geest & Portig.

Fleming, W. L. (1936). *Louisiana state university, 1860–1896*. Baton Rouge: Louisiana State University Press.

Flick, U. (2007). Triangulation in der qualitativen Forschung. In U. Flick, E. v Kardorff, & I. Steinke (Eds.), *Qualitative Forschung Ein Handbuch* (pp. 309–318). Reinbek bei Hamburg: Rowohlt.

Flusty, S. (1999). Building paranoia. In N. Ellin (Ed.), *Architecture of fear* (pp. 48–52). New York: Princeton Architectural Press.

Foner, E. (1995). *Reconstruction. America's unfinished revolution, 1863–1877*. New York: Harper & Row.

Frey, W. H. (1979). Central city white flight: Racial and nonracial causes. *American Sociological Review, 44*(3), 425–448.

Gailing, L. (2014). *Kulturlandschaftspolitik. Die gesellschaftliche Konstituierung von Kulturlandschaft durch Institutionen und Governance* (Planungswissenschaftliche Studien zu Raumordnung und Regionalentwicklung, vol. 4). Detmold: Rohn.

Gailing, L., & Leibenath, M. (2015). The social construction of landscapes: Two theoretical lenses and their empirical applications. *Landscape Research, 40*(2), 123–138. https://doi.org/10.1080/01426397.2013.775233.

Gallagher, L. (2013). *The end of the suburbs. Where the American dream is moving*. New York: Portfolio/Penguin.

Gamerith, W. (2017). USA regional: Der „Alte Süden“. In W. Gamerith, & U. Gerhard (Eds.), *Kulturgeographie der USA. Eine Nation begreifen* (pp. 115–121). Berlin: Springer Spektrum.

Garreau, J. (1992). *Edge city. Life on the new frontier*. New York: Anchor Books.

Gaudet, M., & McDonald, J. C. (Eds.). (2003). *Mardi Gras, Gumbo and Zydeco. Readings in Louisiana Culture*. Jackson: University Press of Mississippi.

Gebhardt, H. (2016). Entwicklungspfade und Perspektiven der Humangeographie im deutschsprachigen Raum – Einige Leitlinien. In J. Aistleitner, M. Coy, & J. Stötter (Eds.), *Die Welt verstehen – eine geographische Herausforderung. Eine Festschrift der Geographie Innsbruck für Axel Borsdorf* (Innsbrucker geographische Studien, vol. 40, pp. 43–59). Innsbruck: Geographie Innsbruck Selbstverlag.

Gebhardt, H., Reuber, P., & Wolkersdorfer, G. (2004). Konzepte und Konstruktionsweisen regionaler Geographien im Wandel der Zeit. *Berichte zur deutschen Landeskunde, 78*(3), 293–312.

Georgi, N. J., & Zafiriadis, K. (2006). The impact of park trees on microclimate in urban areas. *Urban Ecosystems, 9*(3), 195–209.

Gerhard, U. (2017). Reurbanisierung – Mehr als ein neoliberaler Diskurs. In W. Gamerith, & U. Gerhard (Eds.), *Kulturgeographie der USA. Eine Nation begreifen* (pp. 145–152). Berlin: Springer Spektrum.

Gerhard, U., & Gamerith, W. (2017). „From New York to LA“: Aufstieg, Blüte und Krise der US-amerikanischen Stadt. In W. Gamerith, & U. Gerhard (Eds.), *Kulturgeographie der USA. Eine Nation begreifen* (pp. 127–135). Berlin: Springer Spektrum.

Gibson, E. B. (2002). The changing face of HIV/AIDS: An anthropological and epidemiological study of the Baton Rouge area, LSU Master's Theses. 1759. https://digitalcommons.lsu.edu/gradschool_theses/1759/. Accessed 1 July 2019.

Gilbert, A. (1988). The new regional geography in English and French-speaking countries. *Progress in Human Geography, 12*(2), 208–228. https://doi.org/10.1177/030913258801200203.

Gladstone, D. L., & Fainstein, S. S. (2003). The New York and Los Angeles economies. In D. Halle (Ed.), *New York & Los Angeles. Politics, society, and culture: A comparative view* (pp. 79–98). Chicago: University of Chicago Press.

Gläser, J., & Laudel, G. (2010). *Experteninterviews und qualitative Inhaltsanalyse als Instrumente rekonstruierender Untersuchungen* (Lehrbuch, 4. Aufl.). Wiesbaden: VS Verlag.

Glasze, G. (2014). Sozialwissenschaftliche Kartographie-, GIS- und Geoweb-Forschung. *Kartographische Nachrichten, 64*(3), 123–129.

Glasze, G., & Mattissek, A. (2009). Diskursforschung in der Humangeographie: Konzeptionelle Grundlagen und empirische Operationalisierung. In G. Glasze, & A. Mattissek (Eds.), *Handbuch Diskurs und Raum. Theorien und Methoden für die Humangeographie sowie die sozial- und kulturwissenschaftliche Raumforschung* (pp. 11–59). Bielefeld: transcript.

Gleason, D. K., & Brockway, W. R. (1991). *Baton Rouge. Photographs and text.* Baton Rouge: Louisiana State University Press.

Gottlieb, M. S. (1980). Lung cancer and the petroleum industry in Louisiana. *Journal of Occupational medicine, 22*(6), 384–388.

Granitz, E., & Klein, B. (1996). Monopolization by "Raising Rivals' Costs": The standard oil case. *The Journal of Law and Economics, 39*(1), 1–47.

Grant, J. (2006). *Planning the good community. New urbanism in theory and practice* (The RTPI library series, vol. 9). London: Routledge.

Grant, J. L., Nelson, A. C., Forsyth, A., Thompson-Fawcett's, M., Blais, P., & Filion, P. (2013). The future of the suburbs. Suburbs in transition/The resettlement of America's suburbs/Suburbs in global context: The challenges of continued growth and retrofitting/Suburban urbanity: Re-envisioning indigenous settlement practices/Toward a new suburban America: Will we catch the wave?/Optimistic and pessimistic perspectives on the evolution of the North American suburb/Response. *Planning Theory & Practice, 14*(3), 391–415. https://doi.org/10.1080/14649357.2013.808833.

Greider, T., & Garkovich, L. (1994). Landscapes: The social construction of nature and the environment. *Rural Sociology, 59*(1), 1–24. https://doi.org/10.1111/j.1549-0831.1994.tb00519.x.

Ha, K. N. (2006). Die Grenze überqueren? Hybridität als spätkapitalistische Logik der kulturellen Übersetzung und der nationalen Modernisierung. https://eipcp.net/transversal/1206/ha/de. Accessed 29 Nov 2018.

Haas, E. F. (2014). Part Three. In B. H. Wall, & J. C. Rodrigue (Eds.), *Louisiana. A history* (6th ed., pp. 233–323). Chichester: Wiley-Blackwell.

Hair, W. I. (1969). *Bourbonism and Agrarian protest. Louisiana politics 1877–1900*. Baton Rouge: Louisiana State University Press.

Hair, W. I. (1996). *The kingfish and his realm. The life and times of Huey P. Long*. Baton Rouge: Louisiana State University Press.

Hall, T. (1995). 'The second industrial revolution': Cultural reconstructions of industrial regions. *Landscape Research, 20*(3), 112–123. https://doi.org/10.1080/01426399508706465.

Hall, T. (2006). *Urban geography*. London: Routledge.

Hanich, J. (2007). Automobile Einsamkeit. Freeways im Dauerstau – In Los Angeles wird total mobil gemacht. *Lettre International, 76,* 34–37.

Hanlon, B. (2010). *Once the American dream. Inner-ring suburbs of the metropolitan United States*. Philadelphia: Temple University Press.

Hard, G. (1970). „Was ist eine Landschaft?“ Über Etymologie als Denkform in der geographischen Literatur. In D. Bartels (Ed.), *Wirtschafts- und Sozialgeographie* (Neue wissenschaftliche Bibliothek, vol. 35, pp. 66–84). Köln: Kiepenheuer & Witsch.

Hard, G. (1977). Zu den Landschaftsbegriffen der Geographie. In A. Hartlieb von Wallthor, & H. Quirin (Eds.), *„Landschaft“ als interdisziplinäres Forschungsproblem. Vorträge und Diskussionen des Kolloquiums am 7./8. November 1975 in Münster* (pp. 13–24). Münster: Aschendorff.

Hard, G. (1995). *Spuren und Spurenleser. Zur Theorie und Ästhetik des Spurenlesens in der Vegetation und anderswo* (Osnabrücker Studien zur Geographie, vol. 16). Osnabrück: Universitätsverlag Rasch.

Hard, G. (Ed.). (2003). *Dimensionen geographischen Denkens. Aufsätze zur Theorie der Geographie* (Osnabrücker Studien zur Geographie, vol. 23). Göttingen: V & R Unipress.

Hard, G. (2008). Der Spatial Turn, von der Geographie her beobachtet. In J. Döring, & T. Thielmann (Eds.), *Spatial Turn. Das Raumparadigma in den Kultur- und Sozialwissenschaften* (pp. 263–316). Bielefeld: transcript.

Hardinghaus, M. (2004). *Zur amerikanischen Entwicklung der Stadt. Ein Beitrag zur Kulturgenese des City-Suburb-Phänomens unter besonderer Berücksichtigung protestantisch-calvinistischer Leitbilder*. Frankfurt a. M.: Lang.

Harlander, T. (2009). Suburbs, Sun Cities und Gated Communities. Krise und Fragmentierung im Sunbelt der USA. In T. Harlander, & D. Schubert (Eds.), Suburbanisierung und Reurbanisierung. Wohnungsbau und Immobilienkrise in den USA. *Die Alte Stadt 36* (2), 177–198 [Special issue]. Remshalden: BAG-Verlag.

Harris, A. (The Atlantic, Ed.). (2019). The New Secession. Residents of the majority-white southeast corner of Baton Rouge want to make their own city, complete with its own schools, breaking away from the majority-black parts of town. https://www.theatlantic.com/education/archive/2019/05/resegregation-baton-rouge-public-schools/589381/. Accessed 13 Dec 2019.

Hasse, J. (2000). *Die Wunden der Stadt. Für eine neue Ästhetik unserer Städte*. Wien: Passagen-Verlag.

Hasse, J. (2012). *Atmosphären der Stadt. Aufgespürte Räume*. Berlin: Jovis.

Hasse, J. (2017). Die Abwesenheit der Phänomenologie in der deutschen Humangeographie. *Geographica Helvetica, 72*(3), 351–360. https://doi.org/10.5194/gh-72-351-2017.

Hauser, S. (2001). *Metamorphosen des Abfalls. Konzepte für alte Industrieareale*. Frankfurt a. M.: Campus.

Häußermann, H. (1998). "Amerikanisierung" der deutschen Städte. In W. Prigge (Ed.), *Peripherie ist überall* (Edition Bauhaus, vol. 1, pp. 76–83). Frankfurt a. M.: Campus.

Häußermann, H. (2009). Der Suburbanisierung geht das Personal aus. Eine stadtsoziologische Zwischenbilanz. *StadtBauwelt, 181,* 52–57.

Hayden, D. (2004a). *A field guide to sprawl*. New York: W. W. Norton & Company.

Hayden, D. (2004b). *Building Suburbia. Green fields and urban growth, 1820–2000*. New York: Vintage Books.

Hayden, D. (2009). Muster amerikanischer Vorstädte. Ein bauhistorischer Essay. *Bauwelt, 12,* 20–33.

Hebert, M. J. (1999). Beyond black and white: The civil rights movement in Baton Rouge, Louisiana, 1945–1972, LSU Historical Dissertations and Theses. 7045. https://digitalcommons.lsu.edu/gradschool_disstheses/7045/. Accessed 17 June 2019.

Hein, K. (2006). *Hybride Identitäten. Bastelbiografien im Spannungsverhältnis zwischen Lateinamerika und Europa*. Bielefeld: transcript.

Heisler, G. M. (1974). Trees and human comfort in urban areas. *Journal of Forestry, 72*(8), 466–469.

Helfferich, C. (2011). *Die Qualität qualitativer Daten. Manual für die Durchführung qualitativer Interviews* (4th ed.). Wiesbaden: VS Verlag.

Helfferich, C. (2014). Leitfaden- und Experteninterviews. In N. Baur, & J. Blasius (Eds.), *Handbuch Methoden der empirischen Sozialforschung* (pp. 559–574). Wiesbaden: Springer VS.

Hendry, P. M., & Edwards, J. D. (2009). *Old South Baton Rouge. The roots of hope*. Lafayette: University of Louisiana at Lafayette Press.

Hennig, E. (2012). Chicago School. In F. Eckardt (Ed.), *Handbuch Stadtsoziologie* (pp. 95–124). Wiesbaden: VS Verlag.

Henninger, S. (Ed.). (2011). *Stadtökologie. Bausteine des Ökosystems Stadt*. Paderborn: Ferdinand Schöningh.

Henriksson, C., & Friesen, N. (2012). Introduction. In N. Friesen, C. Henriksson, & T. Saevi (Eds.), *Hermeneutic phenomenology in education. Method and practice* (Practice of Research Method, vol. 4, pp. 1–14). Rotterdam: SensePublishers.

Herzog, L. (2013). *Freiheit gehört nicht nur den Reichen – Plädoyer für einen zeitgemäßen Liberalismus* (C.-H.-Beck Paperback, vol. 6127, Orig.-Ausg). München: Beck.

Hesse, M. (2010). Suburbs: The next slum? Explorations into the contested terrain of social construction and political discourse. *Articulo. Journal of Urban Research,* (*Special issue 3*, 43 Absätze). https://articulo.revues.org/1552. Accessed 8 Jan 2020.

Hibbert, C. (1969). *London. The biography of a city*. London: Longmans.

Hibbert, C. (2001). *Rome. The biography of a city*. London: Penguin Books.

Hirsch, J., & Roth, R. (1986). *Das neue Gesicht des Kapitalismus. Vom Fordismus zum Post-Fordismus*. Hamburg: VSA-Verl.

Hochschild, A. R. (2016). *Strangers in their own land. Anger and mourning on the American right*. New York: The New Press.

Hoesterey, I. (2001). *Pastiche. Cultural memory in art, film, literature*. Bloomington: Indiana University Press.

Höfer, W., & Vicenzotti, V. (2013). Post-industrial landscapes: Evolving concepts. In P. Howard, I. Thompson, & E. Waterton (Eds.), *The Routledge companion to landscape studies* (pp. 405–416). London: Routledge.

Hofmeister, B. (1971). *Stadt und Kulturraum Angloamerika*. Braunschweig: Vieweg.

Hofmeister, S., & Kühne, O. (2016). StadtLandschaften: Die neue Hybridität von Stadt und Land. In S. Hofmeister, & O. Kühne (Eds.), *StadtLandschaften. Die neue Hybridität von Stadt und Land* (pp. 1–10). Wiesbaden: Springer VS.

Hokema, D. (2015). Landscape is everywhere. The construction of landscape by US-American laypersons. *Geographische Zeitschrift, 103*(3), 151–170.

Holzner, L. (1996). *Stadtland USA: Die Kulturlandschaft des American Way of Life*. Gotha: Justus Perthes Verlag.

Hoyle, B. (2000). Global and local change on the port-city waterfront. *Geographical Review, 90*(3), 395–417. https://doi.org/10.2307/3250860.

Huebner, W. W., Wojcik, N. C., Rosamilia, K., Jorgensen, G., & Milano, C. A. (2004). Mortality updates (1970–1997) of two refinery/petrochemical plant cohorts at Baton Rouge, Louisiana, and Baytown, Texas. *JOEM – Journal of Occupational and Environmental Medicine, 46*(12), 1229–1245.

Illing, F. (2006). *Kitsch, Kommerz und Kult. Soziologie des schlechten Geschmacks*. Konstanz: UVK Verlagsgesellschaft.

Imbert, D. (2007). The AIAJM: A manifesto for landscape modernity. *Landscape Journal, 26*(2), 219–235.

Internet Archive. (o. J.). End of the Rogue. https://web.archive.org/web/20090808194351/http://archive.salon.com/news/feature/2000/05/12/edwards/index1.html. Accessed 17 June 2019.

Ipsen, D. (2006). *Ort und Landschaft*. Wiesbaden: VS Verlag.

Jackson, J. B. (1984). *Discovering the Vernacular Landscape*. New Haven: Yale University Press.

Jackson, J. B. (2005). Die Zukunft des Vernakulären [1990]. In B. Franzen, & S. Krebs (Eds.), *Landschaftstheorie. Texte der Cultural Landscape Studies* (Kunstwissenschaftliche Bibliothek, vol. 26, pp. 45–56). Köln: König.

Jackson, J. M. (2014). Louisiana cultures. A Creole Mosaic. In R. Orgera, & W. Parent (Eds.), *The Louisiana field guide. Understanding life in the Pelican State* (pp. 36–71). Baton Rouge: Louisiana State University Press.

Jeansonne, G. (1980). Challenge to the New Deal: Huey P. Long and the Redistribution of National Wealth. *Louisiana History: The Journal of the Louisiana Historical Association, 21*(4), 331–339.

Jenal, C. (2019a). (Alt)Industrielandschaften. In O. Kühne, F. Weber, K. Berr, & C. Jenal (Eds.), *Handbuch Landschaft* (pp. 831–841). Wiesbaden: Springer VS.

Jenal, C. (2019b). *„Das ist kein Wald, Ihr Pappnasen!" – Zur sozialen Konstruktion von Wald. Perspektiven von Landschaftstheorie und Landschaftspraxis*. Wiesbaden: Springer VS.

Jenal, C., & Kühne, O. (2018). Postmoderne Stadtlandhybride. In H. Megerle (Ed.), *Bausteine für integrierte und nachhaltige Zukunftsstrategien. Tagungsbeiträge des siebten Hochschultages Ländlicher Raum Baden-Württemberg* (Geographie in Wissenschaft und Praxis, vol. 4, pp. 139–148). Rottenburg am Neckar: Druckerei Maier GmbH.

Jenkins, W. L. & Cowan, B. (2013). Foreword. In *Louisiana State University* (The Campus History Series, p. 6). Charleston: Arcadia Publishing.

Joas, H. (1988). Symbolischer Interaktionismus. Von der Philosophie des Pragmatismus zu einer soziologischen Forschungstradition. *Kölner Zeitschrift für Soziologie und Sozialpsychologie, 40,* 417–446.

Jones, C. (2006). *Paris. The biography of a city*. London: Penguin Books.

Jones, M., & Paasi, A. (2015). Introduction: Regional world(s): Advancing the geography of regions. In M. Jones, & A. Paasi (Eds.), *Regional worlds. Advancing the geography of regions* (pp. 1–5). Abingdon: Routledge.

Jones, T. L. (The Advocate, Ed.). (2019a). Here's why St. George's effort in Baton Rouge will happen, leaders of similar project say. https://www.theadvocate.com/baton_rouge/news/article_1edd239e-2afa-11e9-ac4c-8bf177e0e1a2.html. Accessed 13 Dec 2019.

Jones, T. L. (The Advocate, Ed.). (2019b). What's next for people within the new St. George? As some move forward, others look for way out. https://www.theadvocate.com/baton_rouge/news/article_113c4802-eea2-11e9-b0a2-4b98c655eaa1.html. Accessed 13 Dec 2019.

Jones, L. R., & Bryan, P. W. (1957). *North America. An historical, economic and regional geography*. London: Methuen.

Jorgensen, A., & Tylecote, M. (2007). Ambivalent landscapes – Wilderness in the urban interstices. *Landscape Research, 32*(4), 443–462.

Kane, C. (2018). *Abandoned Baton Rouge. Stories from the ruins*. Mount Pleasant: America Through Time.

Kaplan, D. H., Wheeler, J. O., & Holloway, S. R. (2009). *Urban geography* (2nd ed.). Hoboken: Wiley.

Karl, T. R., Melillo, J. M., & Peterson, T. C. (Eds.). (2009). *Global climate change impacts in the United States. A state of knowledge report from the U.S. global change research program*. Cambridge: Cambridge University Press.

Karlin, S. (2019a). Exxon to move forward with Baton Rouge plant expansion, increase capacity by 450, 000 tons. *The Advocate*. https://www.theadvocate.com/baton_rouge/news/business/article_bf324ffa-3c33-11e9-8a83-8347f3031760.html. Accessed 2 Jan 2020.

Karlin, S. (2019b). Louisiana lawmakers, over objections from Mayor Broome, pass St. George transition bill. *The Advocate*. https://www.theadvocate.com/baton_rouge/news/politics/legislature/article_cee67e14-83c1-11e9-8eaf-3f731f98791a.html. Accessed 2 Jan 2020.

Karson, R. S., & Betsch, C. (2007). *A genius for place. American landscapes of the country place era*. Amherst: University of Massachusetts Press.

Katz, P. (1994). *The new urbanism. Toward an architecture of community*. New York: McGraw-Hill Inc.

Kaufmann, S. (2005). *Soziologie der Landschaft*. Wiesbaden: VS Verlag.

Kazig, R. (2013). Landschaft mit allen Sinnen – Zum Wert des Atmosphärenbegriffs für die Landschaftsforschung. In D. Bruns, & O. Kühne (Eds.), *Landschaften: Theorie, Praxis und internationale Bezüge. Impulse zum Landschaftsbegriff mit seinen ästhetischen, ökonomischen, sozialen und philosophischen Bezügen mit dem Ziel, die Verbindung von Theorie und Planungspraxis zu stärken* (pp. 221–232). Schwerin: Oceano Verlag.

Kazig, R. (2016). Die Bedeutung von Alltagsästhetik im Kontext der Polarisierung und Hybridisierung von Städten. Eine Spurensuche. In F. Weber, & O. Kühne (Eds.), *Fraktale Metropolen. Stadtentwicklung zwischen Devianz, Polarisierung und Hybridisierung* (pp. 215–230). Wiesbaden: Springer VS.

Kazig, R. (2019). Atmosphären und Landschaft. In O. Kühne, F. Weber, K. Berr, & C. Jenal (Eds.), *Handbuch Landschaft* (pp. 453–460). Wiesbaden: Springer VS.

Kazig, R., & Weichhart, P. (2009). Die Neuthematisierung der materiellen Welt in der Humangeographie. *Berichte zur deutschen Landeskunde, 83*(2), 109–128.

Keim, B. D., & Muller, R. A. (1993). Frequency of heavy rainfall events in New Orleans, Louisiana, 1900 to 1991. *Southeastern Geographer, 33*(2), 159–171.

Kelleher Schafer, J. (2014). Part two. In B. H. Wall, & J. C. Rodrigue (Eds.), *Louisiana. A history* (6th ed., pp. 105–232). Chichester: Wiley-Blackwell.

Kindler, H., Weber, F., Kühne, O., & Halder, G. (2019). *Wissenschaftlich Arbeiten in Geographie und Raumwissenschaften. Ein Überblick*. Wiesbaden: Springer VS.

Klohn, W. (1998). Die US-amerikanische Stadt im Wandel. Eine modellhafte Darstellung des „Typischen" der US-amerikanischen Stadt sowie ihrer Entwicklungen und Umstrukturierungen in jüngerer Zeit. *Zeitschrift für den Erdkundeunterricht, 50*(4), 202–208.

Klotz, H. (1985). *Moderne und Postmoderne. Architektur der Gegenwart 1960–1980* (Schriften des Deutschen Architekturmuseums zur Architekturgeschichte und Architekturtheorie, 2. Aufl.). Wiesbaden: Vieweg + Teubner.

Kneer, G. (2009). Akteur-Netzwerk-Theorie. In G. Kneer, & M. Schroer (Eds.), *Handbuch Soziologische Theorien* (pp. 19–39). Wiesbaden: VS Verlag.

Kneer, G., & Nassehi, A. (1997). *Niklas Luhmanns Theorie sozialer Systeme. Eine Einführung.* München: Fink.

Kniffen, F. B., & Hilliard, S. B. (1988). *Louisiana. Its land and people.* Baton Rouge: Louisiana State University Press.

Knox, P., & Pinch, S. (2010). *Urban social geography. An introduction* (6th ed.). Harlow: Pearson Education Limited.

Kolchin, P. (2003). *American slavery. 1619–1877.* New York: Hill and Wang.

Kolen, J. (1995). Recreating (in) nature, visiting history. Second thoughts on landscape reserves and their role in the preservation and experience of the historic environment. *Archaeological Dialogues, 2*(2), 127–159.

Kolen, J., Renes, H., & Hermans, R. (Eds.). (2015). *Landscape Biographies. Geographical, Historical and Archaeological Perspectives on the Production and Transmission of Landscapes.* Amsterdam: Amsterdam University Press.

Kolmar, C. (2019). The 100 largest companies in Louisiana for 2019. https://www.zippia.com/advice/largest-companies-in-louisiana/. Accessed 7 Dec 2019.

Kopytoff, I. (2015 [1986]). The cultural biography of things: Commoditization as process. In A. Appadurai (Ed.), *The social life of things. Commodities in cultural perspective* (12th ed., 64–91). Cambridge: Cambridge University Press.

Körner, S. (2005). Landschaft und Raum im Heimat- und Naturschutz. In M. Weingarten (Ed.), *Strukturierung von Raum und Landschaft. Konzepte in Ökologie und der Theorie gesellschaftlicher Naturverhältnisse* (pp. 107–117). Münster: Westfälisches Dampfboot.

Krätke, S. (2002). *Medienstadt. Urbane Cluster und globale Zentren der Kulturproduktion.* Opladen: Leske + Budrich.

Krisch, R. (2009). Levittown Reloaded. Privater Wohnungsbau "von der Stange" in Long Island/New York. *Die Alte Stadt, 36*(2), 235–246.

Kropp, C. (2015). Regionale StadtLandschaften – Muster der lebensweltlichen Erfahrung postindustrieller Raumproduktion zwischen Homogenisierung und Fragmentierung. *Raumforschung und Raumordnung, 73*(2), 91–106.

Kuai, X., & Zhao, Q. (2017). Examining healthy food accessibility and disparity in Baton Rouge, Louisiana. *Annals of GIS, 23*(2), 103–116. https://doi.org/10.1080/19475683.2017.1304448.

Kühn, M. (2001). Stadt als Kulturlandschaft? Alte Kritik und neue Perspektiven der „Stadtlandschaft". In Akademie für Raumforschung und Landesplanung (Ed.), *Die Zukunft der Kulturlandschaft zwischen Verlust, Bewahrung und Gestaltung* (Forschungs- und Sitzungsberichte, vol. 215, pp. 100–105). Hannover: Selbstverlag.

Kühne, O. (1999). *Die Wetterlagen-, Tages- und Jahreszeitenabhängigkeit der Verteilung von Lufttemperatur, spezifischer Luftfeuchte, Windfeld, Äquivalenttemperatur und anderer bioklimatisch wirksamer Größen im Lokalklima der Stadt Homburg/Saar* (Dissertation zur Erlangung des akademischen Grades eines Doktors der Philosophie, Universität des Saarlandes). Saarbrücken.

Kühne, O. (2006). *Landschaft in der Postmoderne. Das Beispiel des Saarlandes.* Wiesbaden: DUV.

Kühne, O. (2007). Soziale Akzeptanz und Perspektiven der Altindustrielandschaft. Ergebnisse einer empirischen Untersuchung im Saarland. *RaumPlanung, 132*(133), 156–160.

Kühne, O. (2008a). Die Sozialisation von Landschaft – Sozialkonstruktivistische Überlegungen, empirische Befunde und Konsequenzen für den Umgang mit dem Thema Landschaft in Geographie und räumlicher Planung. *Geographische Zeitschrift, 96*(4), 189–206.

Kühne, O. (2008b). *Distinktion – Macht – Landschaft. Zur sozialen Definition von Landschaft.* Wiesbaden: VS Verlag.

Kühne, O. (2008c). Landschaft und Kitsch. Anmerkungen zu impliziten und expliziten Landschaftsvorstellungen. *Naturschutz und Landschaftsplanung, 40*(12), 403–408.

Kühne, O. (2011). Heimat und sozial nachhaltige Landschaftsentwicklung. *Raumforschung und Raumordnung, 69*(5), 291–301. https://doi.org/10.1007/s13147-011-0108-0.

Kühne, O. (2012a). *Stadt – Landschaft – Hybridität. Ästhetische Bezüge im postmodernen Los Angeles mit seinen modernen Persistenzen.* Wiesbaden: Springer VS.

Kühne, O. (2012b). Urban nature between modern and postmodern aesthetics: Reflections based on the social constructivist approach. *Quaestiones Geographicae, 31*(2), 61–70. https://doi.org/10.2478/v10117-012-0019-3.

Kühne, O. (2013a). Landschaft zwischen Objekthaftigkeit und Konstruktion – Überlegungen zur inversen Landschaft. In D. Bruns, & O. Kühne (Eds.), *Landschaften: Theorie, Praxis und internationale Bezüge. Impulse zum Landschaftsbegriff mit seinen ästhetischen, ökonomischen, sozialen und philosophischen Bezügen mit dem Ziel, die Verbindung von Theorie und Planungspraxis zu stärken* (pp. 181–193). Schwerin: Oceano Verlag.

Kühne, O. (2013b). Landschaftsästhetik und regenerative Energien – Grundüberlegungen zu De- und Re-Sensualisierungen und inversen Landschaften. In L. Gailing, & M. Leibenath (Eds.), *Neue Energielandschaften – Neue Perspektiven der Landschaftsforschung* (pp. 101–120). Wiesbaden: Springer VS.

Kühne, O. (2014). Das Konzept der Ökosystemdienstleistungen als Ausdruck ökologischer Kommunikation. Betrachtungen aus der Perspektive Luhmannscher Systemtheorie. *Naturschutz und Landschaftsplanung, 46*(1), 17–22.

Kühne, O. (2015). The streets of Los Angeles: Power and the infrastructure landscape. *Landscape Research, 40*(2), 139–153. https://doi.org/10.1080/01426397.2013.788691.

Kühne, O. (2016a). Die Los Angeles Riots von 1992. Von Postmodernisierungsverlierern und der neuen Bedeutung von ‚Ethnizität'. In F. Weber, & O. Kühne (Eds.), *Fraktale Metropolen. Stadtentwicklung zwischen Devianz, Polarisierung und Hybridisierung* (pp. 331–347). Wiesbaden: Springer VS.

Kühne, O. (2016b). Los Angeles – Machtspezifische Implikationen einer Verkehrsinfrastruktur. In S. Hofmeister, & O. Kühne (Eds.), *StadtLandschaften. Die neue Hybridität von Stadt und Land* (pp. 253–282). Wiesbaden: Springer VS.

Kühne, O. (2016c). Transformation, Hybridisierung, Streben nach Eindeutigkeit und Urbanizing former Suburbs (URFSURBS): Entwicklungen postmoderner Stadtlandhybride in Südkalifornien und in Altindustrieräumen Mitteleuropas – Beobachtungen aus der Perspektive sozialkonstruktivistischer Landschaftsforschung. In S. Hofmeister, & O. Kühne (Eds.), *StadtLandschaften. Die neue Hybridität von Stadt und Land* (pp. 13–36). Wiesbaden: Springer VS.

Kühne, O. (2017). Hybridisierungstendenzen, Raumpastiches und URFSURBs in Südkalifornien als Herausforderung für die Planung. In K. Berr (Ed.), *Architektur- und Planungsethik. Zugänge, Perspektiven, Standpunkte* (pp. 15–32). Wiesbaden: Springer VS.

Kühne, O. (2018a). Die Moralisierung von Landschaft – Überlegungen zu einer problematischen Kommunikation aus Sicht der Luhmannschen Systemtheorie. In S. Hennecke, H. Kegler, K. Klaczynski, & D. Münderlein (Eds.), *Diedrich Bruns wird gelehrt haben. Eine Festschrift* (pp. 115–121). Kassel: Kassel University Press.

Kühne, O. (2018b). *Landscape and power in geographical space as a social-aesthetic construct.* Dordrecht: Springer International Publishing.

Kühne, O. (2018c). *Landschaft und Wandel. Zur Veränderlichkeit von Wahrnehmungen.* Wiesbaden: Springer VS.

Kühne, O. (2018d). *Landschaftstheorie und Landschaftspraxis. Eine Einführung aus sozialkonstruktivistischer Perspektive* (2., aktualisierte und überarbeitete Aufl.). Wiesbaden: Springer VS.

Kühne, O. (2019a). Autopoietische Systemtheorie und Landschaft. In O. Kühne, F. Weber, K. Berr, & C. Jenal (Eds.), *Handbuch Landschaft* (pp. 91–103). Wiesbaden: Springer VS.

Kühne, O. (2019b). Die Sozialisation von Landschaft. In O. Kühne, F. Weber, K. Berr, & C. Jenal (Eds.), *Handbuch Landschaft* (pp. 301–312). Wiesbaden: Springer VS.

Kühne, O. (2019c). *Landscape theories. A brief introduction*. Wiesbaden: Springer VS.

Kühne, O. (2019d). Reboot ‚Regionale Geographie' – Ansätze einer neopragmatischen Rekonfiguration ‚horizontaler Geographien'. *Berichte. Geographie und Landeskunde,* (zur Publikation angenommenes Manuskript).

Kühne, O. (2019d). Sich abzeichnende theoretische Perspektiven für die Landschaftsforschung: Neopragmatismus, Akteur-Netzwerk-Theorie und Assemblage-Theorie. In O. Kühne, F. Weber, K. Berr, & C. Jenal (Eds.), *Handbuch Landschaft* (pp. 153–162). Wiesbaden: Springer VS.

Kühne, O. (2019e). Sozialkonstruktivistische Landschaftstheorie. In O. Kühne, F. Weber, K. Berr, & C. Jenal (Eds.), *Handbuch Landschaft* (pp. 69–79). Wiesbaden: Springer VS.

Kühne, O., & Franke, U. (2010). *Romantische Landschaft. Impulse zur Wiederentdeckung der Romantik in der Landschafts- und Siedlungsgestaltung in der norddeutschen Kulturlandschaft. Ein Plädoyer*. Schwerin: Oceano Verlag.

Kühne, O., & Schönwald, A. (2015a). Biographische Konstruktionen von Mobilität und Landschaft in der Grenzregion San Diego-Tijuana. In J. Scheiner, & C. Holz-Rau (Eds.), *Räumliche Mobilität und Lebenslauf. Studien zu Mobilitätsbiografien und Mobilitätssozialisation* (pp. 221–238). Wiesbaden: Springer VS.

Kühne, O., & Schönwald, A. (2015b). *San Diego Eigenlogiken, Widersprüche und Hybriditäten in und von ‚America's finest city'*. Wiesbaden: Springer VS.

Kühne, O., & Schönwald, A. (2015c). San Diego: *Trouble in Paradise*? Zwischen Stadterneuerung, Reurbanisierung und restriktiver Steuerpolitik. *Geographische Rundschau, 67*(5), 49–54.

Kühne, O., & Schönwald, A. (2018). Hybridisierung und Grenze: das Beispiel San Diego/Tijuana. In M. Heintel, R. Musil, & N. Weixlbaumer (Eds.), *Grenzen. Theoretische, konzeptionelle und praxisbezogene Fragestellungen zu Grenzen und deren Überschreitungen* (pp. 401–417). Wiesbaden: Springer VS.

Kühne, O., & Weber, F. (2015). Der Energienetzausbau in Internetvideos – Eine quantitativ ausgerichtete diskurstheoretisch orientierte Analyse. In S. Kost, & A. Schönwald (Eds.), *Landschaftswandel – Wandel von Machtstrukturen* (pp. 113–126). Wiesbaden: Springer VS.

Kühne, O., & Weber, F. (2019a). *Hybrid California. Annäherungen an den Golden State, seine Entwicklungen, Ästhetisierungen und Inszenierungen*. Wiesbaden: Springer VS.

Kühne, O., & Weber, F. (2019b). Postmoderne Zugriffe und Differenzierungen von Stadt und Land(schaft): Stadtlandhybride, räumliche Pastiches und URFSURBS. In O. Kühne, F. Weber, K. Berr, & C. Jenal (Eds.), *Handbuch Landschaft* (pp. 755–770). Wiesbaden: Springer VS.

Kühne, O., Jenal, C., & Schönwald, A. (2013). Wie konstruiert der Mensch Landschaft? Zum Verhältnis von Landschaft, Heimat und Identität sowie Hinweise für deren Bedeutung für die räumliche Planung. In MULEWF (Ed.), *Landentwicklung und Ländliche Bodenordnung* (Landentwicklung und ländliche Bodenordnung – Nachrichtenblatt, vol. 54, pp. 13–28). Mainz: Selbstverlag.

Kühne, O., Schönwald, A., & Weber, F. (2016). Urban/Rural Hybrids: The Urbanisation of Former Suburbs (URFSURBS). *Quaestiones Geographicae, 35*(4), 23–34. https://doi.org/10.1515/quageo-2016-0032.

Kühne, O., Schönwald, A., & Weber, F. (2017). Die Ästhetik von Stadtlandhybriden: URFSURBS (*Urbanizing former suburbs*) in Südkalifornien und im Großraum Paris. In O. Kühne, H. Megerle, & F. Weber (Eds.), *Landschaftsästhetik und Landschaftswandel* (pp. 177–198). Wiesbaden: Springer VS.

Laclau, E., & Mouffe, C. (1985). *Hegemony and socialist strategy. Towards a radical democratic politics*. London: Verso.

Lane, C. (2013). Taste makers in the "fine-dining" restaurant industry: The attribution of aesthetic and economic value by gastronomic guides. *Poetics, 41*(4), 342–365.

Lang, R. (2003). *Edgeless cities. Exploring the elusive metropolis* (Brookings metro series). Washington, D.C.: Brookings Institution Press.

Lang, R. E., Sanchez, T. W., & Oner, A. C. (2009). Beyond edge city: Office geography in the new metropolis. *Urban Geography, 30*(7), 726–755.

Lankao, P. R., & Qin, H. (2011). Conceptualizing urban vulnerability to global climate and environmental change. *Current Opinion in Environmental Sustainability, 3*(3), 142–149. https://doi.org/10.1016/j.cosust.2010.12.016.

Large, D. C. (2002). *Berlin. Biographie einer Stadt*. München: Beck.

Larsen, L. H., Cottrell, B. J., Dalstrom, H. A., & Dalstrom, K. C. (2007). *Upstream metropolis. An urban biography of Omaha and council bluffs*. Lincoln: University of Nebraska Press.

Larson, S. J., Capel, P. D., Goolsby, D. A., Zaugg, S. D., & Sandstrom, M. W. (1995). Relations between pesticide use and riverine flux in the Mississippi River basin. *Chemosphere, 31*(5), 3305–3321. https://doi.org/10.1016/0045-6535(95)00176-9.

Latour, B. (1998). *Wir sind nie modern gewesen. Versuch einer symmetrischen Anthropologie*. Frankfurt a. M.: Suhrkamp.

Latour, B. (2002 [1999]). *Die Hoffnung der Pandora. Untersuchungen zur Wirklichkeit der Wissenschaft*. Frankfurt a. M.: Suhrkamp.

Lautensach, H. (1973). Über die Erfassung und Abgrenzung von Landschaftsräumen [Erstveröffentlichung 1938]. In K. Paffen (Ed.), *Das Wesen der Landschaft* (Wege der Forschung, vol. 39, pp. 20–38). Darmstadt: WBG.

Lee, M. J. (2006). The populist chameleon: The people's party, Huey Long, George Wallace, and the populist argumentative frame. *Quarterly Journal of Speech, 92*(4), 355–378. https://doi.org/10.1080/00335630601080385.

Lees, L. (2008). Gentrification and social mixing: Towards an inclusive urban renaissance? *Urban Studies, 45*(12), 2449–2470. https://doi.org/10.1177/0042098008097099.

Lee, T., & Black, M. (2017). Cancer alley: Big industry, big problems. https://www.msnbc.com/interactives/geography-of-poverty/se.html. Accessed 10 Jan 2020.

Leibenath, M. (2014). Landschaftsbewertung im Spannungsfeld von Expertenwissen, Politik und Macht. *UVP-report, 28*(2), 44–49. https://www2.ioer.de/recherche/pdf/2014_leibenath_uvp-report.pdf. Accessed 26 Jan 2017.

Lincoln, R. A., Shine, J. P., Chesney, E. J., Vorhees, D. J., Grandjean, P., & Senn, D. B. (2011). Fish consumption and mercury exposure among Louisiana recreational anglers. *Environmental Health Perspectives, 119*(2), 245–251. https://doi.org/10.1289/ehp.1002609.

Lindner, C. (2009). Freiheit und Fairness. In P. Rösler, & C. Lindner (Eds.), *Freiheit: Gefühlt – Gedacht – Gelebt. Liberale Beiträge zu einer Wertediskussion* (pp. 17–28). Wiesbaden: VS Verlag.

Linke, S. (2017). Neue Landschaften und ästhetische Akzeptanzprobleme. In O. Kühne, H. Megerle, & F. Weber (Eds.), *Landschaftsästhetik und Landschaftswandel* (pp. 87–104). Wiesbaden: Springer VS.

Linke, S. (2018). Ästhetik der neuen Energielandschaften – Oder: „Was Schönheit ist, das weiß ich nicht". In O. Kühne, & F. Weber (Eds.), *Bausteine der Energiewende* (pp. 409–429). Wiesbaden: Springer VS.

Linke, S. I. (2019). *Die Ästhetik medialer Landschaftskonstrukte. Theoretische Reflexionen und empirische Befunde*. Wiesbaden: Springer VS.

Lipietz, A. (1991). Zur Zukunft der städtischen Ökologie. In M. Wentz (Ed.), *Stadt-Räume* (pp. 129–136). Frankfurt a. M.: Campus.

LMOGA – Louisiana Mid-Continent Oil And Gas Association. (2019a). History of the industry. https://www.lmoga.com/resources/oil-gas-101/history-of-the-industry/. Accessed 10 Jan 2020.

LMOGA – Louisiana Mid-Continent Oil And Gas Association. (2019b). Industry sectors. Refineries. https://www.lmoga.com/industry-sectors/. Accessed 1 July 2019.

Loew, P. O. (2011). *Danzig. Biographie einer Stadt*. München: Beck.

Logan, W. S. (2000). *Hanoi. Biography of a city*. Sydney: University of New South Wales Press.

Lohr, V. I., Pearson-Mims, C. H., Tarnai, J., & Dillman, D. A. (2004). How urban residents rate and rank the benefits and problems associated with trees in cities. *Journal of Arboriculture, 30*(1), 28–35.

Lomax, T., Schrank, D., & Eisele, B. (Texas A&M Transportation Institute, Ed.). (2019). 2019 urban mobility report and appendices. https://mobility.tamu.edu/umr/report/. Accessed 13 Dec 2019.

Löw, M. (2010). *Soziologie der Städte* (vol. 1976). Frankfurt a. M.: Suhrkamp.

LSU. (2019a). A look back. https://www.lsu.edu/president/history.php. Accessed 2 July 2019.

LSU. (2019b). Fall facts. 2018. https://www.lsu.edu/bgtplan/facts/pdfs/2018fallfacts_final.pdf. Accessed 2 July 2019.

LSU Office of Finance & Administration. (2019). Enrollment by level and gender. https://www.lsu.edu/bgtplan/trend/dashboards/levelgender.php. Accessed 2 July 2019.

Luckett, B. G., Su, L. J., Rood, J. C., & Fontham, E. T. H. (2012). Cadmium exposure and pancreatic cancer in South Louisiana. *Journal of environmental and public health, 12,* 1–11. https://doi.org/10.1155/2012/180186.

Luhmann, N. (1984). *Soziale Systeme. Grundriß einer allgemeinen Theorie*. Frankfurt a. M.: Suhrkamp.

Luhmann, N. (1986). *Ökologische Kommunikation. Kann die moderne Gesellschaft sich auf ökologische Gefährdungen einstellen?* Opladen: Westdeutscher Verlag.

Luhmann, N. (1996). *Die Realität der Massenmedien*. Opladen: Westdeutscher Verlag.

Luhmann, N. (2007[1995]). *The reality of the mass media*. Cambridge: Polity.

Luhmann, N. (2017). *Systemtheorie der Gesellschaft*. Berlin: Suhrkamp.

Lyotard, J.-F. (1979). *La condition postmoderne. Rapport sur le savoir*. Paris: Les Éditions de Minuit.

Makarius, M. (2004). *Ruinen. Die gegenwärtige Vergangenheit*. Paris: Flammarion.

Marcus, J. L., Hurley, L. B., Hare, C. B., Nguyen, D. P., Phengrasamy, T., & Silverberg, M. J., et al. (2016). Preexposure Prophylaxis for HIV Prevention in a Large Integrated Health Care System: Adherence, Renal Safety, and Discontinuation. *JAIDS – Journal of Acquired Immune Deficiency Syndromes (1999), 73*(5), 540–546.

Marston, S. A. (2000). The social construction of scale. *Progress in Human Geography, 24*(2), 219–242. https://doi.org/10.1191/030913200674086272.

Massey, D. (1999). Cities in the world. In D. Massey, J. Allen, & S. Pile (Eds.), *City worlds* (Understanding Cities, pp. 93–150). London: Routledge in association with The Open University.

Mathewson, K. (2009). Carl Sauer and his crititcs. In W. M. Denevan, & K. Mathewson (Eds.), *Carl Sauer on culture and landscape. Readings and commentaries* (pp. 9–28). Baton Rouge: Louisiana State University Press.

Mathewson, K. (2014). The geography of Louisiana. In R. Orgera, & W. Parent (Eds.), *The Louisiana field guide. Understanding life in the Pelican state* (pp. 1–12). Baton Rouge: Louisiana State University Press.

Mathur, A., & da Cunha, D. (2001). *Mississippi floods. Designing a shifting landscape*. New Haven: Yale University Press.

Mattissek, A., & Wiertz, T. (2014). Materialität und Macht im Spiegel der Assemblage-Theorie: Erkundungen am Beispiel der Waldpolitik in Thailand. *Geographica Helvetica, 69*(3), 157–169.

Maturana, H. R., & Varela, F. J. (1987). *The tree of knowledge. The biological roots of human understanding* (Revised edition). Boston: Shambhala.

Mayring, P. (2003). *Qualitative Inhaltsanalyse: Grundlagen und Techniken*. Weinheim: Beltz.

McBride, R. A., Taylor, M. J., & Byrnes, M. R. (2007). Coastal morphodynamics and Chenier-Plain evolution in southwestern Louisiana, USA: A geomorphic model. *Geomorphology, 88*(3–4), 367–422. https://doi.org/10.1016/j.geomorph.2006.11.013.

McCollister, R. (Greater Baton Rouge Business Report, Ed.). (2019). Publisher: St. George could have been avoided. https://www.businessreport.com/opinions/st-george-baton-rouge. Accessed 13 Dec 2019.

McCombs, M. E., & Shaw, D. L. (1972). The Agenda-setting function of mass media. *Public Opinion Quarterly, 36*(2), 176. https://doi.org/10.1086/267990.

McEwen, J. W. (2014). Sense of place, place attachment, and rootedness in Four West Baton Rouge Parish, Louisiana Bars, LSU Doctoral Dissertations. 1500. https://digitalcommons.lsu.edu/gradschool_dissertations/1500/. Accessed 13 June 2019.

McMichael, R. N. (1961). *Plant location factors in the petrochemical industry in Louisiana*. Baton Rouge: Louisiana State University Press.

McMichael, F. A. (2008). *Atlantic loyalties. Americans in Spanish West Florida, 1785–1810*. Athens: University of Georgia Press.

Meijles, E., & van Hoven, B. (2010). Using the *Rural Atelier* as an educational method in landscape studies. *Journal of Geography in Higher Education, 34*(4), 541–560.

Merleau-Ponty, M. (1945). *Phénoménologie de la perception* (Bibliothèque des idées). Paris: Gallimard.

Mettler, S. (2005). *Soldiers to citizens. The G.I. Bill and the making of the greatest generation*. Oxford: Oxford University Press.

Miller, M. (2000). Biography of Hon. John Milliken Parker, Bethel Church, MS., then Orleans Parish, Louisiana. https://files.usgwarchives.net/la/orleans/bios/p-000007.txt. Accessed 2 July 2019.

Misoch, S. (2015). *Qualitative interviews*. Berlin: de Gruyter.

Mitchell, W. J. T. (Ed.). (2002). *Landscape and power* (2nd ed.). Chicago: University of Chicago Press.

Mouffe, C. (2000). Deliberative Democracy or Agonistic Pluralism. Reihe Politikwissenschaft Political Science Series: 72. https://www.ihs.ac.at/publications/pol/pw_72.pdf. Accessed 13 Oct 2017.

Moulaert, F., & Swyngedouw, E. (1989). Survey 15. A regulation approach to the geography of flexible production systems. *Environment and Planning D: Society and Space, 7*(3), 327–345. https://doi.org/10.1068/d070327.

Müller, W., & Rohr-Zänker, R. (2001). Amerikanisierung der „Peripherie" in Deutschland? In K. Brake, J. S. Dangschat, & G. Herfert (Eds.), *Suburbanisierung in Deutschland. Aktuelle Tendenzen* (pp. 27–39). Opladen: Leske + Budrich.

Muzzio, D., & Halper, T. (2002). Pleasantville? The suburb and its representation in American movies. *Urban Affairs Review, 37*(4), 543–574. https://doi.org/10.1177/10780870222185469.

Nagle, A. (2017). *Kill all normies. The online culture wars from Tumblr and 4chan to the alt-right and Trump*. Winchester: Zero Books.

neighborhoodscout.com. (2020). Baton Rouge, LA. Crime Rates. https://www.neighborhoodscout.com/la/baton-rouge/crime. Accessed 28 Jan 2020.

Nicolaou, A. (2019). Amazon becomes fastest-growing music streaming service, Financial Times. https://www.ft.com/content/60633178-a282-11e9-974c-ad1c6ab5efd1. Accessed 21 Oct 2019.

Nogué i Font, J. (1993). Toward a phenomenology of landscape and landscape experience: An example from Catalonia. In D. Seamon (Ed.), *Dwelling, Seeing, and Designing Toward A Phenomenological Ecology* (pp. 159–180). Albany: State University of New York Press.

Nowak, D. J., Civerolo, K. L., Rao, S. T., Sistla, G., Luley, C. J., & Crane, D. E. (2000). A modeling study of the impact of urban trees on ozone. *Atmospheric Environment, 34*(10), 1601–1613. https://doi.org/10.1016/S1352-2310(99)00394-5.

Nussbaum, M. C. (2009). *Frontiers of justice. Disability, nationality, species membership.* Cambridge: Harvard University Press.

Olson Beal, H. K. (2008). Speaking the language of integration: A case study of South Boulevard Foreign Language Academic Immersion Magnet, LSU Doctoral Dissertations. 3169. https://digitalcommons.lsu.edu/gradschool_dissertations/3169/. Accessed 1 July 2019.

Ostrow, D. G., Whitaker, R. E. D., Frasier, K., Cohen, C., Wan, J., Frank, C., et al. (1991). Racial differences in social support and mental health in men with HIV infection: A pilot study. *AIDS care, 3*(1), 55–62.

Otto, A. (2019). Landschaft und der Ausbau der Windenergie. In O. Kühne, F. Weber, K. Berr, & C. Jenal (Eds.), *Handbuch Landschaft* (pp. 859–869). Wiesbaden: Springer VS.

Otto, A., & Leibenath, M. (2013). Windenergielandschaften als Konfliktfeld. Landschaftskonzepte, Argumentationsmuster und Diskurskoalitionen. In L. Gailing, & M. Leibenath (Eds.), *Neue Energielandschaften – Neue Perspektiven der Landschaftsforschung* (pp. 65–75). Wiesbaden: Springer VS.

Otvos, E. G., & Price, W. A. (1979). Problems of chenier genesis and terminology – An overview. *Marine Geology, 31*(3–4), 251–263. https://doi.org/10.1016/0025-3227(79)90036-7.

Paasi, A. (1986). The institutionalization of regions: A theoretical framework for understanding the emergence of regions and the constitution of regional identity. *Fennia – International Journal of Geography, 164*(1), 105–146. https://doi.org/10.11143/9052.

Paasi, A. (1991). Deconstructing regions: Notes on the scales of spatial life. *Environment and Planning A: Economy and Space, 23*(2), 239–256. https://doi.org/10.1068/a230239.

Paasi, A. (2002). Place and region: Regional worlds and words. *Progress in Human Geography, 26*(6), 802–811. https://doi.org/10.1191/0309132502ph404pr.

Paasi, A. (2005). The changing discourses on political boundaries. Mapping the backgrounds, contexts and contents. In H. van Houtum, O. Kramsch, & W. Zierhofer (Eds.), *B/ordering space* (pp. 17–32). Aldershot: Ashgate.

Paasi, A. (2008). Finnish landscape as social practice. Mapping identity and scale. In M. Jones, & K. R. Olwig (Eds.), *Nordic landscapes. Region and belonging on the Northern Edge of Europe* (pp. 511–539). Minneapolis: University of Minnesota Press.

Paasi, A. (2009). Regional geography I. In R. Kitchin, & N. Thrift (Eds.), *International encyclopedia of human geography* (vol. 9, pp. 214–227). Amsterdam: Elsevier.

Paasi, A. (2010). Commentary. Regions are social constructs, but who or what 'constructs' them? Agency in question. *Environment and Planning A, 42*(10), 2296–2301. https://doi.org/10.1068/a42232.

Page, T., Harris, R. H., & Epstein, S. S. (1976). Drinking water and cancer mortality in Louisiana. *Science, 193*(4247), 55–57. https://doi.org/10.1126/science.935854.

Palen, J. J. (1995). *The suburbs.* New York: McGraw-Hill.

Pariser, E. (2011). *The filter bubble. What the internet is hiding from you.* London: Penguin Books.

Parsons, T. (1991 [1951]). *The social system.* London: Routledge.

Paterson, B. (The Advocate, Ed.). (2019). Baton Rouge's bad roads take 4-figure chunk out a driver's bank account each year, study says. https://www.theadvocate.com/baton_rouge/news/article_b308d992-e9f3-11e9-bd73-ef06217d52eb.html. Accessed 13 Dec 2019.

Peck, J. (2005). Struggling with the creative class. *International Journal of Urban and Regional Research, 29*(4), 740–770. https://doi.org/10.1111/j.1468-2427.2005.00620.x.

Penland, S., & Suter, J. R. (1989). The geomorphology of the Mississippi River chenier plain. *Marine Geology, 90*(4), 231–258. https://doi.org/10.1016/0025-3227(89)90127-8.

Pentzold, C. (2007). *Wikipedia. Diskussionsraum und Informationsspeicher im neuen Netz* (Internet Research, vol. 29). Baden-Baden: Nomos.

Perez, A. D., & Hirschman, C. (2009). The changing racial and ethnic composition of the US population: Emerging American identities. *Population and Development Review, 35*(1), 1–51. https://doi.org/10.1111/j.1728-4457.2009.00260.x.

Perry, K. (2004). *The Kingfish in fiction. Huey P. Long and the modern American novel*. Baton Rouge: Louisiana State University Press.

Petrescu, P. (2014). Google organic click-through rates in 2014. https://moz.com/blog/google-organic-click-through-rates-in-2014. Accessed 20 May 2019.

Pietila, A. (2010). *Not in my neighborhood. How bigotry shaped a great American city*. Chicago: Ivan R. Dee.

Popper, K. R. (2011). *The open society and its enemies*. Abingdon: Routledge.

Popper, K. R. (2019[1987]). *Auf der Suche nach einer besseren Welt. Vorträge und Aufsätze aus dreißig Jahren*. München: Piper.

Popper, K. R. (1973). *Objektive Erkenntnis. Ein evolutionärer Entwurf*. Hamburg: Hoffmann und Campe.

Preserve Louisiana. (2019). Our values. https://preserve-louisiana.org/values. Accessed 29 Nov 2019.

Priest, T., & Theriot, J. (2013). Who destroyed the marsh? Oil field canals, coastal ecology, and the debate over Louisiana's shrinking wetlands. In J. Allured, & M. S. Martin (Eds.), *Louisiana legacies. Readings in the history of the Pelican state* (pp. 331–341). Chichester: Wiley-Blackwell.

Prigge, W. (1991). Die Revolution der Städte Lesen. In M. Wentz (Ed.), *Stadt-Räume* (pp. 99–112). Frankfurt a. M.: Campus.

QS World University Rankings. (2019). Top Universities in the US 2018. https://www.topuniversities.com/university-rankings-articles/world-university-rankings/top-universities-us-2018. Accessed 2 July 2019.

Quasten, H. (1997). Grundsätze und Methoden der Erfassung und Bewertung kulturhistorischer Phänomene der Kulturlandschaft. In W. Schenk, K. Fehn, & D. Denecke (Eds.), *Kulturlandschaftspflege Beiträge der Geographie zur räumlichen Planung* (pp. 19–34). Berlin: Borntraeger.

Rapier, G. (2019). The 20 most dangerous US cities for pedestrians. *businessinsider.com*. https://www.businessinsider.com/deadliest-us-cities-for-pedestrians-walking-deaths-2019-01?r=DE&IR=T. Accessed 24 Jan 2020.

Rawls, J. (1971). *A theory of justice*. Cambridge: Harvard University Press.

Rawls, J. (2001). *Justice as fairness. A restatement*. Cambridge: Harvard University Press.

Rehder, J. B. (1971). *Sugar plantation settlements of SouthernLouisiana: A cultural geography* (Dissertation, Louisiana State University). Baton Rouge.

Renzulli, L. A., & Evans, L. (2005). School choice, charter schools, and white flight. *Social Problems, 52*(3), 398–418.

Richard, M. D. (2006). Racial disparities, birth outcomes, and changing demographics of East Baton Rouge Parish, Louisiana, LSU Doctoral Dissertations. 1870. https://digitalcommons.lsu.edu/gradschool_dissertations/1870/. Accessed 1 July 2019.

Richmond Gisolfi, M. (2006). From crop lien to contract farming. The roots of agribusiness in the American South, 1929–1939. *Agricultural History, 80*(2), 167–189.

Riesman, D. (1950). *The lonely crowd*. New Haven: Yale Universaty Press.

Rifkin, J. (2007). *Access – Das Verschwinden des Eigentums. Warum wir weniger besitzen und mehr ausgeben werden* (3., erweiterte Auflage). Frankfurt: Campus.

Rodrigue, S. F., & Phillips, F. (2008). *Baton Rouge*. Charleston: Arcadia Publishing.

Rodrigue, S. F., & Phillips, F. (2011). *Historic Baton Rouge. An illustrated history* (2nd ed.). San Antonio: Historical Publishing Network, a division of Lammert Incorporated.

Roymans, N. (1995). The cultural biography of urnfields and the long-term history of a mythical landscape. *Archaeological Dialogues, 1,* 2–38. https://doi.org/10.1017/S138020380000026X.

Roymans, N., Gerritsen, F., van der Heijden, C., Bosma, K., & Kolen, J. (2009). Landscape biography as research strategy: The case of the South Netherlands project. *Landscape Research, 34*(3), 337–359. https://doi.org/10.1080/01426390802381185.

Ruffin, T. F. (2006). *Under stately oaks. A pictorial history of LSU* (Revised Edition). Baton Rouge: Louisiana State University Press.

Sachs, W. (1989). Die auto-mobile Gesellschaft. Vom Aufstieg und Niedergang einer Utopie. In F.-J. Brüggemeier, & T. Rommelspacher (Eds.), *Besiegte Natur. Geschichte der Umwelt im 19. und 20. Jahrhundert* (pp. 106–123). München: Beck.

Samuels, M. S. (1979). The biography of landscape. Cause and culpability. In D. W. Meinig (Ed.), *The interpretation of ordinary landscapes. Geographical essays* (pp. 51–88). New York: Oxford University Press.

Samuels, D. (NOLA.com – The Times-Picayune, Ed.). (2015). 10 things you hate about Baton Rouge: Your thoughts on the city's biggest problems, NOLA Media Group. https://www.nola.com/news/baton-rouge/2015/01/10_things_you_hate_about_baton.html. Accessed 1 July 2019.

Sassen, S. (2000). Cities in the global economy. In R. Simmonds, & G. Hack (Eds.), *Global city regions. Their emerging forms* (pp. 269–276). London: Spon Press.

Sassen, S. (2001). *The global city. New York, London, Tokyo* (2nd ed.). Princeton: Princeton University Press.

Schach, A., & Lommatzsch, T. (Eds.). (2018). *Influencer Relations. Marketing und PR mit digitalen Meinungsführern.* Wiesbaden: Springer Gabler.

Schaffert, S., Bry, F., Baumeister, J., & Kiesel, M. (2007). Semantic Wiki. *Informatik Spektrum, 30*(6), 434–439. https://doi.org/10.1007/s00287-007-0195-z.

Schein, R. H. (1997). The place of landscape. A conceptual framework for interpreting an American scene. *Annals of the Association of American Geographers, 87*(4), 660–680. https://doi.org/10.1111/1467-8306.00072.

Schenk, W. (2013). Landschaft als zweifache sekundäre Bildung – Historische Aspekte im aktuellen Gebrauch von Landschaft im deutschsprachigen Raum, namentlich in der Geographie. In D. Bruns, & O. Kühne (Eds.), *Landschaften: Theorie, Praxis und internationale Bezüge. Impulse zum Landschaftsbegriff mit seinen ästhetischen, ökonomischen, sozialen und philosophischen Bezügen mit dem Ziel, die Verbindung von Theorie und Planungspraxis zu stärken* (pp. 23–36). Schwerin: Oceano Verlag.

Scherle, N. (2016). *Kulturelle Geographien der Vielfalt. Von der Macht der Differenzen zu einer Logik der Diversität* (Sozial- und Kulturgeographie). Bielefeld: transcript.

Schmaltz, T., & Casserleigh, B. (2019). Annexation becomes latest heated debate after St. George vote, WBRZ 2. https://www.wbrz.com/news/annexation-becomes-latest-heated-debate-after-st-george-vote/. Accessed 13 Dec 2019.

Schmitt-Beck, R. (1990). Über die Bedeutung der Massenmedien für soziale Bewegungen. *Kölner Zeitschrift für Soziologie und Sozialpsychologie, 42*(4), 642–662.

Schneider, P. (2014). *Old man river. The Mississippi river in North American history.* New York: Henry Holt and Company.

Schneider-Sliwa, R. (2005). *USA. Geographie, Geschichte, Wirtschaft, Politik* (Wissenschaftliche Länderkunden). Darmstadt: WBG.

Schönwald, A. (2015). Die Transformation von Altindustrielandschaften. In O. Kühne, K. Gawroński, & J. Hernik (Eds.), *Transformation und Landschaft. Die Folgen sozialer Wandlungsprozesse auf Landschaft* (pp. 63–73). Wiesbaden: Springer VS.

Schönwald, A. (2017). Ästhetik des Hybriden. Mehr Bedeutungsoffenheit für Landschaften durch Hybridisierungen. In O. Kühne, H. Megerle, & F. Weber (Eds.), *Landschaftsästhetik und Landschaftswandel* (pp. 161–175). Wiesbaden: Springer VS.

Schubert, H.-J., Joas, H., & Wenzel, H. (2010). *Pragmatismus zur Einführung. Kreativität, Handlung, Deduktion, Induktion, Abduktion, Chicago School, Sozialreform, symbolische Interaktion* (Zur Einführung, vol. 382). Hamburg: Junius.

Schultz, H.-D. (1997). Räume sind nicht, Räume werden gemacht. Zur Genese "Mitteleuropas" in der deutschen Geographie. *Europa regional, 5*(1), 2–14.

Schulz-Schaeffer, I. (2000). Akteur-Netzwerk-Theorie: Zur Koevolution von Gesellschaft, Natur und Technik. In J. Weyer, & J. Abel (Eds.), *Soziale Netzwerke. Konzepte und Methoden der sozialwissenschaftlichen Netzwerkforschung* (Lehr- und Handbücher der Soziologie, pp. 187–210). München: Oldenbourg.

Schütz, A. (1971 [1962]). *Gesammelte Aufsätze 1. Das Problem der Wirklichkeit.* Den Haag: Martinus Nijhoff.

Schütz, A., & Luckmann, T. (2003 [1975]). *Strukturen der Lebenswelt.* Konstanz: UTB.

Seamon, D. (2019). Whither phenomenology? Thirty years of *EAP*. *Environmental & Architectural Phenomenology, 30*(2), 37–44. (1990–2019: Special 30th-anniversary issue!).

Sedelmeier, T. (2019). Food Deserts – Einblicke in Nahrungslandschaften. In K. Berr, & C. Jenal (Eds.), *Landschaftskonflikte* (pp. 687–698). Wiesbaden: Springer VS.

Sen, A. (2009). *The idea of justice*. Cambridge: Harvard University Press.

Sennett, R. (1991). *Civitas. Die Großstadt und die Kultur des Unterschieds*. Frankfurt: Fischer.

Sentell, W. (2018). Risky riding? Report says Louisiana second most dangerous state for cyclists. *The Advocate*. https://www.theadvocate.com/baton_rouge/news/politics/article_96cce980-c334-11e8-bb33-0b2accfc7d93.html. Accessed 24 Jan 2020.

Seymour, D. (2004). The social construction of taste. In D. Sloan (Ed.), *Culinary taste. Consumer behaviour in the international restaurant sector* (pp. 1–22). Amsterdam: Elsevier Butterworth Heinemann.

Shifman, L. (2013). Memes in a digital world: Reconciling with a conceptual troublemaker. *Journal of Computer-Mediated Communication, 18*(3), 362–377. https://doi.org/10.1111/jcc4.12013.

Siebel, W. (Ed.). (2004). *Die europäische Stadt*. Frankfurt a. M.: Suhrkamp.

Sieber, R. T. (1991). Waterfront revitalization in postindustrial Port Cities of North America. *City & Society, 5*(2), 120–136. https://doi.org/10.1525/city.1991.5.2.120.

Silliman, B. R., van de Koppel, J., McCoy, M. W., Diller, J., Kasozi, G. N., Earl, K., et al. (2012). Degradation and resilience in Louisiana salt marshes after the BP-*Deepwater Horizon* oil spill. *PNAS – Proceedings of the National Academy of Sciences of the United States of America, 109*(28), 11234–11239.

Simmel, G. (1950). The metropolis and mental life. In K. H. Wolff (Ed.), *The sociology of Georg Simmel* (pp. 409–424). Glencoe: The Free Press.

Sivak, M., & Schoettle, B. (2016). Recent decreases in the proportion of persons with a driver's license across all age groups. https://www.umich.edu/~umtriswt/PDF/UMTRI-2016-4.pdf. Accessed 12 Oct 2016.

Smith, N., & DeFilippis, J. (1999). The reassertion of economics: 1990s gentrification in the lower east side. *International Journal of Urban and Regional Research, 23*(4), 638–653. https://doi.org/10.1111/1468-2427.00220.

Soja, E. W. (1989). *Postmodern geographies. The reassertion of space in critical social theory.* London: Verso.

Soja, E. W. (2000). *Postmetropolis. Critical studies of cities and regions*. Oxford: Blackwell.

Sokolowski, R. (2000). *Introduction to phenomenology*. Cambridge: Cambridge University Press.

Song, Y., & Knaap, G.-J. (2003). New urbanism and housing values: A disaggregate assessment. *Journal of Urban Economics, 54*(2), 218–238.

Sorkin, M. (Ed.). (1992). *Variations on a theme park. The New American city and the end of public space*. New York: The Noonday Press.

Southern University and Agricultural & Mechanical College. (2019). President-chancellor's office. https://www.subr.edu/page/supresidentchancellors-office. Accessed 2 July 2019.

Speed, J. G. (1893). Do newspapers now give the News? *The Forum,* (705–711). https://www.unz.com/print/Forum-1893aug-00705. Accessed 20 May 2019.

Speights-Binet, J. (2004). The road to redevelopment: New Urbanism, nostalgia, and the process of downtown revitalization in Baton Rouge, Louisiana, LSU Doctoral Dissertations. 89. https://digitalcommons.lsu.edu/gradschool_dissertations/89/. Accessed 1 July 2019.

statista. (2019a). Marktanteile der meistgenutzten Suchmaschinen auf dem Desktop nach Page Views weltweit in ausgewählten Monaten von August 2015 bis April 2019. https://de.statista.com/statistik/daten/studie/225953/umfrage/die-weltweit-meistgenutzten-suchmaschinen/. Accessed 15 May 2019.

statista. (2019b). Statistiken zur New York Times Company. https://de.statista.com/themen/2554/new-york-times-company/. Accessed 8 July 2019.

Stemmer, B. (2016). *Kooperative Landschaftsbewertung in der räumlichen Planung. Sozialkonstruktivistische Analyse der Landschaftswahrnehmung der Öffentlichkeit.* Wiesbaden: Springer VS.

Stenberg, R. R. (1934). The boundaries of the Louisiana purchase. *The Hispanic American Historical Review, 14*(1), 32–64. https://doi.org/10.2307/2506137.

Stotten, R. (2013). Kulturlandschaft gemeinsam verstehen – Praktische Beispiele der Landschaftssozialisation aus dem Schweizer Alpenraum. *Geographica Helvetica, 68*(2), 117–127. https://doi.org/10.5194/gh-68-117-2013.

Stotten, R. (2019). Kulturlandschaft als Ausdruck von Heimat der bäuerlichen Gesellschaft. In M. Hülz, O. Kühne, & F. Weber (Eds.), *Heimat. Ein vielfältiges Konstrukt* (pp. 149–162). Wiesbaden: Springer VS.

Stringfellow, L., MacLaren, A., Maclean, M., & O'Gorman, K. (2013). Conceptualizing taste: Food, culture and celebrities. *Tourism Management, 37,* 77–85. https://doi.org/10.1016/j.tourman.2012.12.016.

Sudjic, D. (1992). *The 100 mile city.* San Diego: Harcourt Brace & Company.

Sullivan, D. (2016). Google now handles at least 2 trillion searches per year. https://searchengineland.com/google-now-handles-2-999-trillion-searches-per-year-250247. Accessed 20 May 2019.

Sullivan, W. C., & Kuo, F. E. (1996). *Do trees strengthen urban communities, reduce domestic violence?* (Technology Bulletin (Urban & Community Forestry Assistance Program (U.S.), Forestry Report R8-FR 56), vol. 4). Evanston: Northeastern Area State and Private Forestry, Urban Forestry Center for the Midwestern States.

Sumaila, U. R., Cisneros-Montemayor, A. M., Dyck, A., Huang, L., Cheung, W., Jacquet, J., et al. (2012). Impact of the *Deepwater Horizon* well blowout on the economics of US Gulf fisheries. *Canadian Journal of Fisheries and Aquatic Sciences, 69*(3), 499–510.

The City of St. George. (2019). The city of St. George, Louisiana. https://www.stgeorgelouisiana.com/. Accessed 13 Dec 2019.

The Hayride – Southern Politics & Culture. (2019). Isn't it time Baton Rouge's elite start making their peace with St. George? https://thehayride.com/2019/05/isnt-it-time-baton-rouges-elite-start-making-their-peace-with-st-george/. Accessed 13 Dec 2019.

Thieme, G., & Laux, H. D. (2017). Jenseits von „Black and White“: Die Vereinigten Staaten auf dem Weg in eine multiethnische Gesellschaft. In W. Gamerith, & U. Gerhard (Eds.), *Kulturgeographie der USA. Eine Nation begreifen* (pp. 63–72). Berlin: Springer Spektrum.

Thrift, N. (1991). For a new regional geography 2. *Progress in Human Geography, 15*(4), 456–466. https://doi.org/10.1177/030913259101500407.

Tilley, C. (1997). *A phenomenology of landscape. Places, paths and monuments* (Explorations in anthropology). Oxford: Berg.

Together Baton Rouge. (2019). Together Baton Rouge. https://www.togetherbr.org/. Accessed 13 Dec 2019.

Toohey, G., Skene, L., & Gagliano, K. (2019). Despite dip in 2018, Baton Rouge's murder rate still high. How do officials plan to address issue? *The Advocate*. https://www.theadvocate.com/baton_rouge/news/crime_police/article_e14db61a-0fb5-11e9-a7fb-fb14efc3f471.html. Accessed 28 Jan 2020.

Trepanier, C. (1991). The Cajunization of French Louisiana: Forging a regional identity. *The Geographical Journal, 157*(2), 161–171. https://doi.org/10.2307/635273.

Trepl, L. (2012). *Die Idee der Landschaft. Eine Kulturgeschichte von der Aufklärung bis zur Ökologiebewegung*. Bielefeld: transcript.

TRIP. (2019). Louisiana transportation by the numbers. Meeting the state's need for safe, smooth and efficient mobility. https://tripnet.org/wp-content/uploads/2019/10/LA_Transportation_by_the_Numbers_TRIP_Report_October_2019.pdf. Accessed 13 Dec 2019.

TripAdvisor. (2019). Media center. https://tripadvisor.mediaroom.com/us. Accessed 2 July 2019.

Tsai, S. P., Chen, V. W., Fox, E. E., Wendt, J. K., Cheng Wu, X., Foster, D. E., et al. (2004). Cancer incidence among refinery and petrochemical employees in Louisiana, 1983–1999. *Annals of epidemiology, 14*(9), 722–730. https://doi.org/10.1016/j.annepidem.2004.01.005.

Tuan, Y.-F. (1989). Surface phenomena and aesthetic experience. *Annals of the Association of American Geographers, 79*(2), 233–241. https://doi.org/10.1111/j.1467-8306.1989.tb00260.x.

Ungers, O. M. (1990). Die Stadtinseln im Meer der Metropole. Das pluralistische Konzept der „Städtearchipel" – Planung auf historischem Boden. das Neue Berlin (VII). *Frankfurter Allgemeine Zeitung,* p. 37.

United States Census Bureau. (2019a). American FactFinder – Baton Rouge city, Louisiana. https://factfinder.census.gov/faces/nav/jsf/pages/community_facts.xhtml. Accessed 29 Nov 2019.

United States Census Bureau. (2019b). American FactFinder – East Baton Rouge Parish, Louisiana. https://factfinder.census.gov/faces/nav/jsf/pages/community_facts.xhtml. Accessed 1 July 2019.

United States Census Bureau. (2019c). American FactFinder – Louisiana. https://factfinder.census.gov/faces/nav/jsf/pages/community_facts.xhtml. Accessed 1 July 2019.

van Assche, K. (2010). Landscape of the year, or, taking Luhmann to the marshes. Social systems theory and the analysis of ecological and cultural adaptation. *Studia Politica, 42*(3), 110–132.

van Assche, K., & Verschraegen, G. (2008). The limits of planning: Niklas Luhmann's systems theory and the analysis of planning and planning ambitions. *Planning Theory, 7*(3), 263–283. https://doi.org/10.1177/1473095208094824.

van der Valk, A., & van Dijk, T. (Eds.). (2009). *Regional planning for open space*. London: Routledge.

van der Wiel, K., Kapnick, S. B., van Oldenborgh, G. J., Whan, K., Philip, S., Vecchi, G. A., et al. (2017). Rapid attribution of the August 2016 flood-inducing extreme precipitation in south Louisiana to climate change. *Hydrology and Earth System Sciences, 21*(2), 897–921.

van Houtum, H., & Lagendijk, A. (2001). Contextualising regional identity and imagination in the construction of polycentric urban regions: The cases of the Ruhr area and the Basque country. *Urban Studies, 38*(4), 747–767. https://doi.org/10.1080/00420980120035321.

van Kley, J. E. (1999). The vegetation of the high terrace rolling uplands, Louisiana. *Castanea, 64*(4), 318–336.

van Manen, M. (2015). *Writing in the dark. Phenomenological studies in interpretive inquiry*. Walnut Creek: Left Coast Press.

Varela, F. J., Maturana, H. R., & Uribe, R. (1974). Autopoiesis: The organization of living systems, its characterization and a model. *Biosystems, 5*(4), 187–196. https://doi.org/10.1016/0303-2647(74)90031-8.

Vervloet, J. A. J., van Beek, R., & Keunen, L. J. (2010). A biography of the cultural landscape in the eastern Netherlands: Theory and practice of acquisition and propagation of knowledge. In T. Bloemers, H. Kars, A. van der Valk, & M. Wijnen (Eds.), *The Cultural Landscape & Heritage*

Paradox. Protection and Development of the Dutch Archaeological-Historical Landscape and its European Dimension (pp. 131–148). Amsterdam: Amsterdam University Press.

Vester, H.-G. (1993). *Soziologie der Postmoderne*. München: Quintessenz.

Vicenzotti, V. (2008). „Stadt-Wildnis". Bedeutungen, Phänomene und gestalterische Strategien. In Bayerische Akademie für Naturschutz und Landschaftspflege, & Lehrstuhl für Landschaftsbau und Vegetationstechnik (Eds.), *Die Zukunft der Kulturlandschaft – Entwicklungsräume und Handlungsfelder* (pp. 29–37). Laufen.

Waldie, D. J. (2005). *Holy land. A suburban memoir*. New York: W. W. Norton & Company.

Walton, G. M., & Rockoff, H. (2014). *History of the American Economy* (12th ed.). Mason: South-Western, Cengage Learning.

Wang, F., Antipova, A., & Porta, S. (2011). Street centrality and land use intensity in Baton Rouge, Louisiana. *Journal of Transport Geography, 19*(2), 285–293.

Wayand, G. (1998). Pierre Bourdieu: Das Schweigen der Doxa aufbrechen. In P. Imbusch (Ed.), *Macht und Herrschaft. Sozialwissenschaftliche Konzeptionen und Theorien* (pp. 221–237). Opladen: VS Verlag.

Weber, M. (1976 [1922]). *Wirtschaft und Gesellschaft. Grundriß der verstehenden Soziologie*. Tübingen: Mohr.

Weber, F. (2018a). Ein diskurstheoretischer Zugriff auf ‚Landschaft' und ‚Stadtlandhybride' – Annäherungen zwischen Makro- und Mikroperspektive. In S. Hennecke, H. Kegler, K. Klaczynski, & D. Münderlein (Eds.), *Diedrich Bruns wird gelehrt haben. Eine Festschrift* (pp. 122–131). Kassel: Kassel University Press.

Weber, F. (2018b). *Konflikte um die Energiewende. Vom Diskurs zur Praxis*. Wiesbaden: Springer VS.

Weber, F. (2019). Diskurstheoretische Landschaftsforschung. In O. Kühne, F. Weber, K. Berr, & C. Jenal (Eds.), *Handbuch Landschaft* (pp. 105–117). Wiesbaden: Springer VS.

Weber, F., Kühne, O., Jenal, C., Sanio, T., Langer, K., & Igel, M. (2016). Analyse des öffentlichen Diskurses zu gesundheitlichen Auswirkungen von Hochspannungsleitungen – Handlungsempfehlungen für die strahlenschutzbezogene Kommunikation beim Stromnetzausbau. Ressortforschungsbericht. https://doris.bfs.de/jspui/bitstream/urn:nbn:de:0221-2016050414038/3/BfS_2016_3614S80008.pdf. Accessed 17 Oct 2018.

Weber, F., Jenal, C., & Kühne, O. (2017). Die Gewinnung mineralischer Rohstoffe als landschaftsästhetische Herausforderung. Eine Annäherung aus sozialkonstruktivistischer Perspektive. In O. Kühne, H. Megerle, & F. Weber (Eds.), *Landschaftsästhetik und Landschaftswandel* (pp. 245–266). Wiesbaden: Springer VS.

Weber, F., Kühne, O., Jenal, C., Aschenbrand, E., & Artuković, A. (2018). *Sand im Getriebe. Aushandlungsprozesse um die Gewinnung mineralischer Rohstoffe aus konflikttheoretischer Perspektive nach Ralf Dahrendorf*. Wiesbaden: Springer VS.

Weber, F., Kühne, O., & Jenal, C. (2019). Heimat und Landschaft – Zu einem eng relationierten Verhältnis. In O. Kühne, F. Weber, K. Berr, & C. Jenal (Eds.), *Handbuch Landschaft* (pp. 335–349). Wiesbaden: Springer VS.

Webre, S. (1984). The Problem of Indian Slavery in Spanish Louisiana, 1769–1803. *Louisiana History: The Journal of the Louisiana Historical Association, 25*(2), 117–135.

Welsch, W. (1993). Das Ästhetische – Eine Schlüsselkategorie unserer Zeit? In W. Welsch (Ed.), *Die Aktualität des Ästhetischen* (pp. 13–47). München: Fink.

White, L. (Bayou Brief – Nonprofit Journalism for Louisiana, Ed.). (2019). Why three of the five most politically polarized U.S. cities are in Louisiana. And what the failures of suburban sprawl and the wisdom of Lyle Lovett can teach us about fixing a bitterly divided state. https://www.bayoubrief.com/2019/05/21/why-three-of-the-five-most-politically-polarized-u-s-cities-are-in-louisiana/. Accessed 13 Dec 2019.

White, H. K., Hsing, P.-Y., Cho, W., Shank, T. M., Cordes, E. E., Quattrini, A. M., et al. (2012). Impact of the *Deepwater Horizon* oil spill on a deep-water coral community in the Gulf of

Mexico. *PNAS – Proceedings of the National Academy of Sciences of the United States of America, 109*(50), 20303–20308.

Wiegandt, C.-C. (2017). Die Renaissance öffentlicher Räume – Inszeniert und reglementiert. In W. Gamerith, & U. Gerhard (Eds.), *Kulturgeographie der USA. Eine Nation begreifen* (pp. 167–175). Berlin: Springer Spektrum.

Wilds, J., Dufour, C. L., & Cowan, W. G. (1996). *Louisiana, yesterday and today. A historical guide to the state*. Baton Rouge: Louisiana State University Press.

Williams, T. H. (1969). *Huey long*. New York: Alfred A. Knopf.

Wilson, C., & Groth, P. (Eds.). (2003). *Everyday America. Cultural landscape studies after J. B. Jackson*. Berkeley: University of California Press.

Winchester, H. P. M., Kong, L., & Dunn, K. (2003). *Landscapes. Ways of imagining the world*. London: Routledge.

Wöbse, H.-H. (1999). „Kulturlandschaft" und „historische Kulturlandschaft". *Informationen zur Raumentwicklung, 5,* 269–278.

Wöbse, H.-H. (2002). *Landschaftsästhetik. Über das Wesen, die Bedeutung und den Umgang mit landschaftlicher Schönheit*. Stuttgart: Ulmer.

Wojtkiewicz, W. (2015). *Sinn – Bild – Landschaft. Landschaftsverständnisse in der Landschaftsplanung: Eine Untersuchung von Idealvorstellungen und Bedeutungszuweisungen*. Berlin: Technische Universität Berlin.

Wolch, J., Wilson, J. P., & Fehrenbach, J. (2005). Parks and park funding in Los Angeles: An equity-mapping analysis. *Urban Geography, 26*(1), 4–35.

Woodward, C. V. (1981). *Origins of the New South, 1877–1913. A History of the South* (A History of the South, vol. 9). Baton Rouge: LSU Press.

World Population Review. (2019a). Baton Rouge, Louisiana Population 2019. https://worldpopulationreview.com/us-cities/baton-rouge-population/. Accessed 1 July 2019.

World Population Review. (2019b). East Baton Rouge Parish, Louisiana Population 2019. https://worldpopulationreview.com/us-counties/la/east-baton-rouge-parish-population/. Accessed 1 July 2019.

World Population Review. (2019c). Louisiana population 2019. https://worldpopulationreview.com/states/louisiana-population/. Accessed 1 July 2019.

Wright, B. (2005). Living and dying in Louisiana's „Cancer Alley". In R. D. Bullard (Ed.), *The quest for environmental justice. Human rights and the politics of pollution* (pp. 87–107). San Francisco: Sierra Club Books.

Wright, J. D., Donley, A. M., Gualtieri, M. C., & Strickhouser, S. M. (2016). Food deserts: What is the problem? What is the solution? *Society, 53*(2), 171–181. https://doi.org/10.1007/s12115-016-9993-8.

Wylie, J. (2005). A single day's walking: Narrating self and landscape on the South West Coast Path. *Transactions of the Institute of British Geographers, 30*(2), 234–247. https://doi.org/10.1111/j.1475-5661.2005.00163.x.

Wylie, J. (2007). *Landscape*. Abingdon: Routledge.

Wylie, J. (2019). Landscape and phenomenology. In P. Howard, I. Thompson, E. Waterton, & M. Atha (Eds.), *The Routledge companion to landscape studies* (2nd ed., pp. 127–138). London: Routledge.

Xypolia, A. (2003). Children's perception of racial urban boundaries: A case study in Baton Rouge, LSU Master's Theses. 3486. https://digitalcommons.lsu.edu/gradschool_theses/3486/. Accessed 1 July 2019.

Yeates, M. (1990). *The North American city* (4th ed.). New York: Harper & Row.

Yodis, E. G., Colten, C. E., & Hemmerling, S. E. (2016). *Geography of Louisiana* (7th ed.). Boston: McGraw-Hill Learning Solutions.

YouTube. (2019). YouTube-statistik. https://www.youtube.com/yt/press/de/statistics.html. Accessed 3 Nov 2019.

Zapatka, C. (1995). *The American landscape*. New York: Princeton Architectural Press.

Zelinsky, W. (2010). Asserting central authority. In M. P. Conzen (Ed.), *The making of the American landscape* (2nd ed., pp. 329–356). New York: Routledge.

Zierhofer, W. (2003). Natur – Das Andere der Kultur? Konturen einer nicht-essentialistischen Geographie. In H. Gebhardt, P. Reuber, & G. Wolkersdorfer (Eds.), *Kulturgeographie. Aktuelle Ansätze und Entwicklungen* (Spektrum Lehrbuch, pp. 193–212). Heidelberg: Spektrum Akademischer Verlag.

Zierhofer, W. (2008). Das 'Waldsterben' der Informationsgesellschaft. Umweltforschung auf der Grundlage einer sprachpragmatischen Handlungstheorie. In B. Werlen, & K. Gäbler (Eds.), *Geographische Praxis II. Symbolische Aneignungen* (Sozialgeographische Manuskripte, vol. 4, pp 7–23). Jena: Selbstverlag.

Zsilincsar, W. (1972). San Diego – Tijuana. Betrachtungen eines Grenzraumes. *Die Erde, 103,* 128–156.

Zukin, S. (1992). Postmodern urban landscapes: Mapping culture and power. In S. Lash, & J. Friedman (Eds.), *Modernity and identity* (pp. 221–247). Oxford: Blackwell.

Zukin, S. (1993). *Landscapes of power: From Detroit to Disney World.* Berkeley: University of California Press.

Zukin, S. (1995). *The cultures of cities*. Cambridge: Blackwell.

Zukin, S. (1998). Urban lifestyles: Diversity and standardisation in spaces of consumption. *Urban Studies, 35*(5–6), 825–839. https://doi.org/10.1080/0042098984574.